Nuclear Physics

Personal Study Notes

Atomic, **Molecular, And** Nuclear **Physics**

Every Thing I learned In Modern Physics
From High School To Post-Doctorate

By

Mohamed F. El-Hewie

TABLE OF CONTENTS

CONSTANTS

4

Avogadro number .... N_A=6.02296 • 10^{26} 1/kmole
Faraday constant .... F = 9.64914 • 10^7 C/kmole
Elementary charge .... e= $1.60203 10^{-19}$ C
Planck constant .... $h.$ =6.62517 • 10^{-34} J • s
…………………………………………… $\hbar$ = $1.05443.10^{-34}$ J • s
……………………………………………= 6.5817 X 10^{-16} eV • s
Boltzmann constant .... k = $1.3805. 10^{-23}$ J/deg=8.617X X 10^{-5} eV/deg
Velocity of light in vacuo……….c = 2.997925 • 10^8 m/sec
Electric constant ....g_o = 8.854 • 10^{-12} F/m
Weak interaction constant……..g =10^{-62} J.m^3
Bohr magnetos ....μ_B = 9.2732 • 10^{-24} J/T
Electron magnetic moment……. M_e = μ_B
Nuclear magneton .... μ_N = 5.051 • 10^{-27} J/T
Proton magnetic moment……… μ_p =.2.7928 • μ_N
Neutron magnetic moment……. μ_n = 1.913148 • μ_N
Classical electron radius .... r_e = $2.818. 10^{-15}$ m
Radius of action (range) of nuclear forces ....$R \approx 1.5 • 10^{-15}$ rn
Hydrogen atom radius (first Bohr radius) ....$R(H^1)$= $0.529172. 10^{-10}$ m
Mass of mass unit ....m_μ = 1.6603 • 10^{-27} kg
Energy equivalent of mass unit .... $m_\mu c^2$ = .931.4 MeV
Conversion of electron volts into joules ....1 eV= $1.60203 . 10^{-19}$ J
Atomic mass of carbon C^{12} …………..A_r = 12.000000 (standard)
Neutron atomic mass …………..A_r = 1.008665
Proton atomic mass ....A_r =1.007276
Electron atomic mass…………..A_r =0.00054859
Atomic mass of hydrogen atom ………A_r =1.007825
Mass of hydrogen atom . . …………..$M(H^1)$ = $1.67. 10^{-27}$ kg
Unit of measurement of nuclear cross-sections ....1 barn = 10^{-28} m^2
Temperature of Maxwellian distribution at average energy of 1 eV
……………………………………………T= 7737 °K

CHAPTER 14

ARTIFICIAL TRANSMUTATION

Introduction

Transmutation, in general, may be defined as the conversion of the one element into another, *i.e.,* one type of atom into another. *Artificial transmutation* is such a conversion provoked by artificial means, as distinct from the spontaneous transformations witnessed in radioactive substances found in nature. Transmutation is often called also **disintegration,** since the transmuted element disintegrates forming a different new element.

The conversion of the elements into one another has been the dream of the human race for many centuries. Quite early in the Christian era it was current among the Alexandrian Greeks and then spread to the East and West through the nomadic Arabs. In the middle ages it occupied a very prominent place, under the name of **alchemy** whose main purpose was to arrive at what was known as the **philosopher's stone** which would on touch convert base metals into noble ones. The alchemist's heroic search for, this wonderful agent of transmutation resulted in failure, chiefly due to the meager data then available about the ultimate constituents of elements.

In fact, one cannot aspire to transform element into element, if one does not know what elements really are. Hence, before the problem of transmutation could be *seriously* tackled, the ground had to be prepared and this took a long time.

Several centuries of painstaking labor at last brought the scientists close to the fundamental constituents of matter. All matter was first classified into *92 simple elements* and the smallest unit of any one of these, *the atom,* which completely defined the individual nature and properties of that element, was fully recognized. It was next established that the atom was itself composed of two distinct entities, *viz.,* (i) a central positively charged *nucleus* of extremely small dimension, but wherein practically the whole mass of the atom was concentrated, and (ii) very light negatively charged particles, the *electrons,* gyrating round the nucleus at relatively great distances. These peripheral electrons were found to be highly organized and responsible for most of the chemical and physical properties, while the more wonderfully constituted nucleus gave the essential individuality to the element. It was therefore realized that to transform one element into another, the very atom kernel, the nucleus, should be attacked and altered.

The final and decisive stage in this long preparation came with the discovery of radioactivity. The study of the natural radioactive phenomena not only proved beyond any doubt the complexity of nuclear structure but also paved the way for artificial transmutation. For, the occurrence of spontaneous disintegration of the nucleus in the heavy radioactive elements readily suggested the idea whether it might not be possible to break up nuclei of other ordinarily stable elements artificially by bombarding them with high speed particles. Rutherford took up the hint and performed a series of experiments in which the atoms of different stable elements, such as nitrogen, aluminum, phosphorus, etc., were subjected to an intense bombardment by the high velocity α-particles from natural radioactive substances and finally, in 1919, succeeded in transmuting these elements, thus realizing the old dream of the alchemists.

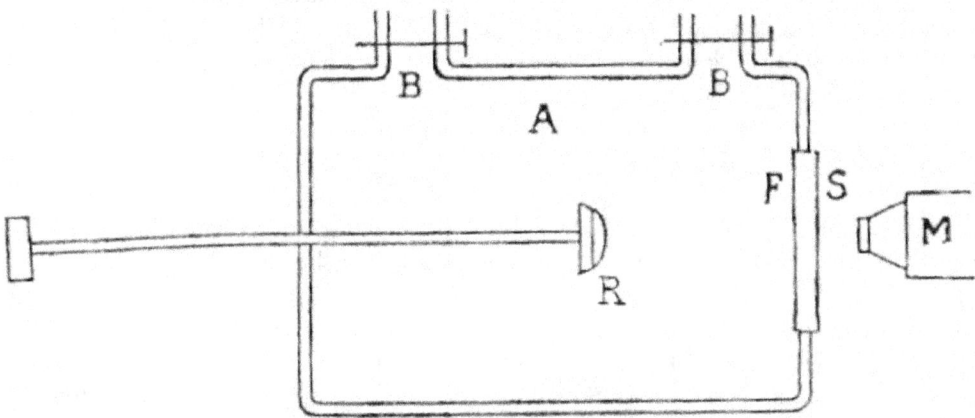

Fig. 269. Rutherford's apparatus used in the first successful transmutation of elements.

Discovery

The very simple apparatus with which Rutherford was able to demonstrate the first artificial transmutation is shown diagrammatically in Fig. 269. One side of a chamber A had an opening which was covered with a thin sheet of silver foil F. A zinc sulphide screen S was placed close to the silver foil and the scintillations on that screen, when they occurred, were observed by a microscope M. The chamber could be filled with various gases through the side tubes BB. The source of α-particles, RaC', was placed at R on a small disc and its distance from F could be varied at will. The range of these α-particles was known to be about 7 cms in air corresponding to an energy of about 8 MeV. As these high speed α-particles shot through various gases in the chamber the resulting effects could be studied by means of the scintillations on S.

When the chamber was filled with oxygen or CO_2, no scintillations were observed, provided the distance between R and F exceeded the range of the α-particles. But, when nitrogen was used in the chamber, scintillations were observed even when the distance between, R and F was as great as 40 cms Since the α-particles themselves could not penetrate such a distance, here was an anomaly which Rutherford interpreted as a real disintegration of nitrogen nuclei resulting from their bombardment by the α-particles and giving rise to long range particles which produced the observed scintillations. Magnetic deflection experiments indicated that these particles were hydrogen nuclei or protons. This was further confirmed by the following two observations:

(i) If the scintillations were produced by the hydrogen nuclei contained as impurity in the nitrogen and projected by elastic collisions with the α-particles, their range would be only about 28 cms, even in the case of head-on collisions, whereas in the experiment the range was as high as 40 cms.

(ii) While hydrogen nuclei are projected chiefly in the forward direction in ordinary collisions of α-particles with hydrogen atoms, in the present case they were found to be emitted in all directions uniformly. These facts showed clearly that the observed protons were emitted by the nitrogen nuclei themselves, when disrupted by the bombarding α-particles, according to the reaction

$$_7N^{14} + _2He^4 \rightarrow {_8O^{17}} + _1H^1,$$

which shows that the new element formed is the rare isotope of oxygen of mass 17. Thus the first transmutation of nitrogen into oxygen was established.

Rutherford, encouraged by this first success, continued his researches and in collaboration with Chadwick, (1921-1924), showed that *many other light elements* from boron to potassium, with the exception of carbon and oxygen, could be transmuted by bombardment with α-particles, resulting in the emission of protons. To perform the experiment with solids, a thin layer or foil of the element under study replaced the silver foil, the source R was pushed close to it and the scintillations produced by the ejected protons were observed. By using various thicknesses of material of known stopping power between F and S, the equivalent ranges in air of the protons emitted in the disintegration of the different substances were determined.

In the case of B, Na, P and Al, the ranges of the emitted protons were 58, 58, 65 and 90 cms respectively, which clearly showed that the energy of the proton was much greater than that of the incident α-particle, the excess of energy being evidently supplied by the nucleus "exploded" under the impact of the α-particle in a veritable transmutation.

Blackett, in 1925, obtained a beautiful confirmation of the disintegration of nitrogen by α-particles with the aid of the cloud chamber Allowing α-particles to shoot through a cloud chamber filled with nitrogen, he photographed the tracks produced. The majority of the tracks thus obtained were straight, typical of α-particles. Several were forked tracks, indicating that an elastic collision had taken place between a nitrogen nucleus and an α-particle. But a very few (8 in 500,000 tracks photographed) presented **a** very unusual feature, as seen in the adjacent photo. Each of them was forked, but one branch was very thick typical of a particle much heavier than the α-particle, while the other very thin, that could have been produced not by an α-particle, but only by a proton. According to the accepted nuclear reaction, the thick track was due to O^{17}.

Disintegration of a nitrogen atom (*Source: Blackett, London*)

This picture shows the tracks of alpha particles (helium-4 nuclei). Near the middle of the picture at the top, two tracks emerge from the end of an alpha particle track. There an alpha particle had caused the disintegration of a nitrogen atom and a proton was emitted (the long track pointing to the left). Earlier Rutherford had used such disintegrations as evidence that protons were a building-block found in all nuclei.

Patrick Blackett (18 November 1897 – 13 July 1974)

Following the first successful experiments of Rutherford, many workers have applied themselves with great ardor to the art of transmutation, devising ingenious methods both for producing fast and intense projectiles required for an effective nuclear bombardment and for observing in detail the products of disintegration, essential for a correct interpretation of results. The progress made in the experimental technique, chiefly during the past century, has been so great that it has now become a relatively easy matter to disintegrate almost any atom; the results obtained are so vast and varied, remarkable and important, both from the practical and theoretical points of view, that the science of artificial transmutation of elements now forms one of the largest and most important branches of nuclear physics. Our aim in this section is to recount briefly the story of this epoch-making achievement of the present century, insisting chiefly on the scientific aspect.

EXPERIMENTAL STUDY OF ARTIFICIAL TRANSMUTATIONS

In order to be able to appreciate fully the great complexity and ingenuity of the experimental technique employed in artificial transmutation,. it is necessary to realize the chief difficulties involved in the process.

Practical difficulties

To effect transmutation, one must necessarily attack the very atom-kernel, the nucleus, wherein resides the essential individuality of the atom and break it up into fragments or at least modify it. But the nucleus is of extremely small dimensions of the order of 10^{-12} cm., which in turn, demands a proportionately small projectile, capable of penetrating sharply the electron groups surrounding the nucleus. Having thus to work with a very tiny target and a very thin missile, the chances of securing proper hits, chiefly "square-hits", must be very small, and in consequence, *transmutation must be a phenomenon of rare occurrence.* This fact calls for devices, which would intensify greatly the number of the projectiles in order to increase the "yield" or occurrence of the phenomenon, on the one hand, and delicate detectors which would record the desired rare events, whenever they happen, on the other.

A second difficulty in transmutation arises from the fact that the *nucleus is well defended by a positive electric field, the potential barrier.* If the projectiles are positively charged, as most of them ordinarily used for this purpose (*e.g.,* α-particles, protons and deuterons) are, many of them will be deflected away and scattered, as they approach the positive field of the nucleus. Even those which are fast enough to surmount the potential barrier will, however, have their speed diminished and consequently lose quickly their efficaciousness. The *discovery of the neutron,* which, having no electric charge, can penetrate the potential barrier unlike the positively charged projectiles *and of the wave mechanical nature* of material particles, which results in a certain probability, however small, of a projectile penetrating into the nucleus, even if its energy is less than what is necessary to surmount the potential barrier, have greatly minimized the difficulties arising under this head.

A third serious problem is *to test whether transmutation has actually taken place, i.e.,* whether the nucleus has been really disrupted and if so, with what results. If the fragments of the broken nucleus are very heavy, they move slowly and remain lost in the mass of the element, and in consequence cannot be detected by any external device. If at least one of the fragments is a light particle, such as a proton, neutron, or α-particle, as is often the case, the process can be more easily checked, since the light particle moves fast enough to get out of the mass of the parent element, and project itself into space.

But another great difficulty arises here, *viz., the possibility of mistaking particles which have nothing to do with the disrupted nucleus* (such as the incident projectiles themselves which might have been scattered, the recoiling nucleus, the protons from hydrogen which may be present as impurity in the element under study, the particles given off by radioactive impurities, etc.,) *as the true fragments of transmutation.* Hence, it must be established that at least some of the particles observed are not of the spurious kind, but really parts of the fractured nuclei. To determine this point with precision, the *nature* of the expelled corpuscles, their *direction* of motion and *energy* must be carefully studied. But this is no easy task, in practice, and a whole complex technique is required to detect accurately the results of the experiments.

The study of transmutation, on its experimental side, may, therefore, be considered to resolve itself into two great practical problems, viz.,

(i) *means of producing powerful sources of projectiles,* and
(ii) *methods of detecting isolated and rare particles which arise from the bombarded nuclei.*

TECHNIQUE FOR PRODUCING SUITABLE AGENTS OF TRANSMUTATION

Different kinds of projectiles
Artificial transmutation of elements was first accomplished by Rutherford using the high speed α-particles from RaC' as projectiles and more than ten years passed before other particles were considered. At present, the projectiles that have been successfully applied to nuclear disintegration are the cosmic rays, α-particles, γ-rays, protons, deuterons, neutrons and electrons. The first three of these may he considered as *natural agents,* being derived from either the mysterious radiations which continuously bombard the earth (cosmic rays) or the natural radioactive substances (α-particles and γ-rays), while the rest as *artificial,* as they are either produced as a result of artificial transmutation (neutrons) or rendered intense and accelerated to a high speed by artificial devices (electrons, protons and deuterons).

Among the **natural agents,** the **cosmic rays** are the most powerful, endowed as they are with fabulous energies of the order of billion electron volts.

The **α-particles** are obtained from many radioactive substances. For quite a long tune they were the sole agents of transmutation and even today they remain very important, capable of effecting very interesting and instructive transmutations.

Of considerable importance has been the use of polonium (RaF), since this can be separated from parent materials which are β- and γ-active and so affords a pure source of α-particles of energy about 5 MeV.

Higher energy α-particles from RaC' (about 8 MeV) are also frequently employed.

The γ-rays obtained from radioactive substances have energies reaching up to 2 to 2.6 MeV. These have been used in a special type of transmutation known as *photo-disintegration.*

Very high energy γ-rays, up to even about 17 MeV, obtained as products of certain disintegration processes, have also been used as bombarding agents.

The general drawback of these natural projectiles is that they are beyond the control of man. It was the desire to control the conditions of his experiment that led the physicist to devise his own artillery. He wanted to be able to choose the kind of projectile and to regulate the speed and intensity with which the projectiles struck the target. Thus it is clear that the special techniques with which we are dealing here, are chiefly directed towards the production of artificial agents.

Of the **artificial agents, neutrons** which have proved to be by far the **best agents of transmutation,** must be considered apart and the artificial accelerators described below do not directly refer to them, on account of their peculiar property of having no electric charge. For one thing, they cannot be accelerated like charged particles, and for another, they turn out to be the more effective agents of transmutation the lower their speed. Hence, in their case artificial means are to be applied, rather to slow down than accelerate them. But they may require the indirect use of artificial accelerators in so far as, for instance, deuterons, accelerated by these devices to high speed and bombarding a lithium target, produce an intense supply of high velocity neutrons (about 15 MeV) which might be needed for effecting certain transmutations.

A *standard and efficient type of neutron source for transmutation purposes* is a small sealed capsule containing a mixture of powdered beryllium and radon gas. The atoms of Be are thus bathed in α-particles emitted by radon; many nuclei of beryllium are thereby disintegrated by the α-particles, producing a great abundance of neutrons (1 m.c.) of radon gives rise to 27,000 neutrons/sec.). For many experiments, the small size of such **a** source makes observations exceedingly simple and effective owing to the large solid angle available for targets.

Techniques of artificial agents
The techniques of producing artificial agents of transmutation are therefore applied chiefly to the two remaining positively charged particles, the protons and the deuterons,

although they have been extended to α-particles in order **to** overcome the low intensity of the α-particles available from natural radioactive sources.

The two main points to be secured in the artificial devices of producing suitable projectiles for transmutation are *controllable acceleration* and *intensification.* According to the wave-mechanical theory, the probability of penetration into a nucleus by a projectile diminishes with the energy. Hence, to disintegrate nuclei with particles of relatively small energy, the number of these particles must be made very large, *i.e.,* their intensity must be increased appreciably. Seventy years of inventive effort have been devoted to the problem and several types of apparatus are now available. Each has its own distinctive feature, but most of them may be grouped in two broad classes:

A. The direct voltage accelerators, in which the projectiles move in straight paths through a long vacuum- tube, impelled by the maximum voltage of the discharge: *e.g., the electrostatic generator of Van de Graaff, the voltage multiplier of Cockcroft and Walton,* etc.

B. The resonance accelerators, in which the projectiles are started at relatively low speeds and by the repeated push of periodic pulses of voltage are brought up to the high energy required for the bombardment of the target: *e.g., the linear accelerator of* Lawrence and Sloan and Thibaud, *the cyclotron* of Lawrence, etc.

We shall now describe the essential features of the more important of these instruments.

THE VAN DE GRAAFF ELECTROSTATIC HIGH TENSION GENERATOR

This high tension generator is simply the old "electrostatic machine" arranged to charge a suitable conductor to very high voltages of the order of several million volts, first devised, in 1931, by Robert Van de Graaff.

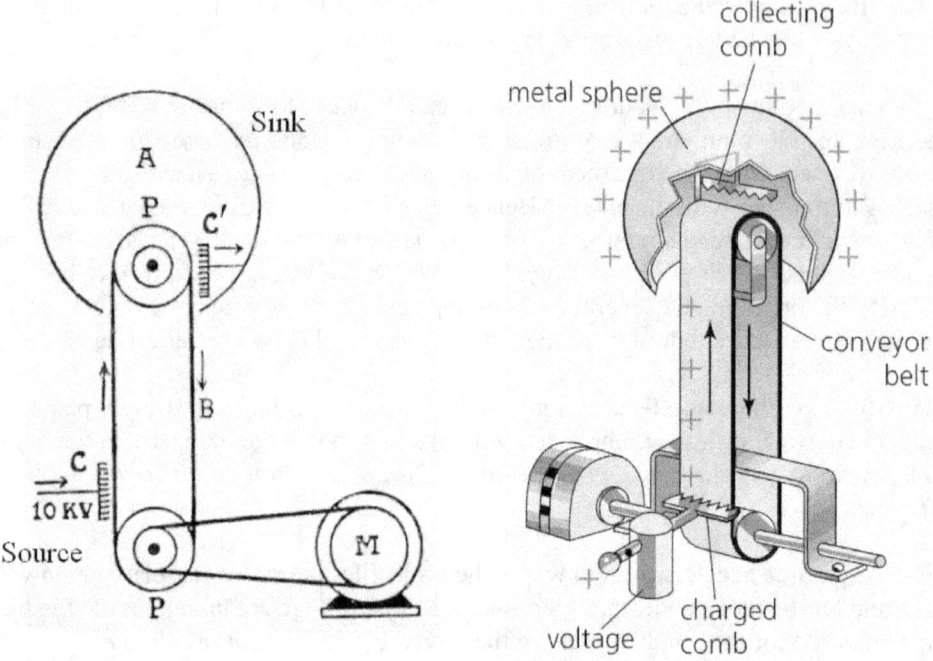

Fig. 270 The principle of Van de Graaff electrostatic generator.

The **principle,** on which the electrostatic generator of Van de Graaff works, is illustrated in Fig. 270. A well insulated endless belt B rapidly rotated between two pulleys PP by means of a motor M, receives electric charges produced by a comparatively low voltage auxiliary generator of about 10,000 volts and sprayed on it by means of a metallic comb C at the bottom. The belt moving upwards carries this charge into a hollow, well insulated metallic sphere A, where the charge is removed from the belt and transferred to the sphere by means of another metallic comb C'. The belt returning to the bottom is charged again and delivers up the charge when inside the sphere.

Charges are thus accumulated and, distributed on the surface of the sphere, while there will be practically no electric field inside it.

With such an arrangement, the sphere can be raised to very high potentials, directly dependent upon the quantity of electricity distributed on its surface.

Technical details
In the practical realization of very high tensions with this arrangement the following points have been carefully attended to:

(i) Since the capacity of the sphere is numerically equal to its radius, the limiting high tension attained by the sphere can be increased, within certain limits, by increasing the *diameter of the sphere.* This increase of diameter increases also the disposable energy on discharging the sphere.

(ii) But the main problem which limits the amount of charge that can be delivered to the sphere and hence also the high tension attainable *is insulation.* With a sphere of a given

14

diameter, insulation will break down beyond a certain limiting potential and heavy sparking will take place to neighboring objects such as the belt, the walls and the ceiling of the laboratory.

This difficulty has been overcome to **a** great extent by the following devices. The sphere is supported on sturdy and long insulating columns; the belt is made with good insulating material which is at the same time flexible enough to pass easily over the pulleys—silk, linen or paper being used for this purpose; the apparatus is lodged in a large and high building. Insulation is further improved by the use of compressed air at high pressure surrounding the apparatus, or better still, by means of Freon gas (a special gas, CCl_2F_2, manufactured for use in refrigerators) which has the additional quality of being non-inflammable. In such cases, the generators enclosed in a huge, thick-walled and pear-shaped steel tank. Generators operating under pressure or in an atmosphere of Freon have the great advantage of being much more compact than those working at atmospheric pressure.

(iii) The characteristics of an electrostatic generator depend also on the arrangement used for the accumulation of charge. For, the *speed of arrival of charges* determines the intensity of the current which the generator will be able to supply. Hence in order to maintain the tension of the generator at a constant value during operation, a constant rate of arrival of charges must be maintained. The intensity of the current can be increased by using wide or multiple belts and by running them at high speed.

(iv) With generators of very high tension; it is often very useful to employ *two identical arrangements,* so that one sphere builds up positive potential, while the other negative, with respect to the earth. Then one can get between the spheres a potential difference whose value is nearly double that of one of them.

(v) Another remarkable improvement made in recent designs is *the dispensing with the auxiliary generator* and *the doubling of the current output.* An initial, small charge on the moving belt, such as is present owing to the friction of the belt with the pulleys, produces by induction a cumulative separation of negative and positive electricity. The belt is made, on its passage out of the sphere, to pick up a charge equal and opposite to the one delivered. Thus the machine is made self-exciting and immediately begins to operate with double the efficiency.

(vi) *The accelerating tube.* Nuclear research requires not only high tension but also a stream of particles which are accelerated to that high voltage. This forms a special technique of its own, known as the "accelerating tube" or "discharge in cascade". The principle first given by Coolidge, in 1926, consists in splitting the very high tension into a series of small tensions which are then applied to the ions to be accelerated in successive regular stages. This has to be done, since the vacuum discharge tube producing the ions will be damaged if such high voltages, as are involved here, are directly applied to it, even if they are sufficiently long to prevent sparking along the outside of the tube.

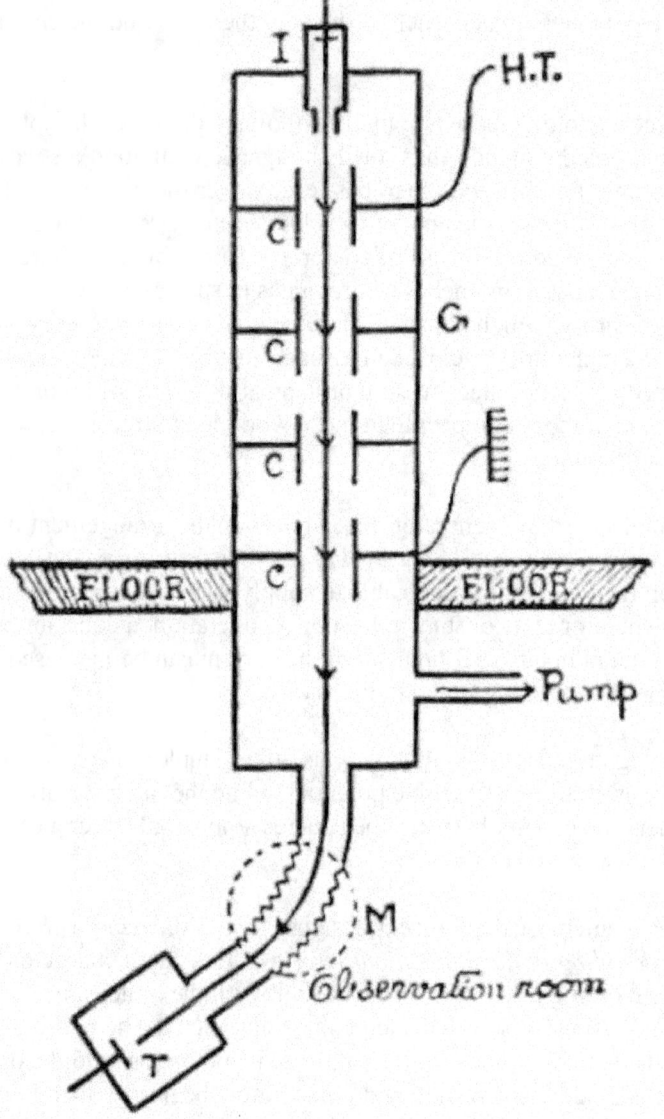

Fig. 271. The accelerating tube.

The accelerating device (Fig. 271) consists of a long, highly evacuated, strong-walled pyrex glass or porcelain tube G which is built up with a great number of sections, containing a series of insulated metal cylinders CC with short gaps between them. The source of ions I, a small discharge tube, is mounted on the top of the accelerating tube. As the ions entering this tube travel along the axis of the cylinders, they are accelerated in the gaps between the cylinders, where there is an intense electric field, which also exerts a focusing effect on the beam when the cylinders are properly separated, so that an intense, well-focused beam of high speed projectiles is obtained. The bottom of the accelerating tube passes through a hole in the floor into the observation room below, where the ions, concentrated and accelerated all along the tube, are deviated, by a suitable

16

and adjustable magnetic field M and made to strike the target T conveniently arranged to observe the results of transmutation. The magnetic field not only gives nearly uniform speed to all the particles that bombard the target but also separates the proper projectiles used from other stray ions of a different kind.

Van de Graaf Generator: An alumnus of the class of 1956 nominates the Van de Graaf generator constructed in 1931 at Round Hill, brought to the Cambridge campus, and donated to the Boston Museum of Science in the 1950s, where it is the featured exhibit in the Thomson Theater of Electricity.

Some of the Van de Graaff generators in use
The first electrostatic generator constructed in 1931 produced only 80,000 volts; but this was perfected and in 1933 Van de Graaff was able to demonstrate at Princeton a generator producing 1.5 million volts with a constant current of 25 microamperes. It consisted of 2 spheres, each 2 ft. in diameter, mounted on upright pyrex insulators 7 ft. long. One sphere was charged positively and the other negatively. A 2.2 inches silk belt run at a speed of 3500 ft./min. was used and a 10 KV supply was employed to spray the belt with electric charges.

The next one to be constructed was at the Department of Terrestrial Magnetism of the Carnegie Institution, Washington, in 1935, under the direction of Tuve, Hafstad and Dahl. This generator using only one sphere, 2 meters in diameter, and containing within itself another concentric sphere of about one meter in diameter is reproduced in the adjacent photo. The globe made of aluminum is supported on three tall stands of insulating material. Through holes pierced on the side of the globe passes the continuous belt, made of special paper, 20 cms broad. This belt is run by means of a motor fixed high on the wall (not shown in the figure) and the speed of rotation is 1200 meters/min. Below the large belt there is a smaller one operated by a dynamo which supplies a modest voltage to the generator of ions, installed in the interior of the globe itself. This machine supplies a tension of 1.2 million volts at a current of 0.75 milliamperes, that concentrates and accelerates the ions in the accelerating tube which is seen extending vertically downward from the bottom of the globe.

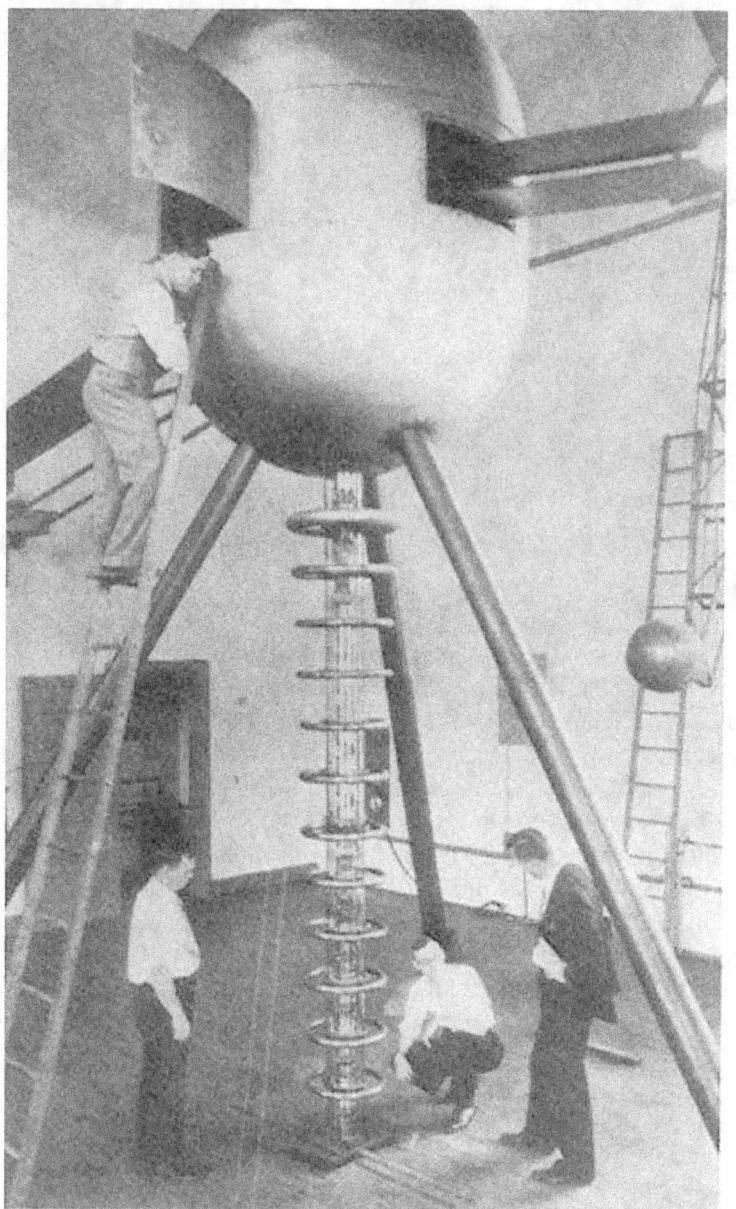

Van de Graaff electrostatic generator- 1.2 million volts (Tuve, Hafstad and Dahl)

In 1937, in the International Exhibition held at Paris, a huge generator, with two identical spheres, each 3 meters in diameter, raised to a high tension of 2.5 million volts, one positively charged and the other negatively charged and hence capable of producing 5 million volts, was demonstrated. It was constructed by A. Lazard under the direction of F. Joliot and Savel.

Van de Graaff's 15-foot generator at MIT's Round Hill Experiment
Station. The spheres stood 43 feet above the ground; their steel trucks
ran on a railroad track to make possible changes in the striking distance.

More recent designs are much larger. In one of the largest of these at Massachusetts
Institute of Technology, the hollow spheres are 15 ft. in diameter; the insulating columns
supporting the spheres are 22 ft. long; the belts made of insulating paper 47 inches wide
are run inside the supporting columns. This generator has actually produced 5.1 million
volts between the two spheres with only 0.1% fluctuation of voltage over a reasonable
time interval. Operating at this potential the apparatus furnishes a current of 1.1
milliamperes.

A similar generator at the Westinghouse Research Laboratory, operating- under air
pressure of 100 lb./sq. in., enclosed in a steel tank, develops 3.7 million volts from a
single metallic sphere to ground, constant to 0.5% for several minutes, yielding a current
of about 3 to 4 milliamperes.

The one at the University of Wisconsin under the direction of Herb which incorporates a number of novel features such as the replacing of the single high potential electrode by three separate concentric ones, operating under pressure with freon-air mixture, etc. develops a steady tension of 4.5 million volts at 5 microamperes.

It may be noted that Van de Graaff generators have been designed to accelerate electrons as well, which have been used for producing high energy X-rays of the order of 1 million electron volts and more. It is interesting to note also what a colossal and complicated machine is required to tackle the extremely minute nucleus.

COCKCROFT AND WALTON TENSION MULTIPLIER

The **basic principle,** on which this high potential device has been developed, is simple. A great number of condensers are charged *in parallel* to a relatively low tension and discharged *in series,* when the tension will evidently be the sum of the tensions of the individual condensers.

Supposing that each of the condensers is charged to 50,000 volts, and that 20 such condensers are discharged in series, the tension of discharge is one million volts. Thus very high voltages can be developed by a suitable choice and large number of condensers.

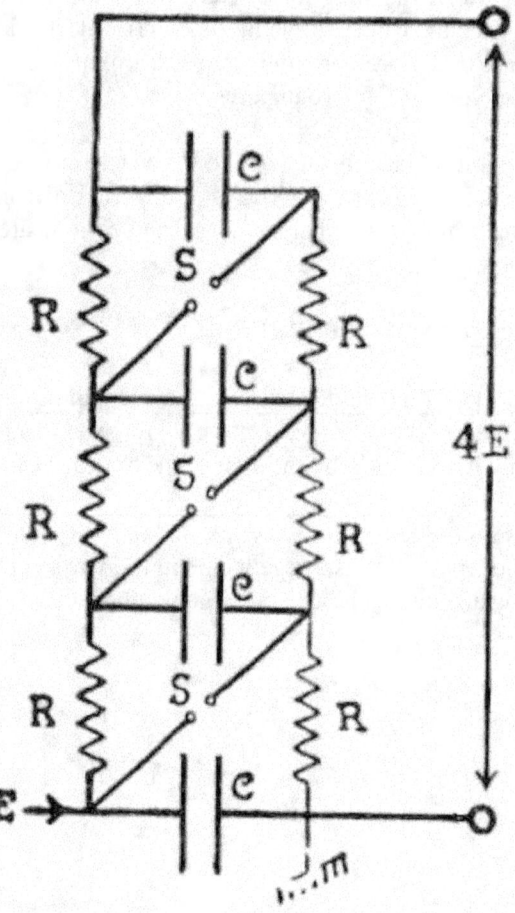

Fig. 272. Impulse generator.

The impulse generator

A first practical application of this principle was realized in what is known as the impulse generator, illustrated in Fig. 272. A number of condensers are arranged in parallel with high resistances included so that they can be charged simultaneously from a relatively low voltage supply E. Each pair of the condensers is fitted with a spark gap S connecting them in series. Each spark gap is in parallel with one high resistance R and one condenser C. When the voltage between the bottom two condensers is high enough to cause a spark across their gap, the spark starts at the bottom and ripples upwards, the breakdown of any one gap with its consequent lowering of resistance being immediately followed by the breakdown of all the other gaps. In this condition, the circuit may be regarded as if the charged condensers are connected in series by the line of conducting spark gaps, so that the voltage difference between the top and the bottom condenser is very high, *viz.,* the initial applied voltage E multiplied by the total number of condensers, *e.g.,* 4E for the arrangement shown in the figure.

In the early days of nuclear research, Brasch and Lange in Austria constructed such an impulse generator capable of developing potentials up to 2.5 million volts, using a full-wave rectifier to supply the charging voltage of 200 kilovolts. In operation they could get about 2 impulses per sec., which they utilized to accelerate particles of transmutation.

The great drawback of this type of generator is that the high voltage is available only for a very short time of about 10^{-5} sec. and hence an uninterrupted current flow for a reasonable time interval cannot be had. The reason for this state of affairs is evidently the fact that the condensers must be discharged simultaneously and there is no way of controlling the spark that travels upwards making the series connection in the, arrangement.

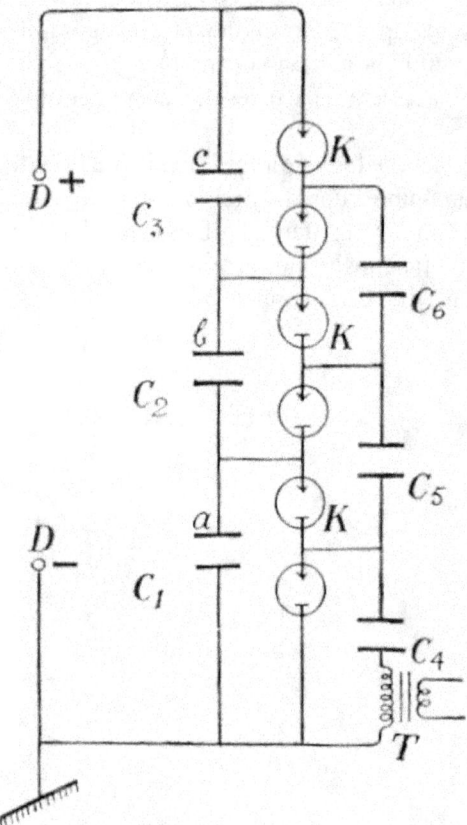

Fig. 273. Principle of Cockcroft and Walton tension multiplier

Cockcroft and Walton tension multiplier

Cockcroft and Walton overcame this difficulty by the use of two columns of condensers interconnected with rectifying vacuum tubes (kenotrons) as shown in Fig. 273. C_1, C_2, C_3 are condensers of equal capacity connected in a series constituting one column, while C_4, C_5, C_6 are similar condensers, also connected in series, forming the second column. The two are cross-connected with diode rectifiers KKK as in the figure. T is a high tension step-up transformer which supplies the voltage for charging the condensers.

The principle of operation may be briefly stated as follows:

Under the combined action of the *oscillatory nature* of the applied voltage from the transformer and the *automatic switching* property of the rectifiers, charges are pumped up to the tops of the columns, C_1 charging C_5, C_5 charging C_2, C_2 charging C_6 and so on while C_6 can never discharge into C_2, nor C_2 into C_5 , etc. In the actual working, the transformer charges a feeding condenser through a pair of rectifiers which are conducting during one half cycle. In the succeeding half cycle these rectifiers are non-conducting and the feeding condenser shares its charge with another through a second set of rectifiers and so on until all the condensers are fully charged. Under these conditions, the total overall tension between the bottom and top of a column is the sum of the individual potentials on the condensers in it. Hence increasing the number of condensers in both the columns any desired high tension can be obtained by a simple multiplication process— hence the name *tension multiplier.* The high voltage thus developed for use between the points DD is evidently not oscillatory but direct. Further, as long as the condensers are kept replenished by the rapid cycle of operations provided by the A.C. supply and the vacuum tube rectifiers, it is possible to produce a small but continuous flow of current between DD at a constant voltage slightly less than the total high tension developed, unlike in the case of the impulse generator.

Technical details

(1) The condensers must be capable of withstanding high tensions of the order of 50 to 100 thousand volts and are mounted on suitable insulating supports.

(2) The vacuum tube rectifiers are of special design, constantly evacuated and the filaments heated by insulated transformers.

(3) The accelerating tube, similar to the one described in connection with the Van do Graaff generator, imparts the high energy to the projectiles in successive steps, thereby removing undue electrical stress inside the tube, as well as securing the required focusing effect.

Three million volt tension multiplier (Cockcroft and Walton) was part of one of the early particle accelerators responsible for development of the atomic bomb. Built in 1937 by Philips of Eindhoven.

Left photo: The two columns on the right side of the picture constitute the tension multiplier, while the smaller column, where a man is standing is the accelerating tube, and the observation room where the projectiles are made to transmute elements, is below him.

Right photo: it is now in the National Science Museum in London, England.

Description of Cockcroft and Walton apparatus

Although Cockcroft and Walton, in their first attempt in 1930, were able to realize an output of only 300,000 volts by this method, a subsequent design in 1932 yielded a potential of about 800,000 volts and their most recent apparatus of gigantic dimensions, represented here, is capable of giving 3 million volts.

These two scientists were the first to succeed in producing high energy projectiles in the laboratory and to show that **transmutation** could be produced by man-made weapon, which is artificial transmutation in the strictest sense.

Sir John Douglas Cockcroft (27 May 1897 – 18 September 1967)

Ernest Thomas Sinton Walton (6 October 1903 – 25 June 1995)

In 1932, working with 300,000 volts and accelerating protons with it, they were able to induce transmutations in lithium and boron. They later showed that, provided sufficient intensity of proton stream was available, occasional disintegrations of lithium could be produced with energies as low as 20,000 electron volts, which thus confirmed the wave mechanical conception of nuclear disintegration.

The intensity of the proton current obtained in the first two designs was low, about 10 microamperes. In 1934, Oliphant and Rutherford were able to obtain an intense beam of projectiles of 100 microamperes at the relatively low voltage of 250,000 volts by an ingenious device, in which were employed a special discharge tube where the ions were produced and a strong and adjustable magnetic field which deflected the projectiles through almost 90° before they struck the target. The latest model of Cockcroft and Walton tension multiplier incorporates all these improvements made by the Cavendish Laboratory workers, so that it is capable of producing intense beams of projectiles at energies even as high as 3 million electron volts.

Ernest Orlando Lawrence (August 8, 1901 – August 27, 1958)

RESONANCE ACCELERATOR—CYCLOTRON

The cyclotron, first devised in 1932 by Prof. E. 0. Lawrence at the Berkeley Institute California, is a powerful weapon in transmutation technique, with great potentialities.

The **basic principle** involved in this ingenious apparatus is known as **resonance acceleration.** Instead of accelerating the projectiles by the application of a constant very high tension in a few big doses, as is done in the two previous methods, they can be very well accelerated to acquire an energy of even several million electron volts by communicating to them periodic pulses of relatively low voltage. This periodic acceleration of the projectiles is achieved by subjecting them to an alternating tension of high frequency. Supposing that the positive ions are accelerated by the alternating potential during a given half-cycle, if they are still under the influence of the applied potential during the next half-cycle in the opposite sense, they would naturally lose the acceleration which they acquired during the first. But, if during the second inverse half-cycle the ions are protected by some device from the influence of the external force and brought under the action of the force only during the third half-cycle, which is in the same sense as the first, then they would receive again an acceleration, of magnitude equal to the first.

Thus subjecting the ions to the alternating potential only during alternate half-periods in the same sense, while shielding them during the intervening half-periods in the opposite sense, their velocity could be steadily increased; and if the alterations are rapid enough, the ions would receive a very high speed within a short time. As the particles have always to be in *step* with the external alternating force in order to be pushed in the right direction, the method is called *resonance acceleration.*

The singular ingenuity of this method lies in the fact that the ions are accelerated to enormous energies not attainable in the other direct methods, and yet, the greatest voltage difference at any moment between any two points of the apparatus is only a few thousands. This brings out also the superiority of the method over the others, as it removes the two great limitations involved in the direct high tension generators, viz., the problems of *insulation* and *separate acceleration.*

> The insulation difficulties set necessarily an. upper limit of about 6 million volts to the Van do Graaff generator and the tension multiplier, while cyclotrons accelerating particles even to 50 MeV and more have already been realized.

A separate accelerating tube with its complex technique that causes so much trouble with other high voltage machines is also eliminated in the remittance accelerator. The method, though simple in principle and superior to others, involves great difficulties in its practical realization, which have, however, been surmounted with great success by Lawrence and his associates as follows:

The first type of the resonance machine, known as the *linear accelerator,* constructed by Lawrence and Sloan in 1931, is shown diagrammatically in Fig. 274.

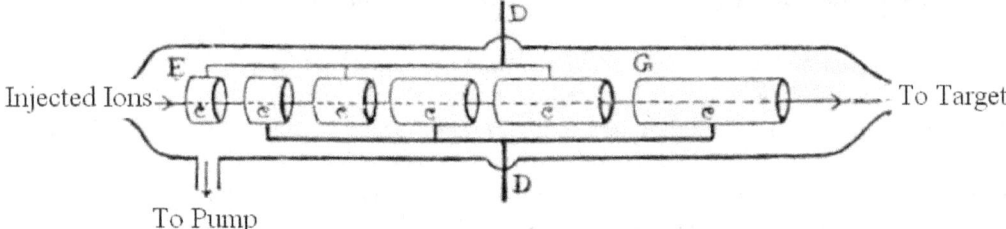

Fig. 274. Linear resonance accelerator

Inside a highly evacuated long glass tube G, a series of metallic cylinders C, C, C, ... of suitably increasing lengths are arranged coaxially with gaps between them. Alternate cylinders are connected together and between the two terminals DD of the two groups thus formed is applied a high voltage high frequency oscillator. Positive ions produced in a discharge tube are made to enter axially this arrangement at one end E. During the proper half-cycle of the applied alternating potential, the ions are accelerated towards the first and shortest cylinder. After entering this cylinder they are in a field-free space and hence continue to move at constant velocity. Now, if the length of this cylinder is such that the ions emerge from it at the moment when the phase of the alternating potential has just reversed, they will be further accelerated towards the second cylinder by the voltage between the two cylinders. The frequency of the applied potential and the lengths of the cylinders are so adjusted that when the ions emerge from each cylinder the electric field has swung round enabling the next gap to give them another accelerating impulse.

It is evident that the lengths of the successive cylinders must be greater and greater in the right proportion in order to allow the more and more rapidly moving ions spend one half-period in each of them, so that they may gain the additional energy at each gap where alone the external field is effective. Using a number of such cylinders the final energy of the ions could be made to correspond to a potential difference (P.D.) many times that used for one acceleration. The concentration of the ions into a narrow beam is effected by the focusing action at the gaps. As the method consists in making the ions proceed along a *straight line* and in *accelerating* them at definite points along that line by voltages produced *in rhythm* by an oscillator, it is called the *linear resonance oscillator.*

Lawrence, in his first apparatus (1931) using 30 cylinders and a short-wave radio generator to supply the high frequency alternating potential of 42,000 volts, was able to accelerate mercury ions to 1.25 MeV with an ion current of 0.1 microampere.

In a later design (1934) with 36 cylinders arranged in a glass tube 185 cms long and an accelerating potential of 80,000 volts, the Hg ions were accelerated to 2.85 MeV at an ionic current of 0.01 microampere.

The *two great drawbacks* of the linear accelerator are:

(i) its *inconvenient length* (being longer the lighter the ions to be accelerated and the greater the maximum energy aimed at), and (ii) the *low, intensity* of the ionic current.

Lawrence overcame these limitations in a simple but ingenious way, *viz.,* by the use of a strong magnetic field which whirled the ions round circular paths, instead of straight, during the half-periods when they had to be protected from the external accelerating field.

The apparatus thus devised is commonly known as the *cyclotron* since the path described by the ions during their acceleration is a series of circles or a spiral. It is sometimes called also the *magnetic resonance accelerator,* since it operates on the principle of resonance between the applied electric and magnetic fields.

The cyclotron consists essentially of a flat, cylindrical evacuated chamber C [Fig. 275 (*a*)] inside which two semi-circular hollow metal boxes DD, called the *"dees",* on account of their shape like the letter D, are arranged with a small gap between them.

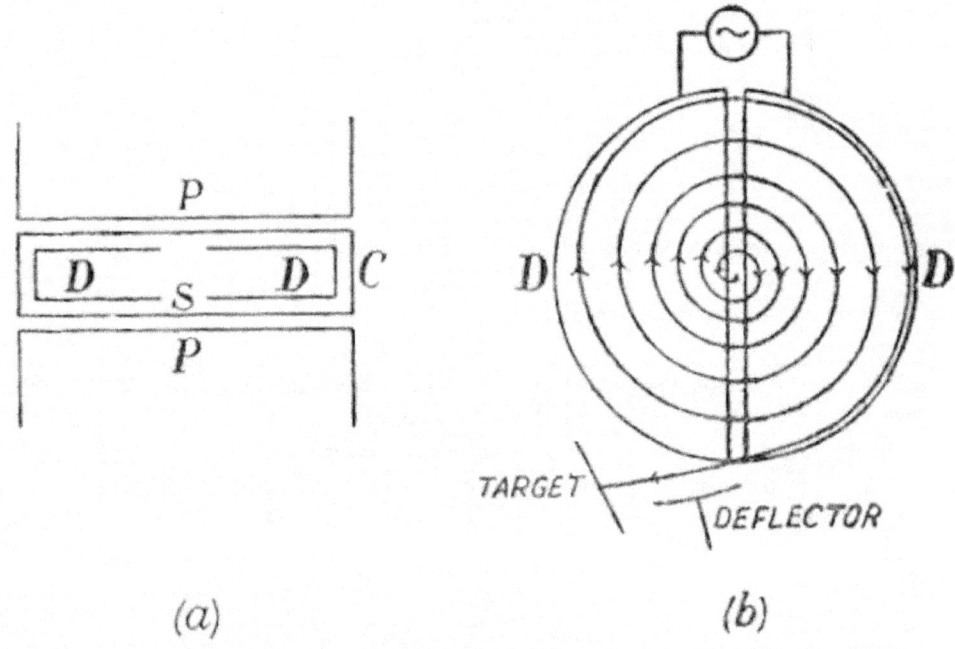

Fig. 275. The cyclotron: (a) sketch, (b) principle of action.

An alternating potential of the order of 10,000 volts and of high frequency (107 cycles) is applied between the dees. The chamber containing the dees is situated between the pole pieces PP of a huge electromagnet capable of producing fields of the order of 15,000 gauss, so that as strong magnetic field can be established perpendicular to the flat face., of the dees. A source S, arranged near the center of the dees, supplies the positive ions to be accelerated.

The **principle of action** of this apparatus is illustrated in Fig. 275 (b). The positive ion produced at S will be drawn into whichever dee happens to be negative at the moment. When once inside the dee the ion being in a field-free space will move with a constant speed. Owing to the magnetic field acting in a perpendicular direction, the ion will describe a small semi-circular path before arriving at the gap between the dees. Now, if the frequency of the applied alternating potential and the magnetic field strength be so chosen that the time required to describe this semi-circle corresponds to one half-cycle, then the ion will reach the gap just at the proper time to be further accelerated by the new reversed electric field across the gap into the opposite dee. Since the particle is now moving with greater velocity, it will describe a semi-circle of greater radius in the second dee. But it is easily seen *that the time taken to describe a semicircle is independent of both the radius of the path and the velocity of the ion.* For, using the general relation representing the action of the magnetic field on the moving ion,

$HEv = Mv^2 / r,$

Where,
H is the strength of the magnetic field,

31

E, M and v are the charge, mass and velocity of the ion, and r the radius of the circular path, we get

$r = Mv / HE$;

hence $r \propto v$.

Now the time t required to traverse a semi-circle is given by

$t = \pi / \omega = \pi / (v/r) = \pi r / v = \pi M / HE$

> The time taken to describe a semi-circle is, therefore, independent both of r and v and is the same for all ions with the same mass and charge.

Hence, the ion describes all semi-circles, whatever be their radii, in exactly the same time, the larger velocity of the ion compensating for the longer path to be traversed. This would be no longer true, if the ion is moving so fast that the relativistic change in mass cannot be neglected, but the desired results are attained without forcing the speed to such heights.

Thus it is clear that the ion will arrive each time at the gap between the dees exactly at the moment when the alternating potential can accelerate it further. By this means several hundred separate accelerating impulses can be given to each ion which in consequence will describe a series of semi-circles of ever-increasing radius in the two dees, gradually spiraling outwards and finally emerging at the outer rim of the dees. For each complete turn of the spiral the ion gains an energy equal to twice the tension applied between the dees. It will emerge, therefore, with an energy corresponding to a potential very much higher than that used in the accelerating process (given by $2n$V, where n is the number of revolutions executed by the ion and **V** the applied potential). At the periphery of the dees an auxiliary negative electrode deflects the accelerated ions on to the target to be bombarded. Although there may be ions which cross the gap before and after the peak value of the alternating potential is reached and hence acquire less energy each time they cross the gap, yet on account of the fact that these ions make more revolutions before reaching the periphery of the dees and in so doing acquire the same total amount of energy as the ions which cross the gap exactly at the peak value, *all the ions emerging from the dees will be homogeneous in energy.*

The final energy attainable depends on the magnetic field strength and the radius of the final path since the energy of the ion is given by

$W = (1/2) M v^2 = (H E r)^2 / 2M.$

Hence a powerful magnet giving a large field, uniform over a considerable area, has to be used in this device. It may be thought that the voltage can be increased indefinitely, provided the above conditions are satisfied. But it is not so, since a limit to the maximum energy that can be attained is set, ultimately, by the relativistic variation of mass of the ion at very high velocities which tends to throw the ion out of step with the alternating potential.

Technical details

The cyclotron, though simple enough in principle, is a complicated and expensive apparatus where a good many accessories are required and several delicate technical details must be attended to:

(1) The *dees* are usually made by cutting a flat circular copper-spun box, a few inches in height and 30 inches and more in diameter, into two halves. They are then arranged inside the vacuum chamber by means of insulating supports, separated from each other by a gap of a few inches.

(2) The *vacuum chamber* is made of brass with heavy steel for its top and bottom, which renders the magnetic field stronger inside the chamber. In the big cyclotrons in use, the vacuum chamber is so heavy a crane is needed to lift the top off when it has to be removed from between the poles of the magnet for inspection or repair. High vacuum in the chamber is required to prevent collisions resulting in the scattering of the ions from their normal paths and thereby reducing the efficiency of the machine. It is no easy task to evacuate a large chamber of diameter 3 feet and more. The heavy steel top of the chamber is clamped down on a rubber tape so tightly that no air can creep in, and all other inlets around the wall of the chamber are also closed by rubber cords. Even with all

these precautions, the chamber sometimes springs a leak; then the operations have to be suspended until the leak is found out and repaired. The probability of collisions between ions even in a good vacuum chamber becomes greater, the greater the number of revolutions, because both the path and the velocity of ions increase. In practice, it has been found that an ion making more than 150 revolutions will surely meet with a collision in a cyclotron of medium size. This also limits the maximum velocity that can be attained by a projectile in the cyclotron.

(3) The *powerful electromagnet,* callable of producing 15 to 20 thousand gauss with pole-pieces as large as the vacuum chamber in extent, weighs some hundred tons and more, a good portion of which is iron, used for the core and pole-pieces, while copper wire used for winding and tube for water cooling purposes amount to about ten tons. This accounts not only for the large and massive size but also for the high cost of cyclotrons. The consumption of electric power in such electromagnets is bound to be very high.

(4) The *alternating potential* of some tens of thousands of volts and of high frequencies of 10 to 15 megacycles, used for accelerating the projectile, is by itself a feat of modern radiotechnique, which involves the construction of powerful short-wave transmitters operating at about 20 meters and 50 kilowatts. The proper frequency of the transmitter is determined by the strength of the magnetic field and the charge and mass of the ions. For, the cyclotron resonance condition requires a frequency

$$f = \omega / 2\pi = v / 2\pi r = (1/2\pi)(E/M) H.$$

This means that the cyclotron frequency of an ion, though independent of r, is directly proportional to H and E /M. Hence also the same frequency oscillator would produce resonance for both deuterons and α-particles, but twice this frequency is required for protons. Great care is taken to stabilize the frequency of the transmitter. This powerful electric oscillator is connected through a step-up transformer to the two dees by leads passing through the insulating supports of the dues.

(5) The *positive ions* to the accelerated are produced by ionization of gas. After thoroughly evacuating, the chamber containing the dees is filled with the gas whose positive ions are to be used (*e.g.,* hydrogen, deuterium, helium) at a pressure of about 10^{-4} cm. A stream of electrons shot into the gas at the center of the dees from a hot filament placed just above the dees at the negative potential relative to the mean potential of the dees, ionizes the gas and the resulting positive ions are drawn into the negative dee. Recently, special capillary ion sources, which give a more copious supply than the hot filament, have been used. This avoids also electrical discharges inside the dees so that narrower dees and smaller air gaps between the poles can be secured.

(6) The *ionic beam to be effective must be focused.* It may be asked how the ion beam could at all be focused as the ions execute the irregular spiral towards the periphery of the dees. The electric field across the gap and the magnetic field perpendicular to the dees do this work quite efficiently, the former in the inner regions, while the latter towards the periphery of the dees. Matching these two effects, so that the final beam is neither too

wide nor too narrow, is a difficult task. 'Shims," *i.e.,* iron sheets inserted between the poles and vacuum chamber are used for this purpose.

(7) Even a very small change in the mass of the ion due to *relativistic variation,* and *small inhomogeneity of the magnetic field* due to slight falling off in intensity at the edge of the pole-piece, affect both the focusing of the beam and the resonance conditions necessary to produce very high energy particles. It is not easy to secure the optimum conditions for these two factors together, since the "shimming" required to produce resonance is not just that required for the best focusing. On account of this practical difficulty limitations are set on the maximum energy obtainable with the cyclotron for a given alternating potential.

Calculations show that for 50,000 volts on the dees, the highest practical energies are 15 MeV for protons, 21 MeV for deuterons and 42 MeV for helium.

Only through a great sacrifice in intensity can the maximum energy be further increased for the given potential on the dees.

(8) The *deflector electrode,* used to pull away the projectiles from their normal trajectory and direct them to the target beyond the dees, requires a large voltage, sometimes as high as 60,000 volts and more, chiefly when the target is placed outside the vacuum chamber. Even with this high voltage the deflector cannot pull all the projectiles through; many of them hit the dee near the window and are lost from the beam. The heat generated by these impacts is very great, so that a thin molybdenum plate is to be substituted for the copper just where the projectiles hit the dee.

(9) The *ionic beam,* when allowed to emerge from the cyclotron through a thin window into air, can be observed visually having a violet ionization glow for a range of 2 to 3 feet, *as* seen in the adjacent picture. The *ion current* obtained with a cyclotron is, however, rather low, reaching a maximum value of about 100 microamperes. This method does not serve, therefore, for the production of a steady current at a constant voltage as do the Van de Graaff generator and the tension multiplier. But it is a very powerful weapon available today for nuclear transmutation and its great importance as a producer of high intensity neutrons and of large quantities of artificial radioelements which are being used in the study of many interesting and difficult problems in biology and medicine is widely recognized.

A beam of 10 MeV deuterons about two feet in length emerging from a cyclotron into air.

The first external cyclotron beam, obtained on March 26, 1936. The glow arises from the ionization of the air by the 5.8 MeV deuterons.

(10) The *target*. In the first designs the target was placed inside the chamber itself. In the later machines it is arranged outside, resulting in the great advantage that many different elements can be bombarded in rapid succession without impairing the main vacuum inside the chamber.

Some of the cyclotrons in use

Although it is not easy to construct a cyclotron on account of the many technical difficulties there are several cyclotrons in use today. After WWII, the majority of constructed cyclotrons were found in the United States of America. In England there were three, one at each of the laboratories in Cambridge, Birmingham and Liverpool. Paris, Copenhagen, Stockholm, Leningrad and Tokyo had one each. In India there was one in Calcutta.

Lawrence, who was awarded the Nobel Prize for this outstanding invention, had supervised the building of several cyclotrons at California. His first cyclotron (1932) having pole pieces 12" in diameter, operated with a magnetic field of 14,000 gauss and an alternating potential of 5,000 volts on the dees, was capable of producing 1.2 MeV protons.

The second cyclotron built by him (1934-1936), the general features of which are shown here, is very large and massive. A huge electromagnet weighing 85 tons, with pole-pieces 37 inches in diameter produces a field of 15,000 gauss between the pole faces 3 ½ inches apart.

In the picture here:

AA represents the core of the magnet,

BB the winding of the magnet, weighing 9 tons,

PP the pole-pieces

and

C the vacuum chamber 35 inches in diameter and 3 inches in height.

In the picture above are shown the two dees marked 1 and 2 inside the vacuum chamber, the ionic source at the center marked 3, and at the periphery various accessories, such as the insulating supports deflecting electrode, etc. A vacuum tube oscillator capable of furnishing 20 kilowatts power supplies the alternating potential of 15,000 volts at **a** frequency of l0 megacycles. Huge water tanks 3 ft. thick surround the cyclotron to protect the experimenter from the harmful biological effects of the very intense neutron radiation produced by the impact of the fast ions on the copper dees. This machine produces deuterons with energies of 8 MeV and α-particles of 16 MeV. Most of the cyclotrons in various parts of the world have been built on this model.

Lawrence and his collaborators, still unsatisfied, constructed in 1939, a much larger instrument, with a magnet weighing some 200 tons provided with pole-pieces 60 inches in diameter. The high frequency oscillator supplying voltage to the dees operates at 60 kilowatts. With this apparatus, Lawrence has produced 8 MeV protons. 16 MeV deuterons and 38 MeV α-particles.

MORE POWERFUL PARTICLE-ACCELERATING MACHINES

In recent years, accelerating machines much more powerful than the cyclotron, and capable of delivering energies in hundreds and even thousands of million electron volts, have been conceived and already constructed or are being built. Though based on the fundamental principle of resonance or synchronism, common with the cyclotron, these instruments involve additional new techniques which enable them to produce extremely powerful projectiles comparable to those produced by cosmic rays. They are called the *synchro-cyclotron*, the *betatron*, the *proton-synchrotron (betatron)* and the *cavity and guided wave accelerators*.

The synchro-cyclotron or the relativistic ion cyclotron

After building several cyclotrons of progressively larger power and bigger size, Lawrence, in 1946, devised this new machine, where an additional new principle of "phase-stable orbits" obtained by suitable frequency modulation was utilized to compensate for the relativistic increase of mass of particles accelerated to very high velocities.

The conventional fixed frequency cyclotron can accelerate ions to energies of about 30 MeV. This energy limit is due to the relativistic increase in mass with energy. The frequency of revolution of ions in a magnetic field is

$$\omega = (E / M) (H / c)$$

and their energy is

$$W = M c^2 = HE c / \omega$$

Hence, when relativistic variation of mass is taken into account, as an ion's energy \V increases, its cyclotron frequency ω decreases, so that it lags in phase behind the accelerating radio-frequency potential. The phase-lag is eliminated in the synchro-cyclotron by frequency modulation, *i.e.*, by suitably altering the frequency of the applied alternating potential to correspond to the lagging rate of the more massive ions, so that those ions move in phase stable orbits, *i.e.,* they move in such a way that they are caught at the right moment in every revolution to be accelerated.

Both theory and experiment indicate that about only one per cent of the ions leaving the source are captured in phase-stable orbits, so that ions ejected from the machine after having been accelerated are grouped in small pulses. Hence this device produces high-energy ions at the sacrifice of beam intensity.

The Berkeley synchro-cyclotron, represented schematically in Fig. 276 has a huge magnet with a pole diameter of 181 inches and weighing 4,000 tons. The modulation frequency is 120 cycles/sec., and is produced by a large rotating condenser that forms part of the capacitance in the *r-f.* oscillator circuit. The *r-f* peak voltage is 15,000 volts. To reduce scattering of the ions by the residual gas molecules, the pressure in the vacuum chamber is reduced to about 10^{-6} cm. of mercury. Each ion makes 10^4 revolutions before attaining the maximum energy. On account of this great number of revolutions, the electrostatic focusing, which is so important in the fixed frequency cyclotron, is net essential for the synchro-cyclotron and hence only one dee is employed. This feature makes the ion beam much more accessible for experiments.

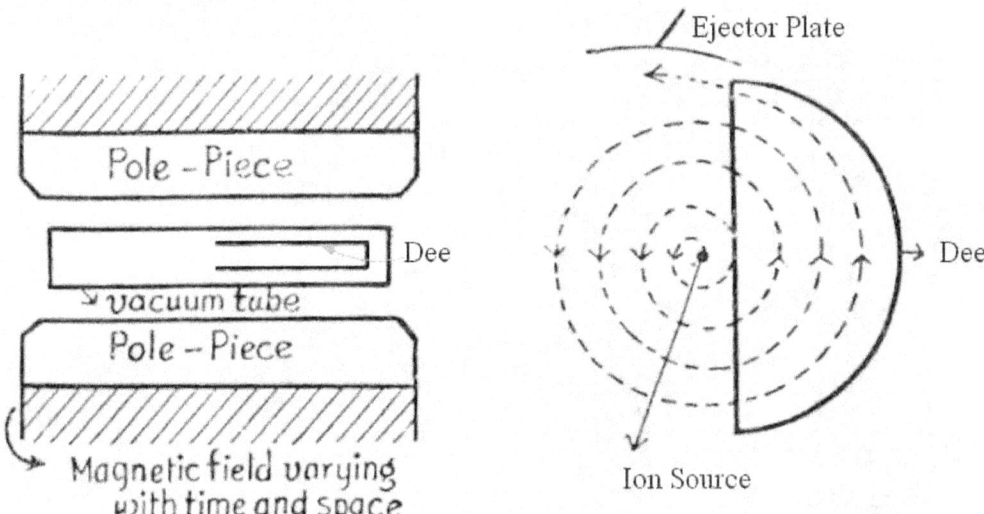

Fig. 276. Synchro-cyclotron.

The ejector plate with a huge positive voltage pulse deflects the ion groups on the target. This synchro-cyclotron can accelerate deuterons to 200 MeV, α-particles to 400 MeV and protons to 350 MeV. It has already been used in the production of mesons (very high energy particles discovered in cosmic rays) in the laboratory. It is now being rebuilt to accelerate protons to 750 MeV. There exist a few other synchro-cyclotrons in the United States of America.

The betatron or the induction accelerator

The principle on which this electron accelerator operates, has already been stated.

- It involves a device that is free from the difficulty of relativistic variation of mass, which prevents the use of the cyclotron for accelerating the electrons.

- It employs no electric field; but the magnetic field itself is alternating and produces an inductive electromotive force on the circular motion of the electrons by which they are accelerated—hence the name "induction accelerator".

Betatrons accelerating electrons up to 100 MeV are in use since 1941, while the upper limit attainable with them is about 500 MeV.

Some very *important results* have already been obtained *with the betatron, e.g.,*

(i) *transmutation* of Cu into Ni and Ag into Cd and Pd;
(ii) *mesons* have been produced by bombarding atoms with the 100 MeV electrons;
(iii) *production of visible light* in the betatron, which confirms the postulate of the classical electromagnetic theory that in large scale phenomena an electron revolving in a circular orbit should radiate energy, which has been contradicted by the quantum theory referring to the small scale phenomena.

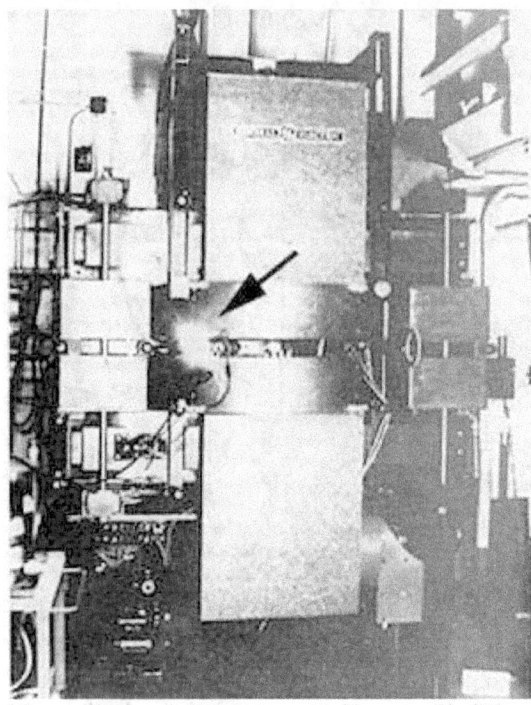

The radiation is seen as a small spot of brilliant white light by an observer looking in to the vacuum tube

The GE team. From left to right: Langmuir, Elder, Gurewitsch, Charlton and Pollock

Elder, Gurewitsch, Langmuir and Pollock, working with G. E. C. betatron, found that above 30 MeV visible light was produced by the rapidly circulating electrons in accordance with the classical theory. As the energy of the electrons was increased, more and more was radiated in this way and eventually the machine ceased to function efficiently.

The **proton synchrotron,** an apparatus which accelerates protons through hundreds of thousands of small impulses, is very similar in principle to the betatron. The protons are made to circulate in a ring-shaped vacuum tube under the influence of a magnetic field and are accelerated once in each revolution when they pass between electrodes connected to a high frequency generator (Fig. 277).

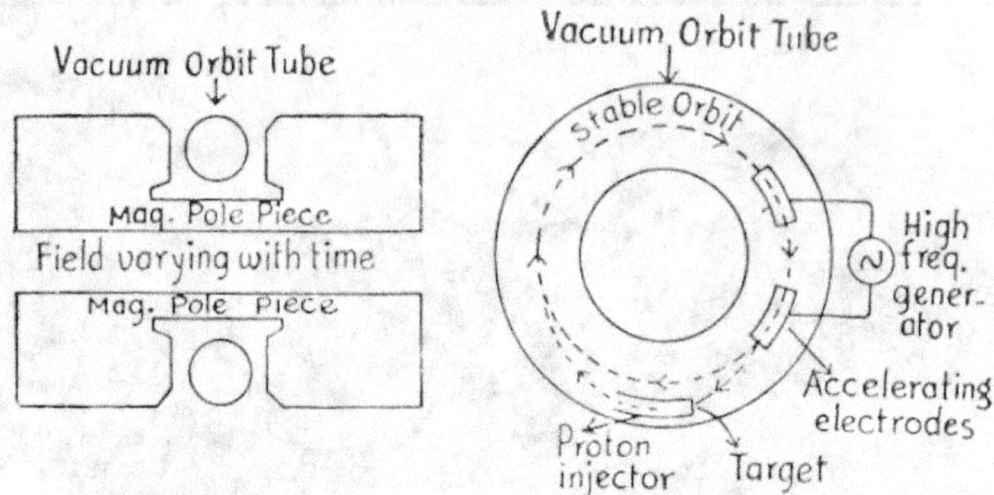

Fig. 277. Proton synchrotron.

The field strength and the frequency of the oscillator are simultaneously varied in such a way that the particles travel in a constant, stable orbit and arrive always at the electrodes when the applied voltage is in the right phase for acceleration. The vacuum chamber, instead of being of glass, as in most betatrons or of metal as in cyclotrons, is sealed by stretching sheets of non-porous rubber over corrugated strips of stainless steel.

Whereas synchro-cyclotrons appear to have an upper limit of about 750 MeV, the proton-synchrotrons are able to build up voltages in the billions (1000 millions) without disturbing the stability of the paths traveled by the particles, or impairing the intensity of the beam. First suggested by McMillan of California (1945), a 1300 MeV synchrotron has been constructed at the University of Birmingham. In U.S.A. one of 3000 MeV (3 BeV) the Brookhaven **cosmotron** is in operation at the Brookhaven National Laboratory in Long Island, and another of 6 BeV, the **bevatron,** has started functioning early in 1955 at the University of California. It is believed that this machine would enable experimenters to create protons and neutrons.

Similarly high energy mesons, found so far only in cosmic rays, could be produced in pairs when protons having energies of the order of 600 MeV are made to bombard nuclei.

Thus scientists are getting very near at producing inside the laboratory such energetic particles that are bound to lead to a rapid advance in the knowledge of the nucleus.

General view of the Brookhaven cosmotron

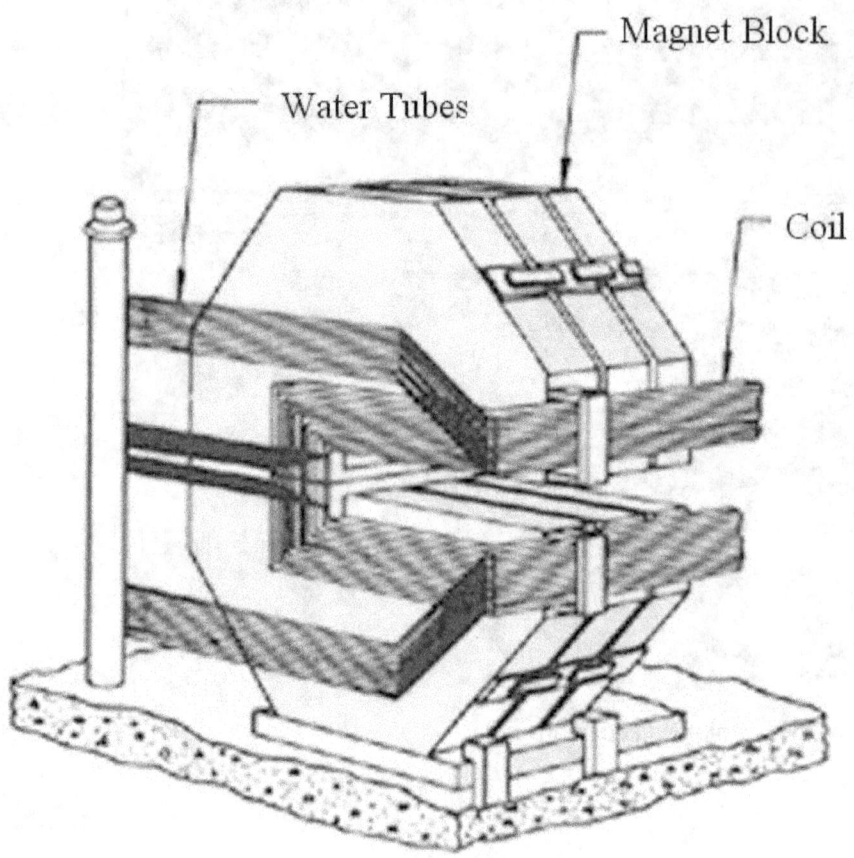

Water Tubes
Magnet Block
Coil

The U.S.A.E.C. has approved, in 1954, the design and construction at the Brookhaven National Laboratory of a *25 BeV alternating gradient synchrotron,* capable of producing beams of protons of energies up to 25 billion electron volts, at an estimated cost of 20 million dollars. This new ultra-high energy particle accelerator will use a series of alternate strongly converging and diverging magnetic fields to confine the proton beam in a tube of relatively small cross-section. This focusing effect allows the production of high-energy beams with smaller electromagnets and related equipment than would otherwise be possible. The construction of the new machine was completed in 6 years. Then, the most powerful proton synchrotron in the world was in Russia, where a 10 BeV machine was built in the Physical Laboratory of the Academy of Sciences of the U.S.S.R. and put in operation in April 1957.

The cavity and guided wave linear accelerators
Although the original linear accelerator was soon superseded by the more energetic circular accelerators, such as the cyclotron and its still more powerful modifications described above, yet, of late, the linear accelerator with ingenious improvements has been reintroduced to take its place in the class of proton synchrotrons and betatrons. These modern linear machines are able to deliver a well-collimated and intense beam of protons of accurately regulated energy, though they cannot touch the very high values of energy

developed in the proton synchrotrons. As for accelerating electrons, they are found to be far superior to the betatron, nay more, they can reach higher energies and deliver large currents than any other machine.

Two forms of the linear accelerator have been developed, the one suited for accelerating protons, known as the *"cavity"* accelerator and the other designed chiefly for accelerating electrons, called the *"guided wave"* accelerator.

In the cavity type, the intensity of the ionic current is greatly increased by reducing the power losses, with the help of the cavity resonator technique (first utilized in radar) of generating high radio-frequency voltages. The metallic cylinders of the linear accelerator known as the "drift-tubes" are encased in a cavity resonator, to which power is supplied by a number of high-power radio transmitters. The cavity transforms this radio energy into an alternating electric field, which makes currents rush back and forth across the gaps between the ends of the drift tubes. At any instant the direction of the accelerating force is the same in all the gaps. If the structure is so designed that a group of particles takes the time of one full cycle of the alternating voltage to travel through each drift-tube, then the particles will always find themselves in an accelerating field when they cross a gap.

The Berkeley linear accelerator which has been in operation since 1954, is constructed on this principle. It increases the energy of protons from a 4 MeV Van de Graaff injector to a final energy of 31.5 MeV. The accelerator consists of a cavity 40 feet long and 39 inches in diameter, excited at resonance in a longitudinal electric mode with a radio-frequency power of about 2.1×10^6 watts peak at 202.5 Mc. Acceleration is made possible by the introduction of 46 axial ``drift tubes'' into the cavity, which is designed so that the particles traverse the distance between the centers of successive tubes in one cycle of the rf power. The proton bunches are longitudinally stable as in the synchrotron, and are stabilized transversely by the action of converging fields produced by focusing grids. The electrical cavity is constructed like an inverted airplane fuselage and is supported in a vacuum tank. Power is supplied by 9 high-powered oscillators fed from a pulse generator of the artificial transmission line type. Output currents are 3×10^{-7} ampere, average, and 60 μa, peak. The beam has a diameter of 1 cm and an angular divergence of 10^{-3} radian.

A 70 MeV machine of similar design is in operation at the University of Minnesota. In England, the British Atomic Energy Research Establishment at Harwell is building a 600 MeV proton machine.

In the "guided wave" type which is particularly applicable to electrons, an electromagnetic wave is made to travel along a long continuous cylindrical pipe. The electric field generating the wave is produced by a high-power (20 million watts) and very high frequency (3000 Mc) klystron generator. Of the different kinds of traveling waves used, a very interesting one is the so-called TM wave, in which the electric field has always a component parallel to the axis of the pipe. In order to synchronies the speed of the traveling wave with that of the particles, carefully designed discs of conducting material, known as "corrugated wave guides" are arranged at regular intervals along the pipe. The wave thus suitably slowed down carries along with it the particles that happen to be in its path to steadily higher velocities. The linear accelerator built at Stanford University, America, is of this type. The cylindrical pipe is 220 feet long; 20 million watts of power from a klystron generator is injected at regular feed points, 10 feet apart along the pipe, so that the total power input is about 400 million watts. It is capable of accelerating electrons up to 600 MeV and more.

TECHNIQUES USED IN THE ANALYSIS OF TRANSMUTATION

The second great problem in the experimental study of artificial transmutation, which is more important than even the production of powerful atomic projectiles, consists in devising means of detecting the fragments which fly forth from the bombarded nuclei and of studying their nature, energy and direction. For, it is by this process alone that one can decide whether transmutation has really taken place and even understand the inner mechanism of transmutations by the knowledge obtained of the energy liberated or absorbed during the reaction, of the nature of the nuclei formed, etc.

There is no *one single technique* which is capable of furnishing at one stroke all the required data. Several different methods must be taken together to arrive at sure conclusions. Of these methods, some analyze the statistical effect of transmutation, while others indicate single elementary processes. Some measure the energy alone without giving any indication about the nature or direction of the particles of transmutation. Others permit the quantitative counting of the number of particles set free. Still others furnish photographs of the tracks of these particles with the possibility of detecting their nature, energy and direction. Hence, the methods that are described below are not exclusive of one another, but complementary.

Fortunately the particles of transmutation to be detected are limited in kind: protons, deuterons, α-particles, electrons, positive and negative, (charged particles), neutrons (uncharged particles) and γ-rays (electromagnetic radiation). The methods devised concern themselves chiefly with the study of charged particles. Neutrons are detected indirectly by means of the charged particles, such as protons or more massive nuclei which they chase before them by impact, while γ-rays by their secondary effects, such as photoelectrons, Compton electrons and pair production.

The different techniques developed for analyzing the results of nuclear bombardment may be classified under the following heads:

(1) Scintillation method.
(2) Chemical separation method.
(3) Magnetic spectrograph method.
(4) Ionization chamber method.
(5) Counter method.
(6) Cloud chamber method.
(7) Photographic emulsion method.

As most of these methods were devised even before artificial transmutation was discovered, chiefly in connection with the study of natural radioactivity, we have already dealt with several of them in their appropriate places. But many of them have been modified and improved under the impetus of modern alchemy to such an extent that they bear little resemblance to their simple and primitive forerunners. Hence, in the present account of these methods, we shall limit ourselves to such of their special features as have reference to their employment in artificial transmutation.

METHOD OF SCINTILLATIONS

The principle of this method has been given earlier in this book. It is a simple and direct method which enables charged particles to be counted individually. It has also an historic interest, as the one by which the first proof of artificial transmutation was obtained. The fluorescent materials commonly used for the screen on which the scintillations are observed are barium platino-cyanide, calcium tungstate and zinc sulphide. The duration of a scintillation is of the order of 10^{-4} sec.

Although Rutherford was able to establish the accuracy of this method by comparing the results obtained with it and those given by more refined methods, such as the Geiger counter, and an experienced observer can count from 90 to 95% of the impinging particles, yet it is a long, tedious and nerve-racking process to count thousands of dim flashes for hours in dark rooms. Further, though the method may be well suited for counting α-particles it can hardly be used in the case of protons and it is almost impossible to use it for counting electrons.

If the fluorescent screen is bombarded with the γ-rays, often produced during transmutations, electrons are liberated from the fluorescent material and the surrounding matter, which make the screen shine all over with a feeble glow, against which the flashes made by the particles to be counted are difficult to discern. Such are the drawbacks of the method; yet, on account of the extreme simplicity of the technique which has also the unique advantage that no barrier, not even a gas, need intervene between the detecting screen and the source of the fragments, it is still used, but in a highly refined form, known as the *scintillation counter,* as we shall see presently.

METHOD OF CHEMICAL SEPARATIONS

Methods of chemical analysis, though delicate, are not fine enough to detect directly the particles with which we are concerned here. The use of certain standard chemical separation techniques, however, has materially assisted in the interpretation of the results of transmutation, chiefly in the case of induced radioactivity. A good guess as to the nature of a nuclear reaction makes it possible to separate the possible products and by observing in which of the separated residues the activity lies, the *nature* of the element produced in transmutation can be determined. This is usually accomplished by employing a radioactively inert element of the kind to be separated, as a carrier, in order to have a sufficient quantity to work with. This method was introduced in 1934 for α-induced radioactivity by Curie and Joliot and for neutron processes by Fermi. Positron activity obtained by deuteron bombardment has been studied in a like manner by McMillan and Livingstone.

A valuable modification was initiated by Szilard and Chalmers, when they realized that chemical bonds could be broken by the recoil of the activated atom and so the free element could be released from a chemical compound. If the element in question is bombarded in the form of a compound which, once dissociated by recoil action in the process, is not reproduced spontaneously, the activated atoms remain in a chemical state different from the bulk of the inactive substance and can be separated by ordinary chemical methods. For example, by bombarding chlorine in the form of an organic compound, chloroform, or as a chlorate, the radioactive chlorine produced remains in an atomic or ionic state and can be separated by the addition of a trace of chlorine ion and precipitation with silver nitrate.

Precipitation of the radioactive material from a large sample used as a target concentrates the activity into a small sample which can be brought close to the recording instrument and so increases the observable intensity. This separation method is particularly important in transmutations by slow neutrons, where the substance to be bombarded can be dissolved in the water which is used to slow down the neutrons.

MAGNETIC SPECTROGRAPH METHOD

This method, which has already been described in detail hi several places, with its highly developed technique, is used to measure the *energies* of charged particles as well as of γ-rays emitted by disintegrated nuclei to a high degree of accuracy.

IONIZATION CHAMBER METHOD

The ionization chamber method in the present highly improved form is a very sensitive electrical method of studying the *statistical effect* of nuclear disintegration, from which the *nature and energy* and even the *number* of the fragments flying off from the disrupted nuclei can be determined.

The **basic principle** that underlies the method is that charged particles in motion produce, in a gas ionization varying with their nature and velocity. Hence, an accurate estimate of the number of ions produced under given conditions, for instance, in air at N.T.P., enables one to know the nature of the ionizing particle as well as its energy. Thus for example, from independent experiments on ionization current produced in an ionization chamber by different charged particles, *the number of ion-pairs formed in air at N.T.P. per cm. path is obtained as follows:*

1. *Electrons*	Fast	(2 Mev)	45 pairs about	
	Less fast	(60 Kev)	200 ,, ,,	
2. *Protons*	Fast (from transmutations)		3,000 ,, ,,	
	At the end of their range		15,000 ,, ,,	
3. *α-particles*	Fast	(8 MeV)	25,000 ,, ,,	
	At the end of their range		75,000 ,, ,,	

From these data, we see that α-particles produced in air at N.T.P. ion-pairs varying from 25,000 to 75,000-per cm. range, according to their speed, the faster the particle the less being the ionization produced.

Protons produce four to ten times less of ion-pairs, while electrons much less still, as compared with the others; but they also form a relatively great quantity of ion-pairs at the end of their range. Now, in the disintegration experiments, if we measure the number of ions produced by the fragments of the disrupted nuclei, we can easily conclude to the nature of the fragments, whether α-particle, proton or electron.

Again, it is known from other independent experiments that

a fast-flying charged particle loses energy, on an average, of 30 to 35 electron volts for every ion-pair it produces.

Hence, the number of ion-pairs formed by a fast particle going through a gas is about a thirtieth of the number of electron volts which it loses in its transit. Here we have a means of estimating the energy of the particle from the ion-pairs it produces. It is, however, important to measure the ionization at a well-defined position of the range, on account of the peculiar nature of the ionization curve (Bragg type) indicating an increase in ionization at lower speeds with a sudden drop at the end.

Technical details

The ionization chamber, as regards its essential features, has already been described. Mention will be made here, therefore, only of the additional special technique involved in its use with transmutation experiments. The improvements made consist chiefly in the following points:

(i) *Maintenance of a constant high potential difference* (*P.D.*)

The potential difference must be high enough, so that the electric field between the electrodes reaches approximately saturation, in which state most of the ions formed reach the electrode before they could recombine.

(ii) *The dimensions of the chamber as well as the nature and pressure of the gas used in it* are determined by the type of radiation to be measured. Thus, for example, in the case of α-particles which are completely absorbed by a few cms of air, the chamber is usually filled with air at atmospheric pressure and its linear dimensions need not be greater than the range of the particles, since large size would not increase the ionization produced. For β- and γ-rays, which are much more penetrating, larger chambers have to be used. In order to increase the ionization in the case of γ-rays, it is found useful to replace air with a more intensely absorbing gas, *e.g.,* methyl iodide, or to increase the pressure to 30 to 50 atmospheres.

(iii) *The sensitivity or quick collective power of the chamber* can, in general, be increased by the use of highly absorbing gases at high pressures. It must be noted, however, that it does not increase in proportion to the density, since at high pressures there is an increasing degree of recombination of the ions. This can be avoided to a large extent by the use of very pure rare gases, such as argon, in which the recombination effect is small up to very high pressures.

(iv) *The sensitivity of the electrical device* which measures the feeble ionization current has been progressively improved so that at present it is possible to detect the primary ionization due to even single particles.

Electroscopes and electrometers

In the initial stages, the gold-leaf electroscope of the tilted type of great sensitivity or the quadrant electrometer, whose sensitivity was increased to a high degree either by increasing its voltage sensitivity or by decreasing its capacity, was used. They were able to detect and even measure the total charge of a few thousands of ions without amplification.

String Electrometer

The next step in the development was a modification of the electroscope, designed by Lauritsen, where the gold leaf was replaced by a tiny metal-coated quartz fiber, about 5 microns in diameter and 6 mm. long, the movements of which could be observed by a microscope, or better still, recorded by photography on a moving film. This arrangement known as the string electrometer possesses the advantages of *small size* which enables the fiber to return to its normal position very quickly after each displacement *high sensitivity* due to the low capacity of the small fiber and *special ruggedness* lending itself to any orientation. The primary ionization produced by single particles can be readily measured with this device. Each ionizing article is revealed by a jump on the photographic film, the size of the jump being a measure of the number of ions produced by a particle.

But the *major difficulty* in this method, as in the older ones, though minimized to a good extent, is the *lack of quickness of response to particles that follow one another in rapid succession.*

Linear amplifier

With the advent of radio-technique of amplification by thermionic valves, the above difficulty was overcome to a great extent by a device, known as the linear amplifier, which augmented without distortion the feeble ionization current from a single particle to a value which would easily register on an oscillograph or an electrical counter. Originally developed by Greimacher, this method has been brought to a high degree of perfection by Ward, Wynn-Williams and Cave of the Cavendish Laboratory, by Dunning of Columbia and by Maurice de Broglie of Paris. The apparatus consists of two parts:

(1) The ionization chamber with one or two suitable vacuum tubes destined to receive faithfully the very small electric pulse produced by the primary ions in the chamber.

(2) The amplifier proper, with three to five stages of resistance-capacity coupled valves of great power, and an oscillograph or counter at the end.

The ionization chamber is ordinarily shallow, of the order of a centimeter, well protected from noise and vibrations. Likewise, the valves in the first amplification stages are shielded from outside disturbances. The overall voltage amplification is of the order of 10^6 or 10^7. The amplification is linear, *i.e.,* the final amplified pulse registered on the oscillograph is strictly proportional to the charge due to primary ionization. Under these conditions, the oscillograph record indicates the passage of each particle by a *linear jump,* the length of which is proportional to the number of ions which the particle produced in crossing the ionization chamber. Since this number depends on the nature and mass of the ionizing particle as well as on its velocity and consequently its energy, the method is capable of giving valuable information about the *nature* and *energy* of the particles.

With proper amplification, particles that produce a primary ionization of 500 and more ions can be detected. Those that produce less number of ions, such as fast electrons, cannot be detected. Hence, the method is well suited for the detection of heavy ionizing particles such as α-particles, protons, recoil nuclei, even in the presence of β- and γ-rays, since these latter produce not single measurable pulses but only a background disturbance through the fluctuations in their ionization. The effect of this disturbance is smaller when the resolving power is higher and consequently an intense electric field is applied to the ionization chamber to ensure quick collection of the ions.

Further, different groups of the same kind of particles emitted in a transmutation can be readily detected, since in the oscillograph record, the length of the linear trace gives immediately the speed of each particle at the moment of its passage through the ionization chamber. Thus, for example, in a disintegration which gives rise to several groups of protons, when the ionization chamber is placed at a certain distance from the target bombarded, the protons emitted are recorded by the pulses on the oscillogram. Now protons of the same group give pulses of the same amplitude. Thus, on a single band one can read the total number of particles which has traversed the chamber and the proportion of each of the groups. On increasing the distance of the chamber from the target little by little one finds on each band the different groups with their amount and one can thus come to a correct conclusion by a comparative study of the different bands.

In the research of nuclei projected by the collision of neutrons with atoms, this method furnishes very precious information, since each recoil nucleus is recorded with indication of its energy. If the chamber walls are coated with Li or B, energetic α-particles are emitted in neutron disintegration and will be recorded by the amplifier.

(v) *Exclusion* of *parasitical effects*
Background effect due to electrical leakage, cosmic rays and contamination of radioactive materials should be minimized, as far as possible, since they cannot be completely eliminated. The use of high pressure in the ionization chamber, lead screens surrounding the chamber, etc. serve this purpose to a certain extent. The background ionization still remaining can be estimated by an independent experiment and deducted from observations to get accurate results.

(vi) *Use of the cathode ray oscillograph*
If a modern cathode ray oscillograph be substituted in a place of the older type of oscillographs, the results are very much enhanced. The sharp kicks in the electronic beam on the screen record very faithfully the passage of particles in the ionization chamber even when they follow each other in rapid succession. The phenomena that occur on the screen can be photographed and analyzed at leisure.

The ionization chamber method was first applied to the analysis of transmutation by G. Hoffman and his colleague Pose, The latter was able to observe fragments from aluminum nuclei bombarded by α-particles, which produced as few as three thousand ion-pairs, thanks to the very sensitive electrometer devised by the former. Fermi and his co-workers used a rather large ionization chamber, filled with CO_2 at a high pressure of several atmospheres and provided with a thin aluminum window, in the measurement of neutron-induced radioactivity in various elements. The ionization chamber with the string electrometer has been used for the detection of γ-rays and neutrons, where the interchangeable walls of lead and cellophane of the chamber make it differentially sensitive to the gamma and neutron radiations. The linear amplifier has been used by Dunning, Maurice de Broglie and others in transmutations where α-particles, protons and neutrons are involved, as already stated. Using the Wynn-Williams 'scale of eight" thyratron system to operate a mechanical counter at the final stage of amplification of the linear amplifier, it has been possible to count particles which enter the ionization chamber at the rate of about 5000 per minute.

COUNTER METHOD

This automatic electric device, as it was used today, with its auxiliary circuits of multi-electrode vacuum tubes, and latter semiconductor devices, of high amplifying and resolving powers, is perhaps the *most efficient and accurate method of quantitative counting of individual particles occurring in extremely rapid succession.* There are, in actual use, three types of counters, built on three different basic principles, *viz.*

(i) the *Geiger counter* depending on the ionization of gases,

(ii) the *scintillation counter* working on the scintillations produced by fluorescent materials,
And
(iii) *crystal counter* where photoconductivity is utilized.

I. THE GEIGER COUNTER

This type of counter was first devised by Rutherford and Geiger, in 1908, for counting the number of α-particles emitted in a given time by a known quantity of a radioactive substance in connection with their experiment on the determination of the charge carried by a single α-particle. In 1928, it was developed further by Geiger and Muller, which resulted in a great increase of speed of counting as well as in the adaptability of the instrument to detect all kinds of radiation.

Principle
The Geiger counter is essentially a simple ionization chamber, very compact on account of its small dimensions and very efficient due to the automatic amplification, inside the chamber itself, of the weak ionization current produced by the passage of a single charged particle. This internal amplification is secured by the application of a carefully adjusted strong electric field between the two electrodes of the chamber. In an ordinary ionization chamber, the voltage on the electrodes is just large enough to collect the ions produced by the passage of a charged particle.

Following a suggestion made by Gorton, in 1905, for X-ray ionization, Rutherford and Geiger increased the voltage on the chamber to such a high value that the electrodes in the chamber just did not spark over, so that even a single charged particle passing through the chamber was enough to set the chamber off and cause a discharge. Under these conditions, the relatively few ions produced by the particle, finding themselves in a strong electric field, are immediately and violently accelerated towards the electrodes. In their passage they produce more ions by collisions with neutral molecules, until the cumulative effect results in a discharge, large enough to be measurable.

It is essential that the discharge caused by the passage of the particle should not become permanent, but must be quickly and automatically extinguished. This is achieved either by the gas in the chamber or by a suitable adjustment of the accompanying circuit or by both. For, then only the system will be reset in proper condition for responding effectively to successive particles. The counter acts somewhat like an *automatic* rifle which resets itself every time it has been fired by the 'trigger' action of the passing particle.

There are a number of empirical rules (some partially understood, some not at all) about the size and the shape of the chamber, the proportioning and conditioning of the electrodes, the nature, purity and density of the gas and the intensity of the electric field, which makes the problem of the inner mechanism of the counter operation difficult for a complete solution. Given, however, the great value and the frequent use of these counters, we shall state briefly some of the important practical details about their construction and operation.

Construction

There are two kinds of Geiger counters in use, called the *point counter* and the *tube* or *Geiger-Midler counter*. They are differentiated chiefly by the shape of the electrodes.

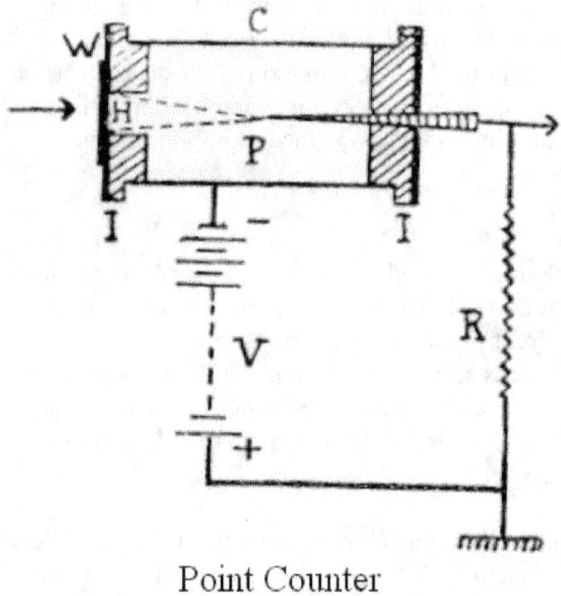

Point Counter

Fig. 278. The Geiger point counter.

In the **point counter** (Fig. 278) the outer electrode is a hollow metallic cylinder C provided with two insulating plugs I's. In one of these plugs there is a circular hole through which the particles are allowed to enter the counter. Through the other plug passes the inner electrode which consists of either a sharp needle or a small sphere at the end of a thin wire P and is held along the axis of the cylinder. The apparatus can be operated at atmospheric pressure which makes it valuable for certain experiments. If it is, however, to be worked at reduced pressure, the circular hole H is closed by a thin foil W. The outer electrode is raised to a high positive or negative potential (from 1000 to 3000 volts). The inner electrode is grounded through a very high leak resistance R (from 10 to 5000 MΩ) which is in series with the source of the high tension. In the strong electric field thus established between the electrodes, a very relatively few ions formed by the passage of a single charged particle through the counter produce, by multiple impacts and secondary effects near the surface of the electrodes, what is essentially a cumulative ionization resulting in a discharge. Although the reasons why the discharge is interrupted have not been clearly ascertained, yet it may be said that due to the presence of the high resistance in the circuit, the discharge lowers the voltage so that the current ceases or is "quenched". Had there been no resistance, the high voltage used would have been enough to run a steady discharge after the passage of the first ionizing particle. Without any further ionization, the voltage then builds up again only gradually to its original value through the high resistance and the counter is ready for the next firing.

Under good working conditions, almost every particle that produces ions in the conical space defined by the point of the inner electrode and the circular window gives rise to a discharge. The maximum current intensity varies rapidly with the applied voltage and can be raised to a value of 10^{-5} ampere.

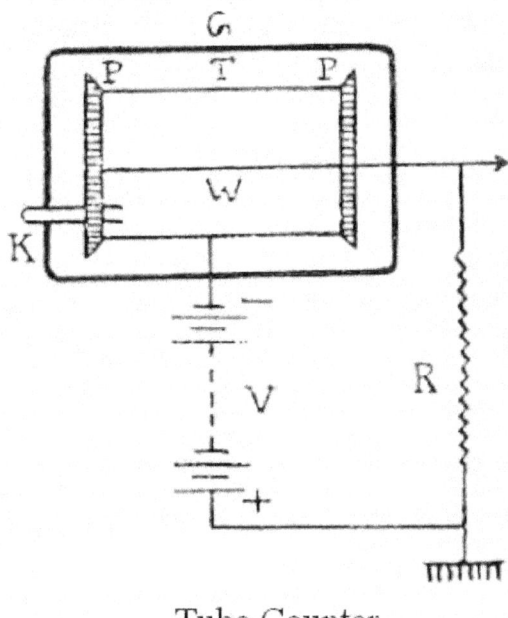

Tube Counter

Fig. 279. The Geiger-Műller counter.

In the **tube** or **Geiger- Műller counter** (Fig. 279) the outer electrode is a metal tube T (brass or nickel), 1 to 5 cms in diameter, 10 to 50 cms long and the inner electrode a very fine wire W, 0.1 to 0.5 mm. thick, usually of tungsten, stretched along the axis of the tube and well insulated from it by means of the ebonite plug PP. The electrodes are usually sealed in a thin-walled glass tube G. The gas contained in the counter may be either air, hydrogen, argon, or even a mixture, at a reduced pressure of 2 to 10 cms of mercury. The counter is evacuated and filled with the desired gas at the chosen pressure with the help of a glass tube K, which is then sealed off. The introduction of a gas mixture in the counter not only lowers the values of the high voltage to be applied but also aids in quenching the discharge. A strong electric field is established between the two electrodes by the application of a steady high potential. The external circuit contains a high leak resistance R as in the previous case.

The design of the outer tube electrode as well as the nature of the gas employed vary according to the different uses of the counter. Thus, for instance, to detect γ-rays, the tube has pretty thick walls (1 to 3 mm.), although the practice is to keep the amount of material at a minimum in order to reduce the contamination background Secondary electrons released chiefly from the *walls* of the tube produce the discharge.

When intended for the study *of β-particles* a thin aluminum window, about 0.1 mm. thick, is used to admit the radiation. Heavy particles such as *protons* or *α-particles* are readily detected with a thin-walled tube counter operating at atmospheric pressure.

For the detection of *neutrons,* the counter is filled with hydrogen, the protons chased by the impact of neutrons with the hydrogen atoms constituting the 'triggering' particles; or better still, the tube is made of silver, easily activated by the neutrons, in which case, the electrons of disintegration of the radioactive substance artificially produced by the neutrons will operate the counter.

The tube counter has several advantages over the point counter: viz.,

(1) the duration of the discharge is reduced,
(2) the sensitive region extends throughout the total space between the electrodes,
(3) a wider range of voltage is obtained for the sensitive condition
and
(4) it is not so readily disturbed by discharge.

Operation
For the proper and effective functioning of the counter the following points are of great importance:

(a) *The high tension required to set the counter in the sensitive state*

> The voltage applied across the counter must lie within a definite range, often pretty narrow; if it is lower, the particles do not produce discharge; if higher, the discharge becomes continuous.

In order to fix the correct margin of the high tension, it is important to study the behavior of a given counter when the applied potential is steadily increased and to obtain its characteristic curve. For this purpose, some standard source in used and the number of output impulses is noted for different values of the voltage. When the data are plotted a curve as shown in Fig. 280 is obtained.

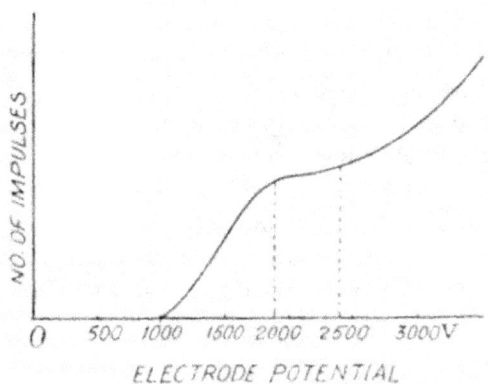

Fig. 280. Sensitivity curve of a counter.

It is seen that no impulses occur until a certain minimum voltage of about 1000 volts is applied, which means that there is no secondary ionization until a pretty high value of voltage is reached. Then, there is a region, where the number of impulses increases almost *linearly* with the voltage. It is known as the *multiplication region* where the total number of ions produced is proportional to the number of original ions. At this stage, the impulses are still very small, the *multiplication factor, i.e.,* the ratio. between the total current produced in the discharge and the primary ionization current due to the particle is between 10^3 and 10^4. Under these conditions, the counter can function as a *proportional counter* and is very useful to count heavy particles, such as α-particles and protons, even in the presence of *γ*- or *β-radi*ation.

As the applied potential is further increased the number of impulses remains constant over a certain region indicated by the *fiat part of the curve*. It is essential that such a plateau exists if accurate quantitative results are desired. In this region, the magnitude of the impulses becomes independent of the amount of original ionization and is a function only of the potential, nature of the gas, the resistance R and the geometrical conditions of the apparatus. The impulses due to strongly or weakly ionizing particles (α or β) cannot be distinguished in this region. Hence, the *counter set in this portion responds to all kinds of radiation, independent of their nature and becomes an efficient device for quantitative counting of particles,* since the existence of the saturation region which may extend through several hundreds of volts, indicates the *number of impulses per unit time remains constant for a given source, and any observed change in that number can he due solely to a change in the intensity of the source.* A further increase in the voltage produces spontaneous discharges which must be avoided by all means.

(b) *Devices employed for high speed counting*

As the *chief use of* a counter is *to register separately particles which enter it in very quick succession, means* were sought to achieve this purpose. In general, the *resolving power* of a counter depends upon both the *duration of the discharge* and *the time of recovery;* the greater will be the resolving power, the smaller the value of these two time factors. These depend, among other things, also upon the capacity and leak resistance which therefore must be as small as compatible with -satisfactory operation of the counter.

In the first stages of development of the counter technique, the counting of particles was done, by connecting the inner wire electrode to a short period string electrometer and registering the displacements of the string caused by the discharges on a moving film with a photographic arrangement. The number of counts thus recorded was not high.

With the advent of radio technic, rapid progress in the rate of counting has been made. Vacuum tube circuits have been designed which, when used with the Geiger counter, can resolve and count individual particles, even though they occur in extremely quick succession.

The conventional circuit

One of the first of such auxiliary devices is illustrated in Fig. 281.

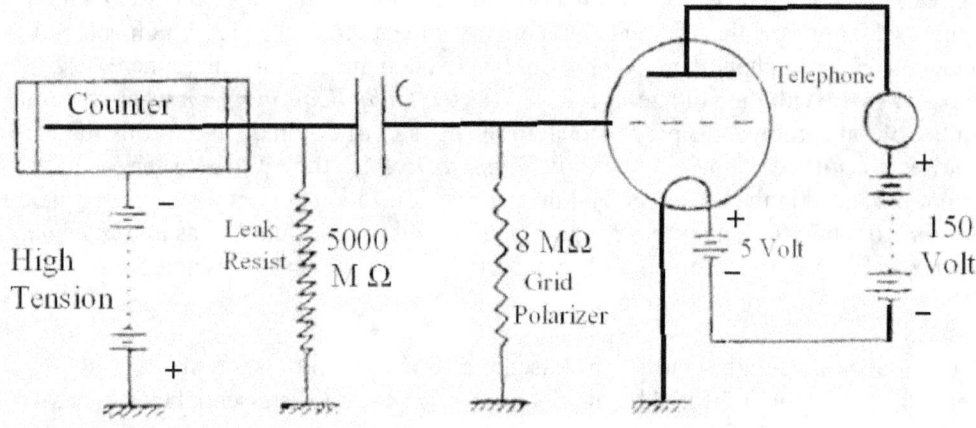

Fig. 281. Conventional circuit of a G.M. counter.

The central wire of the counter which is positive with respect to the outer tube electrode raised to a high negative tension is connected through a capacity C to the grid of a tube amplifier. The leak resistance is of the order of 5000 MΩ. A smaller resistance of about 8 MΩ is put between the capacity and the grid to polarize the latter. The momentary current produced at each discharge of the counter is amplified in the vacuum tube so that it is capable of operating a telephone in the plate circuit, and an audible "click" is heard at each discharge. Thus one can hear the passage of each particle through the counter. But the most practical method of counting consists in amplifying further the discharge current with a second stage tube amplifier so that it is strong enough to operate a mechanical counter. With such an arrangement, it is possible to count at the rate of about 3000 impulses per minute. It is evident that the high value of the leak resistance makes the time constant of the circuit large and in consequence diminishes the resolving power which, ultimately, sets the upper limit to the counting rate.

The Neher-Harper circuit

In 1936, Neher and Harper devised a type of amplifier circuit which besides amplifying the impulses of the Geiger counter, improved its resolving power to a great extent. In their arrangement, shown in Fig. 282, the inner wire electrode of the counter is connected not to the grid, as in the previous case, but to the plate of the amplifier tube, a pentode, and raised to a high positive potential. The outer cylinder electrode is connected to the grid. The high tension for both the counter wire and the plate of the amplifier is applied at E_p, which can be measured with an electrostatic voltmeter V_2. The value of the grid resistance R_g, in general, depends upon the physical constants of the counter and may vary from 1 to 10 MΩ. A resistance of 3 MΩ is suitable for use at R_p. The control grid potential E_g, which becomes critical under certain conditions, is regulated by the potentiometer P and the voltmeter V_1. The screen-grid potential E_g, is not critical and its value is kept constant at 45 volts. A coupling condenser C is introduced in the output circuit; the impulses obtained at the output may be further amplified by a second tube before passing them on to the recording circuit.

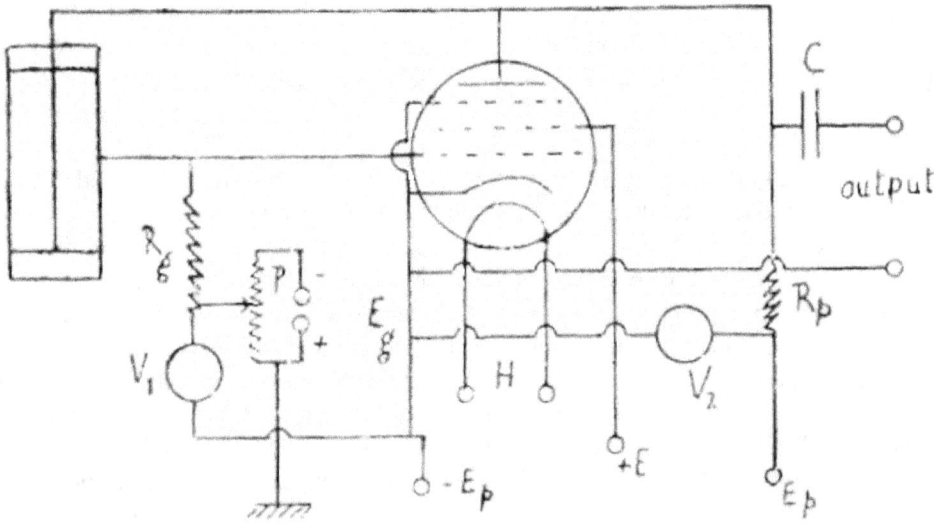

Fig. 252. Neher-Harper circuit.

Under normal conditions, *i.e.,* prior to the passage of an ionizing particle, the voltage E_g maintains the control grid sufficiently negative to prevent the flow of any appreciable. plate current through R_p. The potential E_p is adjusted to such a value that the effective voltage across the counter is just above the threshold value, i.e. the counter is assumed to be in the 'sensitive' state. The passage of an ionizing particle through the counter initiates a discharge between its electrodes. This discharge makes the potential of the control grid less negative, thereby increasing the plate current through R_p. If the change in the grid potential is of sufficient magnitude, the voltage drop in R_p will result in the effective counter voltage dropping below the threshold value. Thus the counter discharge is extinguished and the circuit recovers to normal conditions in a period determined by the time constant (CR).

The amplifying action of the pentode permits satisfactory operation with a total circuit resistance far below the value ordinarily employed, viz. of the order of only 1 to 10 MΩ, as contrasted with the 5,000 MΩ of the conventional circuit. This low value in the resistance results in a time constant small enough to permit counting at rates as high as 200,000 particles per minute. It is to be noted, however, that such high counting rates are never realized in practice. Several factors such as the physical constants of the counter, the characteristics of the recording circuit, etc. limit it to about 10,000 per minute, which shows still the tremendous progress made over the five per minute of Rutherford's first counter.

Wynn Williams scaling circuit

No mechanical meter, due to its inherent inertia, can achieve the fast counting rates realized with the Neher-Harper circuit. But this difficulty has been overcome by an ingenious arrangement of vacuum tubes known as the *scaling circuit,* first realized in practice by Wynn-Williams. By means of these circuits, only one impulse out of two, four, eight or more is registered on the mechanical meter. The probability that two impulses will come in such quick succession as not to be registered separately is thus much reduced,

In the scaling circuit devised by Wynn Williams, thyratrons are used as "inertialess relays" between the fast incoming particles and the comparatively slow moving mechanical recorder, so that counting speed is made independent of the inertia of the recording device and dependent only on the electrical characteristics of the circuit. Several units of two thyratrons each are arranged *in cascade,* so as to reduce the rate of counting by a factor of two per unit, until the final counting rate is sufficiently slow to enable a mechanical meter to be used. For most purposes, three such units are sufficient to give a speed reduction of 2^3 or 8 times and, in consequence, the arrangement is called a **"scale of eight".** In general, the scaling factor is given by 2^n, where n is the number the "scale of two" units used. For exceedingly high rates of counting, "a **scale of 256"** can be used, which has eight units.

(c) *Background effect*

The universal sensitivity of the Geiger counter, recording, as it does, identically all sorts of radiations, constitutes a source of difficulty which has to be eliminated or at least minimized in practice. The counter responds to cosmic rays and radioactive contaminations, so that the natural background is always present. Some idea of this background effect can be had from the following:

A tube counter 1 cm. in diameter and 4 cms in length gives a minimum of 10 to 20 impulses per minute, even when no ionizing source is used. This background effect can be reduced to about one-half by screening the counter with a few cms of lead.

Point counters have a lower background effect, because of the smaller extension of the sensitive region. They are therefore sometimes preferred for the measurement of α- or β-particles, while the tube counters, on account of their higher sensitivity, are generally used for measuring -γ-rays.

The Geiger counter, thus refined and fortified by auxiliary devices, has become an important and valuable instrument in nuclear research—a veritable "watch dog" of the atomic world.

2. THE SCINTILLATION COUNTER

This super-speed counting device is a modern and highly advanced version of the simple zinc sulphide scintillation screen first employed by Rutherford and his co-workers. By the use of suitable electronic devices, it has become possible to remove the great limiting factor imposed by the visual observation in the original scintillation method and the new technique has a capacity for counting that surpasses not only the comparatively slow rate of its prototype, but also the much more efficient Geiger tube counters. We shall first describe briefly the chief components of the modern scintillation counter and then state some of its important uses.

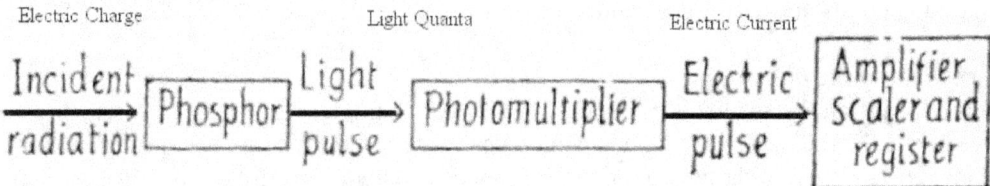

Fig. 283. The scintillation counter.

Components of the scintillation counter

The main parts are shown diagrammatically in Fig. 283. They are:

(1) A *phosphor, i.e.,* a suitable scintillating material. In addition to a few natural mineral (inorganic) crystals, such as zinc sulphide, calcium tungstate, etc., there are now available a great number of organic crystals, certain organic liquids and even plastics which scintillate when exposed to radiations and are called by the common name 'phosphors'.

The organic phosphors are all highly transparent and can often be produced as large single crystals; liquid phosphors can he had in large volume, which thus extends the field of application of the scintillation counter; plastic phosphors are non-fragile, optically clear and can be obtained in any required size or shape. The most important characteristic of these phosphors is that *the flash of light emitted by them is extremely short, the resolution time within the phosphor being of the order* 10^{-10} *sec.* This is one of the reasons which make the scintillation counter more efficient than even the tube counters.

(2) A *photomultiplier* which is an *electron multiplier* has already been described. The output pulse of a photomultiplier is very narrow, since all the electrons do not arrive together and in consequence the resolving time of the tube, *i.e.,* the shortest interval between two successive pulses which are registered separately, is of the order of 10^{-10}

sec. This is about ten thousand times shorter than the resolving time of a gas-filled counter, which is of the order of a microsecond. Hence, for work involving high counting rates or measurement of short times between the fast arrival of particles the photomultiplier is much superior to the Geiger counter. Of the two types of photomultiplier, the R.C.A. has a resolving time of 10^{-9} sec., while the E.M.I. 10^{-8} sec. The lower efficiency of the latter is due to the somewhat long and variable time of transit from one stage to the next; hence it is not so sensitive of the influence of stray magnetic field as the former, which is an advantage under certain circumstances of operation, *e.g.*, in the vicinity of transformers.

(3) *Electronic valve equipment* functioning as amplifier, scatter and register. Most parts of this component are common with those used with counters of other types, *e.g.*, Geiger counter. But, on account of the short resolution time of both phosphor and photomultiplier, the amplifier must have correspondingly 'fast' circuits which have been readily secured by the recent development of *transistors*.

In the actual set up of a scintillation counter with the above described components the following points are carefully attended to:

(*a*) *Choice of the phosphor*
This depends on the special uses of the counter. For example, when it is necessary to investigate the radioactive content of small weak sources, or of sources emitting radiation of low energy, a pretty large size crystal phosphor is taken, a small hole is drilled in it and the radioactive sample is placed in that hole. A more recent and better method is to dissolve the sample in a liquid phosphor, in which case practically all the emitted radiation is absorbed within the phosphor, thereby giving a high detection efficiency.

(*b*) *Mounting the phosphor* demands the following conditions: (*i*) Light other than that produced in the phosphor has to be excluded; (*ii*) the scintillations produced. have to be efficiently directed to the photocathode surface of the photomultiplier; (*iii*) the phosphor must be protected from mechanical damage and if hygroscopic, from moisture; (*iv*) the phosphor must be screened from extraneous radiation. Numerous types of phosphor-holder have been developed to meet these requirements, each depending upon the purpose for which the counter is designed. The problem of maximum light-collection from the phosphor has been solved by *either* silvering or better still, frosting all the surfaces of the phosphor, except that through which light passes to the photomultiplier.

(*c*) *Light guide* is a device to make the light produced by the phosphor reach the photomultiplier with as little loss as possible, chiefly when the photomultiplier has to be kept separated by a good distance from the phosphor, as in the case when a scintillation counter is used in conjunction with particle accelerators like the cyclotron, employing huge magnets. It is usually a glass tube containing water or a glass rod several feet long, at one end of which the phosphor is mounted and at the other the photomultiplier. The light from the phosphor is carried along the guide without much loss to the photomultiplier, thanks to the phenomenon of total internal reflection. The light guide is usually enclosed in a light-proof outer covering.

(*d*) *Voltage supply to the photomultiplier.* A constant voltage produced by a stabilized 'power pack' has to be supplied to the chain of resistors which divides the P.D. into the necessary steps to supply each dynode with its correct tension.

(*e*) *Discriminator in the amplifier-Scaler circuit.* When counting the pulses caused by the incident radiation, it is necessary to exclude the many smaller pulses due to the spontaneous emission of electrons from the dynode surfaces. For this purpose, a 'discriminator circuit' is introduced between the amplifier and Scaler circuits. It is an electronic device for sorting out the required size of pulse from all those passed on by the amplifier. Such discriminators are used chiefly in the analysis of a heterogeneous incident radiation, when the components are to be separated and their relative intensities are to be measured.

(*f*) *Scaler and recorder.* The scaling circuit is usually so arranged that the sorted pulses received by it are passed on 'in powers of ten' to the recorder. Different kinds of recording are in use: the total number of pulses may be indicated visually or recorded graphically; the count rate, *i.e.,* the number of pulses in unit time may be directly given by the recorder; in some cases the cathode ray oscillograph is used as the recording device.

Thus, the scintillation counter is not a simple apparatus, but an assembly of several contrivances which can be varied to suit the need. The overall resolving time attainable by an actual scintillation counter is about 10^{-9} sec., being limited chiefly by fluctuations in the transit times of electrons in the photomultiplier and by the decay time of the phosphor.

Some important uses of the scintillation counter
The scintillation counter fitted with any of the known inorganic phosphors has proved a very efficient *detector of protons, deuterons and α-particles,* even in the presence of a heavy background of β-.or γ-radiation, since the above-mentioned particles produce much stronger flashes than the electrons or gamma rays.

Likewise, *in the detection of γ-rays,* the scintillation counter with an inorganic phosphor, such as calcium tungstate, is much more efficient than other types of counter. This makes it well suited for measurements on weak sources, chiefly when the sources are a good distance away. For example, geological surveys for radioactive ores can now be made with this extremely sensitive counter from aeroplanes, unlike the usual tedious and difficult prospecting on the ground. It has also proved an asset in reducing the 'tracer' dose used in medicine to a minimum, on account of its high sensitivity.

For the *detection of β-particles* an organic phosphor counter is commonly preferred. When the source is weak, it can be put within the phosphor itself with best results, as mentioned earlier.

The scintillation counter has also proved quite good in the *detection of neutrons.* Since the organic phosphors used in the counter are rich in hydrogen, the neutrons entering the counter readily release by collision (chiefly when they are fast) protons which give rise to scintillations. An important technique used in this connection is known as the *'scintillation telescope'* in which a number of phosphors are arranged in a line to form a telescope that records only when a particle passes through all the phosphors according to. a 'coincidence technique'. The scintillation telescope has been found extremely helpful in the measurement of *extremely short periods* of *nuclear isomers,* of the order of 10^{-9} sec.

Recently, many types of scintillation counters have been devised in the *study* of *cosmic rays* for the detection of mesons and other unstable particles of very high energy. This has been possible due to the ability of the scintillation counter to respond specifically to events separated by extremely short intervals of time. The scintillation telescope forms naturally an important technique in these investigations, since the axis along which the phosphors are mounted can be rotated, so that the particles arriving in any direction relative to that of the primary beam can be studied.

But the most interesting and efficient type is the *liquid scintillation counter.* With the advent of liquid organic phosphors, it has become possible to use large volume liquid counters which are bound to throw light on problems not approachable by other methods. For example, sure counters have been used in the measurement of the energy spectrum of extremely fast moving charged particles met with in cosmic rays or produced by the modern powerful particle accelerators, such as the synchro-cyclotron. The energy of the primary particles analyzed may go up to several billion volts and the scintillation counter technique involving a novel type of radiation is quite ingenious.

In 1934, Cerenkov, a Russian scientist, discovered that very fast electrons moving through a transparent medium emitted light at a definite angle from the direction of motion of the electrons. Such an effect was called **Cerenkov radiation** and was interpreted as due to the electrons moving with velocities greater than that of light in the medium where the electric field of the electron is strongly perturbed and an optical "shock wave" is generated. Can a particle travel with a velocity greater than that of light?

According to the theory of relativity no particle can travel with a velocity greater than that of light *in vacuum.* In the present case, we are dealing with particles traveling in a *medium* with velocities greater than light and giving rise to a new type of radiation.

That such a phenomenon really occurs can be readily understood by an example from acoustics. When an aeroplane travels with velocities smaller than that of sound in air, the sound emitted is due only to the oscillatory motions of the engines and other moving parts of the plane. But

when supersonic velocities are reached, as in the case of modern jet planes, then even the uniform motion of the aeroplane produces an additional radiation of very intense sound waves called the Mach's waves.

On this analogy one may call the Cerenkov effect as "singing electrons". Cerenkov used β-rays and Compton electrons.

With electrons of 2 MeV fairly strong source of radiation could be had. The radiation was found to be bluish white in color and was identical in appearance for all solids and liquids. Its spectrum was continuous. Its chief peculiarity was asymmetric emission, *i.e.,* it was emitted in directions making a definite angle with the direction of motion of the electrons and thus forming a cone whose axis was the path of the electrons.

If the Cerenkov radiation propagates outward from the source in directions forming a cone of half-angle θ whose axis is the path of the particle, it can be readily shown that $\cos \theta = c / n v,$
where
c is the velocity of light,
n the refractive index of the medium and
v the velocity of the primary particle.

Putting $v/c = \beta$, $\cos \theta = 1 / \beta n$. Since $\cos \theta$ is always less than or equal to unity, the condition for the emission of Cerenkov radiation is $\beta n > 1$. It can be shown that the total energy W thus radiated is given by

$$W = \frac{e^2 l}{c^2} \int v dv \left(1 - \frac{1}{\beta^2 n^2} \right)$$

where *l* is the path of the particle and v the frequency of the emitted radiation. By observation of θ in a medium of known refractive index, θ can be estimated.

Cerenkov radiation is produced in the atmosphere by cosmic rays and can be observed as large light pulses directly correlated with the cosmic rays.

It can also be produced artificially by allowing the high-speed charged particles, obtained from the modern powerful particle accelerators, such as the synchrocyclotron, traverse a transparent medium.

The Cerenkov effect can be used to detect high-speed particles and estimate their energy. Thus, R.L. Mather, in 1951, allowing the proton beam from the 184-inch Berkeley cyclotron to pass through dense flint glass ($n = 1.925$) or amorphous silver chloride ($n = 2.07$) was able to observe the Cerenkov radiation as a cone of visible light. Measuring the angle $\theta,$ he estimated the energy of the proton beam as 345 MeV.

It may be noted that the Cerenkov effect can also be utilized to detect high-energy electrons in beams containing heavier charged particles such as mesons. For example, a beam of particles brought out from the target of the Chicago University synchro - cyclotron by magnetic deflection contained π^- mesons of 124 MeV, μ^- mesons of 144 MeV and electrons of 220 MeV all having the same momentum. By allowing the beam to pass through 7 inches of aluminum, the speeds of both kinds of mesons were sufficiently reduced to prevent them from exciting Cerenkov radiation in a phosphor. But the electrons still had enough speed to produce the Cerenkov effect and so they could be observed independently of the charged mesons in the same beam.

Coming back to the liquid scintillation counter, with which we are dealing now, its efficiency can be understood by the following example.

When a primary electron of very high energy is allowed to fall on a liquid scintillation counter, it first forms what is known as 'electron showers' of lower energy. These 'shower' electrons are still fast and in their flight through the transparent liquid phosphor of the counter emit a small amount of Cerenkov radiations. Thus, it is possible to capture the whole shower within a single phosphor and collect the Cerenkov radiations in the photo-multiplier. Then the size of the output pulses in the counter will be proportional to the energy of the particle initiating the shower and the analysis of such super-energy particles can be made. The scintillation counter used in these investigations is known as the *large-volume Cerenkov counter.*

It may be noted also that a scintillation counter having 300 litres of liquid phosphor and 90 photomultipliers has been employed recently in *the detection of the most elusive of all nuclear particles, the neutrino.* Hence, the importance of this type of counter in modern researches is readily understood; it is now one of the most valuable of all modern counter techniques.

3. THE CRYSTAL COUNTER

The initiation of this type of counter started with the discovery by Sir R. Robertson, in 1934, that certain kinds of diamond, now known as type 2 diamonds, exhibited marked photoelectric or photoconductive effect when exposed to light.

In 1945, P. J. Van Heerden made the first serious investigation of crystals as radiation counters. He was very successful with silver chloride crystals, but failed completely with diamond, probably due to the fact that he could not get at the type 2 variety.

In 1947, E. W. Wooldridge, A.J. Ahearn and J. A. Burton in the Bell Telephone Laboratories obtained positive results with diamond as α- particle counter and in the following year, L. F. Curtis and B.W. Brown of the National Bureau of Standards succeeded to build a diamond counter that detected γ-rays. Thus the crystal counter was ushered into the experimental arena of nuclear physics.

A crystal counter (Fig. 284) consists of a natural crystal of diamond, silver chloride or lithium fluoride mounted between two metal electrodes maintained at a high voltage. Exposure to radiation causes a pulse of current to flow across the crystal and this pulse is then amplified. The mechanism of the observed current may be construed as follows:

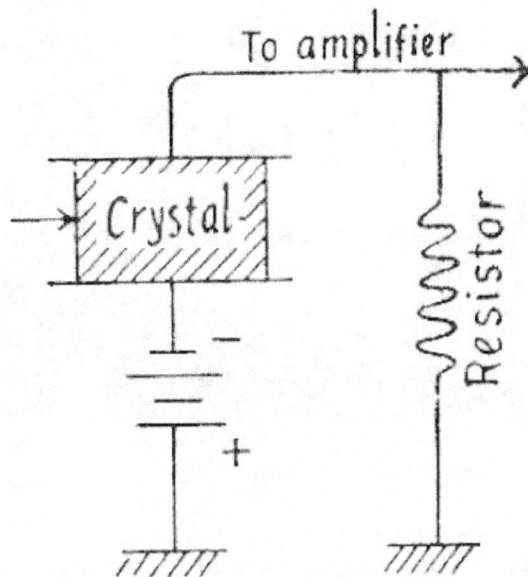

Fig. 284. Crystal counter.

The impact of the incident radiation or particle knocks out a single electron from one of the atoms of the crystal lattice. The electron thus released is attracted towards the positive electrode and is accelerated as it moves towards it. On the way, it hits other atoms and knocks more electrons from them, so that by the time the positive electrode is reached enough electrons have been made free to produce a measurable pulse.

This type of counter is found to be faster than the Geiger counter, due evidently to the rapid rise of pulses and the short recovery time in the crystals used. The technique, however, is still in the initial stages of development, since suitable crystals have to be selected by trial and error. It is found that while almost all diamonds can be used to detect α- particles, only one in a hundred or so can be used for counting γ-rays.

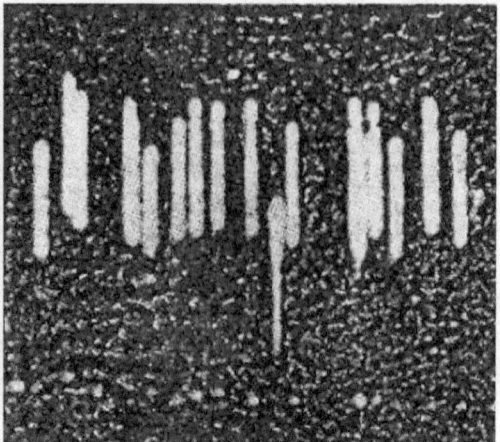

Oscilloscope pulses obtained with crystal counter: (*a*) *d*iamond, (*b*) germanium using mono-energetic alpha particles.

Recently, semi-conductors, like germanium and silicon crystals, have been tried with better results. A typical oscilloscope record of pulses obtained from the bombardment of the crystal counter by .mono-energetic alpha particles is reproduced here: (a) refers to diamond, while (b) to germanium. It is seen that the pulse height distribution as well as the signal to noise ratio are decidedly better with germanium than with diamond. The germanium crystal counter holds the bright promise of being a compact, high-speed device with a time of resolution of the order of 10^{-9} sec. It requires only an electrical connection, unlike the complicated light path linkage of the scintillation counter; it responds to each of the basic nuclear particles, if the appropriate shape and composition are used.

CLOUD CHAMBER METHOD

The cloud chamber is one of the most powerful tools available today for the study of artificial transmutation on account of the spectacular and many-sided results it is capable of producing. It enables the investigator to take photos of the atoms he splits and thereby secure direct visual evidence of the track of an atomic projectile hitting another atom and the tracks of the fragments of transmutation. With the counter technique one can hear the particles of transmutation pass by; with the cloud chamber one can see them. This fact contributes, in no small measure, to the convincing reality of the nuclear processes analyzed.

Furthermore, this wonderful apparatus, unlike the others, provides at one stroke all the different important information required in the analysis of transmutation. A *single cloud chamber photograph,* can demonstrate with compelling evidence *how many fragments* are formed in a single process, what are the *directions* in which they fly away and *how far* they are able to travel through the gas in the chamber the aspect and size of the track can indicate the *nature* of the particle and the curvature of the track in an applied magnetic field supplies the value of the *energy* of the particle that made the track. Thus the cloud chamber, in the hands of a skilful operator, is bound to give very precious data in the analysis of individual nuclear processes.

The *fundamental principle* of the cloud chamber, discovered in 1897 by C.T.R. Wilson, as well as its first application in the determination of the electronic charge, have already been given. Careful research as regards constructional and operational details has now made the cloud chamber a very valuable tool in many a modern research.

The first cloud chamber

In 1912, Wilson designed a cloud chamber which made it possible for the first time to see and photograph the tracks of individual particles. This prototype of all modern cloud chambers is shown diagrammatically in Fig. 285. A is a cylindrical expansion cloud chamber with walls and roof of glass, containing dust-free air saturated with water vapor. Directly below A is a movable piston P whose rapid descent produces the adiabatic expansion. A large evacuated vessel V is connected to the piston through a tube provided with a valve C. When C is opened, the air under the piston rushes into the vacuum chamber, thereby causing the piston to drop suddenly. The wooden blocks WW are to reduce the air space below the piston. Water at the bottom of the apparatus ensures saturation in the chamber. The expansion ratio can be adjusted by altering the height of the piston. An electric field E is maintained in A to sweep out the spontaneously produced ions that would create a diffuse fog. This electric field is usually cut off just before expansion.

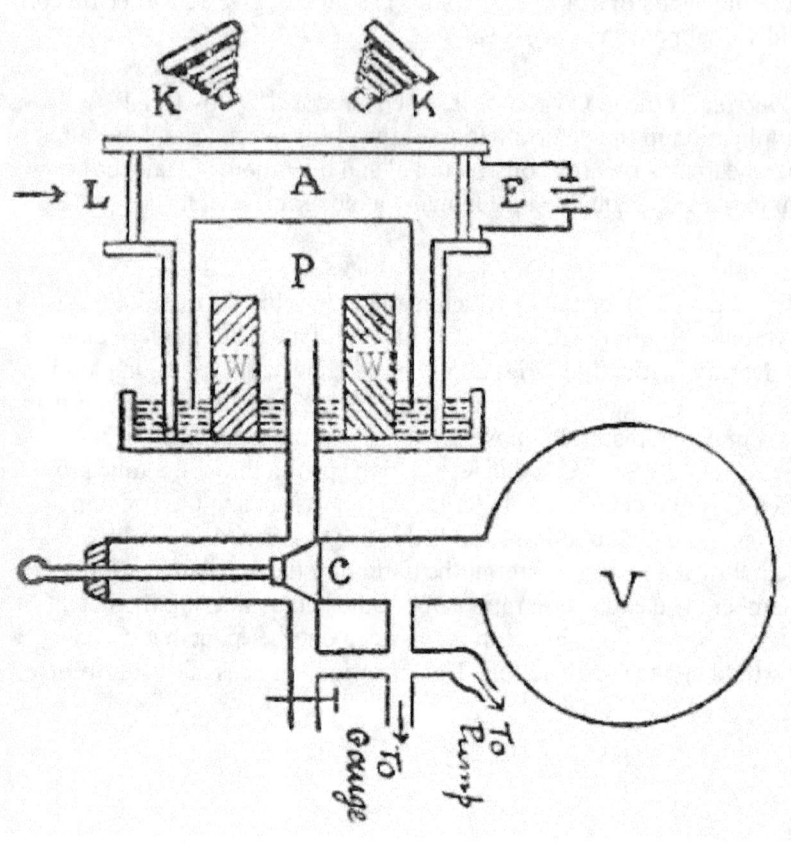

Fig. 285. Wilson's 1911 cloud chamber.

Immediately after expansion, the particles are allowed to enter the chamber through a side window; droplets are formed on all the ions produced by the particle, so that they mark the track of the particle. At this precise moment, if the chamber is strongly illumined with a horizontal beam of light L from a mercury vapor discharge lamp, the track of the particle can be visually observed as a white line of fog against the blackened top surface of the piston. The tracks of individual particles can be photographed by a camera, or better still, by two cameras KK focused on the interior of the chamber through the glass roof. The various operations, *viz.,* expansion, admission of ionizing particles into the chamber, illumination and photographing must be performed in very rapid succession to obtain good results. For this purpose, Wilson used an ingenious arrangement with a falling weight, which first opened the valve C and allowed the expansion to take place, then let the particles into the chamber and finally made a contact which flashed on the arc light and clicked the camera for taking the picture. As the chamber heats up very quickly, in about 1/50th of a second, the observations must be completed before the chamber has become so warm that the condensation droplets disappear.

With such a relatively simple apparatus, Wilson and others were able to obtain beautiful photographs of individual ionizing particles, such as α- particles, β-particles, etc. The α-particle tracks are thick and straight, while the electron tracks very fine, evidently due to the widely different ionizing powers in the two cases. Photons do not produce by themselves any tracks but manifest themselves by the tracks of their secondary electrons.

Hence it becomes possible to distinguish the particles from the different aspects of the tracks they produce. By setting up a magnetic field of known intensity at right angles to the path of the particles photographed, the tracks obtained will be curved into circular arcs. Measuring the radii of curvature of these circular tracks, the energies of the particles which made them can be determined.

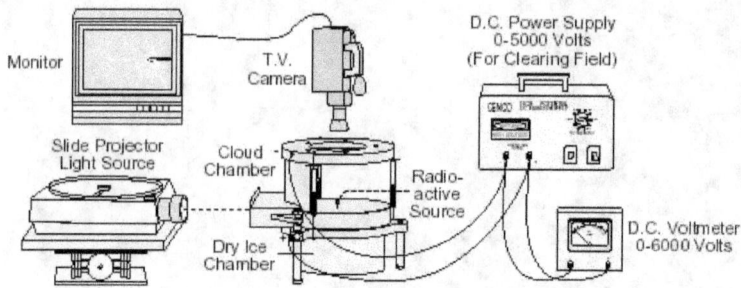

This Wilson continuously sensitive cloud chamber can work for up to several hours, displaying tracks of various radioactive decay products.

Modern expansion cloud chamber

In the adjacent picture are shown two types of modern expansion cloud chamber. The one marked A (at the top) is a *horizontal chamber* illumined by the arc lamp B. It is well suited for transmutation experiments. The other marked C (at the lower half) is a *vertical chamber* using a rubber diaphragm instead of the piston, designed for the study of cosmic rays.

In the adaptation of the expansion cloud chamber to modern requirements of nuclear research the following improvements have been made:

(*a*) *The expansion technique*
Mechanically the expansion is achieved in a variety of ways. A reciprocating piston to compress the gas slowly and expand it rapidly can be operated by a cam, by an electromagnet or by changing the pressure under the piston. Rubber diaphragm with similar pressure regulation provides another satisfactory method. Excessive turbulence is avoided by maintaining a geometrical simplicity in design and with mesh screens just below the active portion of the chamber. The operation of the mechanism of expansion is usually carried out by an *electrical liming system* so that expansions are produced at regular intervals of about 30 secs., in order to maintain temperature equilibrium. The same electrical system resets the chamber and controls the illumination and the camera, which thus renders the operation of the cloud chamber *automatic*.

(*b*) *The expansion ratio*
This can be estimated from the known constants of the gas and vapor in the chamber which produce supersaturation. It is, however, usually found experimentally, to get the sharpest definition of the tracks. For water *vapor in air* this ratio is in the region of 1.25 to 1.38, but is much lower for other vapors such as alcohol or for other gases as helium or

argon. For *a mixture of water and alcohol vapors in argon,* the ratio is about 1.12 and even less.

(*c*) *Nature and pressure of the gas in the chamber*

Hydrogen, instead of air, is used along with saturated water or alcohol vapor in order to minimize the scattering effects which might distort the paths of the particles and thereby introduce error in the measurement of the radii of curvature of the tracks when the energies of the particles are to be estimated. Chambers are constructed to function at low and high pressures. With high pressures of the gas and large diameters of the chamber, the particle range observable in the chamber can be increased to include the whole or even very fast proton tracks. The use of low pressures (down to a few cms. of Hg) of the gas enables particles of small energy, such as heavy recoil atoms from disintegrations or scattering processes, to be observed.

(*d*) *Illumination and photographic techniques* play a great part in the perfect recording of the tracks. As a source of illumination, the *carbon arc is* used with an arrangement whereby the power line is momentarily short-circuited to give a flash of intense illumination. *Capillary mercury arcs, burning aluminum foil in a current of oxygen* are also found satisfactory.

Wilson used two cameras set at an angle in order to obtain simultaneously *two stereoscopic pictures of* the tracks. This method which is frequently used enables the reconstruction of the paths of the particles in space, so that their ranges and directions can be easily measured. A modification by Kurie (1932) replaces the two cameras with a single camera and lens system, equipped with mirrors and a prism, to take the two stereoscopic pictures on a movie film. With the automatic device to control the illumination and a cinematograph camera, once the apparatus has been carefully regulated, thousands of photographs can be taken in a short time, as is necessary when rare phenomena, like artificial transmutations, are to be recorded.

(*e*) *Source of the ionizing particles*

For α- particles the source is usually placed inside the cloud chamber. The source of β-particles is arranged outside the chamber and the particles allowed into the chamber through a thin window. γ-rays are studied by observing the secondary electrons produced in the chamber by Compton, photoelectric or pair production effects.

For the Compton effect, a thin sheet of carbon is placed within the chamber and irradiated with γ-rays. For the other two effects, a thin sheet of lead is used. In the case of neutrons, the chamber is filled with hydrogen gas and water vapor, and the recoil protons chased by the neutrons that collide with the hydrogen atoms can be observed and photographed.

Use of the expansion cloud chamber in artificial transmutation

The expansion cloud chamber method lends itself most specially to the case where the element to be transmuted is in the gaseous state. Thus, for instance, Blackett employed it successfully in the transmutation of nitrogen by α- particles. A study of the forked tracks obtained in such cases, by reconstructing them in space, measuring the angles between them and their ranges, gives very valuable indication on the nature of the collision, whether it is a question of a mere elastic collision or of a real transmutation of the nucleus track. Two expansion chamber photographs of elastic collisions of α- particles with helium and hydrogen nuclei are reproduced here. Feather discovered transmutation with neutrons by the same method.

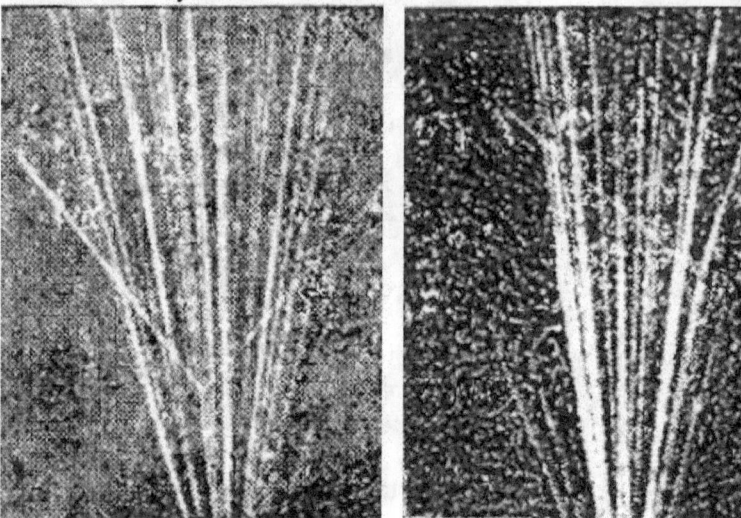

Cloud chamber photographs of elastic collisions (a) a-particle with a He nucleus (b) a-particle with a H nucleus (Rutherford, Chadwick and Ellis)

If the element to be transmuted is in the solid state, it is suitably arranged in the form of a thin target inside the chamber so that the projectiles can effectively bombard it and the particles of transmutation can be conveniently photographed. Thus Cockcroft, Walton, Dee and Gilbert were able to obtain splendid photographs of

α-particle tracks from the disintegration of Li and B by proton bombardment, as illustrated in the adjacent picture. In general, this method is very precious in the study of transmutations, chiefly when the phenomenon is not very rare.

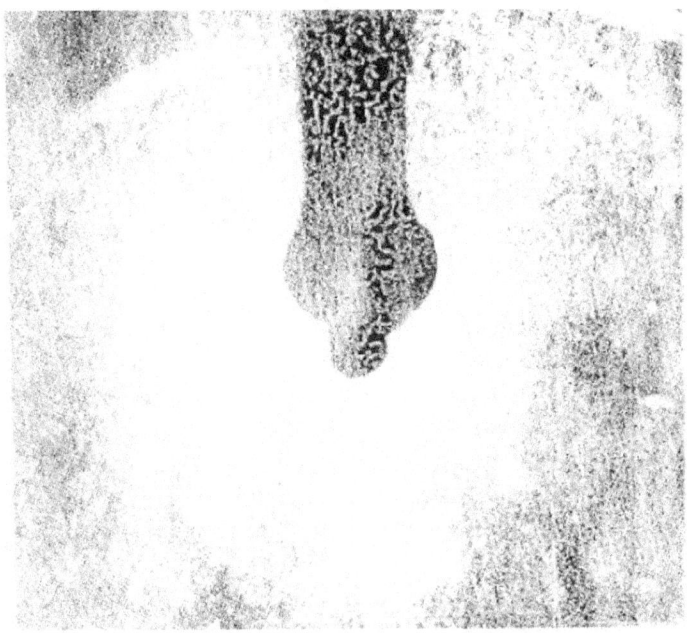

$_5B^{11} + {}_1H^1 \rightarrow 3\,{}_2He^4$ (P. I. Dee and C. W. Gilbert)

Limitations of the method

Although the expansion cloud chamber is extensively used in transmutation experiments and yields results in harmony with those obtained by other methods, a number of difficulties limit the extent of its utility. For instance,

(i) one is not always sure of the sense of the track photographed;
(ii) the range of the particle may exceed the dimensions of the chamber so that the whole track is not photographed;
(iii) there remains a certain amount of uncertainty about the nature of the nuclei constituting the arms of the forked tracks, chiefly when the agent of disintegration is neutron whose track is mot registered by the chamber, as it is not electrically charged;
(iv) as the chamber contains several gases, it is at times difficult to know which of them has suffered transmutation;
(v) the knowledge acquired about the range-energy relations for the recoil nuclei is still insufficient and this handicaps the accuracy of the observations.

Another major difficulty of the method is that the apparatus records only the particles which fly off during a hundredth of a second about and then lies idle for several seconds and more, while it is being prepared for its next brief interval of effectiveness. Conditions suited to good track formation last for about one-tenth of a second, which is called the 'sensitive time' of the chamber.

The expansion results in a turbulent motion of the gas and the temperature differences give rise to convection currents. The swirling of the gas must die down and the temperature return to normal before another expansion can be made and this may require about a minute which is called the 'dead time' of the chamber.

Consequently, one is obliged in the study of a rare phenomenon to make a great number of expansions and photograph almost endlessly, before one meets with the event looked for. Thus only a few in hundred and even thousands of photos taken may become useful. Further, for the same reason, it is hardly possible to use this method for counting the number of times a given phenomenon repeats itself in a given time.

This difficulty of the ineffectiveness of the apparatus over a considerable time has been overcome by two ingenious devices, one, first conceived and realized in practice by Blackett, in 1932, known as the *counter-controlled cloud chamber* and the other initiated by Langsdorf, in 1939, called the *diffusion cloud chamber*. The former is suited exclusively for the study of cosmic rays, while the latter can be employed in the artificial disintegration process as well. We shall, therefore, deal here only with the diffusion cloud chamber postponing the treatment of the other technique to the chapter on cosmic rays.

Diffusion cloud chamber

The first description of a working model of a *continuously sensitive* diffusion cloud chamber was given by Langsdorf as early as 1939. But it is only recently, *i.e.,* after WWII that several chambers of this type have been constructed and successfully operated in different laboratories, engaged in nuclear and cosmic ray researches, such as the California Institute of Technology, University of Glasgow, University of Liverpool, etc.

Principle

In diffusion cloud chambers, the supersaturation is obtained by the diffusion of a vapor from the warm roof of the chamber downward through an inert gas to the cold floor. Condensation occurs on ions in a sensitive layer near the bottom. Since the vapor supply is constantly replenished by diffusion from the roof, the chamber is almost continuously sensitive to ion tracks. When the temperature and vapor distribution have settled down to a steady state, tracks may be seen forming in the sensitive region and falling under gravity to the bottom. It may take half an hour to reach this condition, during which if the air is dusty, a dense cloud of droplets, with the dust particles as nuclei, will be seen to form and gradually disappear. By proper adjustment of the operating conditions, the depth of the sensitive region can be increased to several centimeters. The diffusion cloud chamber is normally simpler in design than the expansion cloud chamber, as it has no moving parts, valves and resetting mechanisms.

Technical details

There are many variable quantities in the actual construction of a diffusion chamber and much of the experimental work has been devoted to finding how the operation is affected by increasing or reducing the pressure in the chamber, changing the gas in the chamber or the condensant vapor, altering the temperatures at the top and the bottom as well as the size of the chamber, etc. The man experiments and the theory of the instrument which R. P. Shutt published in 1951 show that a wide variety of conditions are compatible with track formation. Hence we shall give only the essential details involved in the construction and successful operation of a chamber of this kind.

(1) *The chamber,* shown diagrammatically in Fig. 286, is usually made with glass rings about one foot in diameter and half a foot in height, strengthened by a brass cylinder with perspex (brand name for Poly(methyl methacrylate, PMMA))windows. The bottom of the chamber is a metal plate (brass or aluminum) 1/4 inch thick blackened with stove polish or black bakelite. When the bottom surface becomes covered with liquid, it forms an excellent background for photography. The roof of the chamber is a metallic plate, with a central glass window for observation and photography. The top and bottom plates are sealed to the glass ring with rubber gaskets to have the chamber gas-tight.

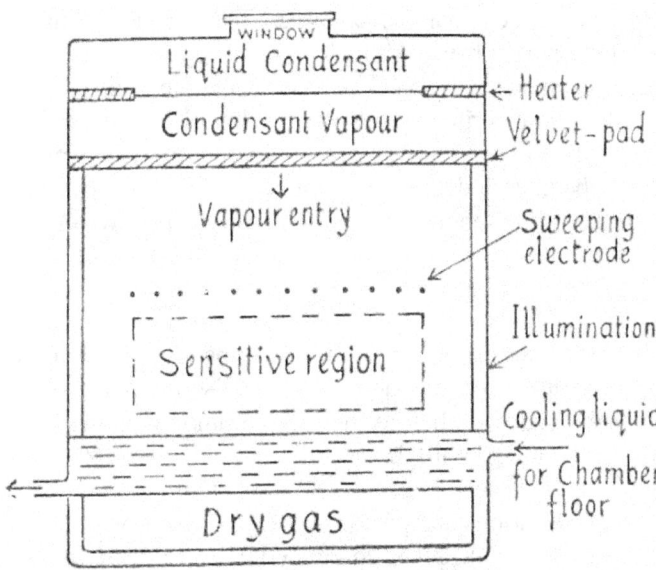

Fig. 286. Diffusion cloud chamber.

(2) *The condensant liquid* is contained in a ring-shaped metal trough just below the top-plate and an electric heater is used to promote evaporation. The bottom of the trough has small perforations and is covered with a velvet pad which thus gets uniformly soaked with the liquid. It *is* essential for the maintenance of stable operation that the condensant vapor should be light as compared with the gas used in the chamber. The deepest sensitive layers have been obtained with mixtures of methyl alcohol, ethyl alcohol and water. However, since the composition of the mixture changes with evaporation, pure methyl alcohol is usually employed.

(3) *The gas in the chamber.* Air may be used as the inert gas in the chamber; although argon appears to give better results. Methyl alcohol diffusing through air at atmospheric pressure is found quite satisfactory. Lighter gases, like hydrogen, and helium, have also been used under high pressures.

(4) *Temperature conditions of the bottom and top of the chamber.* The bottom of the chamber is usually maintained at about —70°C by a block of dry ice pressed firmly against the metal base-plate. Another method of cooling is by partially submerging the base plate in a bath of alcohol and dry ice. Although the temperature of the floor of the chamber is not very critical, it is desirable that it should be below —40°C to obtain a suitable high degree of supersaturation. The top of the chamber is maintained in the whereabouts of 40°C, depending on the size of the chamber and the nature of the gas and vapor used. As the temperature of the top is increased, the depth of the sensitive layer is increased until an optimum temperature is reached above which the sensitive region breaks up into different layers, which is evidently to be avoided.

(5) *The electric sweep field* applied between the top and bottom plates of the chamber is found necessary here also as in the case of the expansion chamber. The purpose of this is to remove from the chamber any ions left over from old tracks which may otherwise act as centers of condensation and cause droplet formation. This is desirable not only to prevent unwanted background droplets from confusing the pictures but also to avoid depleting the vapor supply in the sensitive region. The sweep field voltage depends on the actual working conditions of the chamber and may vary from 20 to 100 volts per cm. It is usually switched off shortly before taking photos, so that the tracks themselves, consisting of a double row of positive and negative ions, are not pulled apart and thereby lose their sharpness.

It is to be noted that a number of other components, such as flash lamps of high intensity, cameras capable of taking stereoscopic photos of the tracks, etc., are also necessary to complete the construction of a diffusion chamber, but, as they are common with the expansion chamber technique, we shall not deal with them.

As an example of the actual operating conditions and performance of a diffusion chamber, we shall mention briefly one of the researches conducted by R. P. Shutt and his associates in 1951. They used the diffusion chamber to investigate the collisions of negative pi-mesons with protons. The pi-mesons were artificially produced with a particle accelerator. The chamber was filled with hydrogen at a pressure of 21 atmospheres to supply the protons. With the top of the chamber at 20°C and the bottom at -65°C and using methyl alcohol vapor, they obtained a track-sensitive layer 6 cm. deep. The cyclotron ion source was pulsed every 4 to 6 secs, so as to produce about 20 meson tracks in the chamber which were photographed stereoscopically. Between pulses a sweeping field of about 1,000 volts was applied while tracks settled down and vapor was replenished.

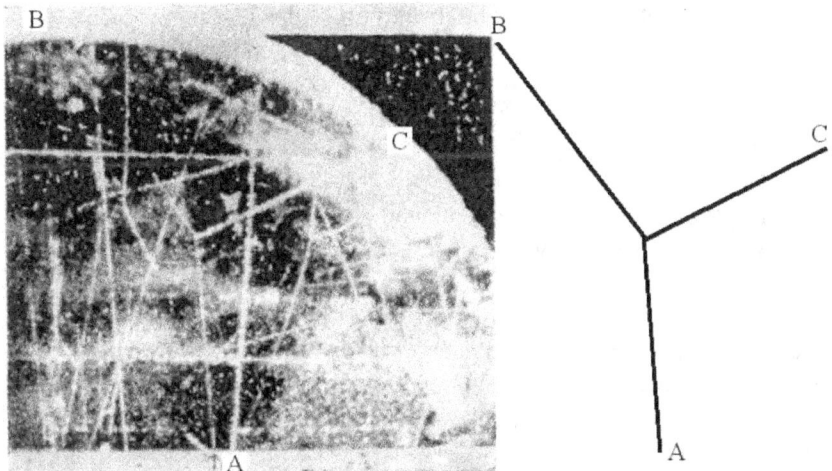

Photo of tracks in a diffusion cloud chamber (H.P. Shutt)

They were able to take 5,600 photos in a day, one of which is reproduced here. In it, among other events, we see the following case quite clearly. A negative pi-meson entering the chamber at A, strikes a proton and leaves at B. The struck proton recoils towards C.

The chief advance of the diffusion chamber over the expansion type is its *nearly continuous operation*, when once the apparatus settles down to working conditions. Although. there is a 'dead time' with this chamber also (during which the sensitive region has to build up to the required concentration of vapor after the passage of each burst of ionizing particles) it is considerably shorter than that of an expansion chamber of comparable size and gas pressure. Other points in its favor are comparative ease and cheapness of construction, lack of moving parts such as diaphragms and valves, absence of swirls and eddies which distort the tracks, etc. The technique is still in the initial stages of development, but promises a bright future.

PHOTOGRAPHIC EMULSION METHOD

Although the basic discovery of the photographic emulsion method was made as early as 1911 by S. Kinoshita and M. Beingmanns who observed that α- particles, moving through the sensitive layer (emulsion), of a photographic plate, produced, in some cases, developable tracks that could be rendered visible under a microscope as rows of dark silver granules, this technique as a research tool for the detection of charged particles was left undeveloped for quite a long time, on account of certain inherent difficulties, such as small penetration of the particles in the emulsion, complex and variable nature of the emulsion, etc., which were thought almost insuperable.

It is only of late that the method has received the precision and versatility required in instruments employed in the detection and analysis of the particles of transmutation. The two scientists .mainly responsible for such a development are Dr. C. F. Powell of the University of Bristol, England, and his collaborator, G.P.S. Occhialini, who, since 1938, have devoted themselves with untiring zeal to the resolution of the several problems involved in the technique and have already achieved remarkable success. To these scientists must be added the names of the two well-known photographic companies, Ilford Ltd., in the first place and then Kodak Ltd., which have produced the special plates required.

The **fundamental principle** of the method is simple and almost the same as that of ordinary photography. Charged particles, when allowed to fall on a photographic plate, will have the same effect on the sensitive layer of the plate as would exposure to light. Just as the latent image of the subject photographed is preserved by the silver halide grains until the plate is developed, so also the tracks of the charged particles in the emulsion, which will, in consequence, appear, on developing the plate, as so many trails of black grains of metallic silver. These can be viewed directly under a microscope or projected on a screen and photographed.

Technical details
In the practical application of this principle to the detection and analysis of particles of transmutation, improvements had to be made with the photographic emulsions to render them sensitive enough to react effectively to the particles under study and several difficulties had to be overcome in the analysis of the photographic records obtained. These various technical details may be summarized as follows:

(i) *Preparation of special photographic emulsions*
In the ordinary photographic plate, the number of grains that are "exposed" by the passage of a particle through the emulsion is not great enough to make interpretation of the plates easy. As early as 1939, Dr. Powell had been in touch with Ilford Co., with the hope of their preparing plates in which the number of sensitive grains would be greater than normal. The first attempts met with little success and it was only in 1947, that a series of new plates was produced with special emulsions which contained a *very high concentration of sensitive grains,* about ten times as much silver halide for a given quantity of gelatin as an ordinary emulsion. This close packing of the grains would enable the track lengths of the particles to be measured accurately and hence energy determinations based-ion track length's become more reliable.

In these special emulsions, *the size of the sensitize grains vary from 0.5 μ to 0.1μ* according to which the different grades of plates are named A, B, C, D, etc. By varying the average size of the grains it is possible to control the sensitivity towards different types of radiation. Thus a plate can be prepared which will permit a somewhat selective detection of protons, α-particles or fission fragments.

These specially prepared plates have also much *thicker coatings of emulsion than in ordinary photography,* a thickness commonly employed being 100μ (1/10 mm). This would be the longest distance of any track in a direction at right-angles to the surface of the plate. Since the range of particles in the emulsion runs from 10μ to a full mm., many tracks could be followed only when they run parallel to the plate surface. The photographic record of a nuclear disintegration will thus be a *star-shaped explosion* within the emulsion, with the tracks of the resulting particles scattering in a variety of directions. Some might leave the plate before their course is, completed and others come to rest within the emulsion, in which case their ranges can be measured. For the recording of particles which enter the emulsion at large angles with regard to the plane of the emulsion, very thickly coated plates up to 300μ, have been used.

(ii) *The experimental set-up*
The arrangements of the specially prepared plates to record nuclear phenomena depends to a great extent on the particular problems to be solved; but two simple methods have been used so far, in connection with artificial transmutation,

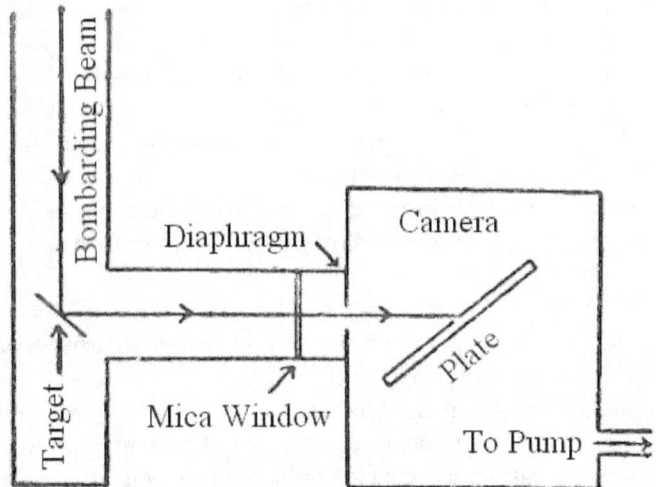

Fig. 287. Apparatus for exposing photographic emulsion to particles of disintegration.

One of them, applicable to cases where there are only two products of disintegration, *viz.,* the recoiling nucleus and a fast light particle, such as a proton, is shown diagrammatically in Fig. 287. The bombarding beam, say of deuterons, accelerated by a Cockcroft and Walton tension multiplier impinges upon a sample of the element, say lithium. The resulting particles, protons, may be stopped to a known extent by a mica window and then passed through a diaphragm into a camera and on to the photographic emulsion, the angular position of which with regard to the direction of the incident protons may be varied: and measured. If necessary, the camera chamber can be evacuated. The protons falling on the emulsion will produce their characteristic tracks in it, which, after having been developed and fixed, can be observed and studied. The application of the law of conservation of momentum allows the analysis of the transmutation process from observations on the one track recorded in the emulsion in such cases.

The second method, well suited to cases where several particles are emitted in a given disintegration, consists in incorporating the elements to be transmuted in the emulsion. The emulsion of a photographic plate normally contains a number of elements, notably silver, bromine, iodine, carbon, hydrogen, nitrogen and oxygen. Hence, when ordinary plates are used, one can expect to find only transmutations of the nuclei of this limited number of atoms. But Powell has developed methods of introducing into the emulsion almost any other element which one desires to transmute. This can be done by simply mixing a salt of the material with the other ingredients of the emulsion during manufacture, or by "sandwiching" a thin layer of the material between two layers of the emulsion, or better still, introducing fine, insoluble particles of the desired element into the emulsion, of such a type that they could be distinguished from developed silver grains, in the microscopic examination of the processed plate, by their size or color, in which case the observed disintegration can be identified with certainty as due to a particular types of nucleus.

(iii) *Photographing the tracks of particles on a magnified scale*
In order to conveniently observe and study the minute tracks of the particles in the sensitive layer of the plate, they are magnified considerably (1,000 to 2,000 times) by a microscope with an oil immersion objective of high aperture and adequate sub-stage illumination in good adjustment. The magnified image of the tracks is projected on a screen and photographed. If a track happens to lie in a plane parallel to the surface of the emulsion, it can be photographed with a single exposure. If, on the other hand, the track dips into the emulsion, it is necessary to focus on different parts of the track in turn and take a series of microphotographs. These are afterwards carefully arranged to form a *mosaic.*

Dr. Occhialini has designed and constructed an ingenious form of projection microscope, called the *telephanto* which is of great value to this part of the work. It is capable of automatically scanning the complete area of any plate which is put into it for projection. It also varies its focus periodically so that all sections of the plate may be examined, in depth as well as in area. When it is in operation, it creates the illusion that individual tracks and "stars", which may have been left weeks before in the plate, are alive and moving.

Apart from its value in speeding up examination, the same equipment can be used with manual control to photograph the projection screen, with the microscope focused at different depths, so that complete pictures can be taken showing tracks in three dimensions, in one flat plane.

(iv) *Analysis of the tracks*
Generally speaking, the nature of the particle can be deduced from the appearance of its track, which depends upon the number of ionizations the particle causes per unit length. The chance that a silver halide grain is made developable depends of the number of ionizations it receives.

An α- particle with its twofold charge and mass of four units causes many more ionizations per unit length of path than, for instance, a proton of the same energy with its single charge and unit mass. Hence the intensity of ionization produced by an α- *particle* is more than sufficient to render developable all the silver halide grains it traverses.

A proton having the same initial energy as an α- particle but a greater speed will cause so few ionizations in individual grains that not all the grains at the beginning of the track will be made developable. It is this characteristic which enables the investigator to distinguish between α- particle and proton tracks in the photographic emulsion.

For the same reason, a *deuteron* track will be roughly intermediate between those of the proton and α- particle.

Although *electrons* cause ionization and undoubtedly bring about a blackening of the emulsion, actual electron tracks are not easily observed. This is partly due to their very light mass, which causes them to be scattered in zig-zag paths in the emulsion.

In addition, the ionization per unit length is much weaker than in the case of protons. Another difficulty is the background of fog grains, (i.e., those which are developable without exposure) which makes it impossible to the observer to distinguish the grains made developable by electrons. Recently, however, (1948) Kodak Ltd. has produced an emulsion suitable for the detection of electrons.

The energy of the particle can be determined from the *range* of *its track*. But in this connection there are a number of points which have to be carefully solved. In the first place, there is an appreciable shrinkage of the photographic emulsion in processing, amounting to as much as 43%. Dr. Powell's greatest achievement, accomplished by long and patient investigation, is that he developed a method of measuring the particle tracks accurately in spite of this shrinkage. He was able to establish that the vertical shrinkage of the emulsion was not only uniform over the central area of the plate but was also constant for all plates of the same type.

Secondly, since the emulsion is a mixed material, it is of equal importance to obtain a direct and experimental comparison between the ranges of particles of different types and speeds passing through air and through a photographic emulsion. This comparison work is carried out as follows:

Firing a given kind of particles, whose range in air, is known, into the photographic emulsion, a range calibration for the particular emulsion in use can be made. Thus, for instance, using the α-particles from polonium which is nearly a pure α-emitter, if-the source is placed at increasing distances from the photographic plate and the corresponding- lengths of the tracks recorded in the emulsion are measured, it is found that (a) the number of silver grains produced in a track is directly proportional to the length of the track and (b) the range (3.8 cms) in air of α- particles emitted by Po corresponds roughly to 20 to 25 μ in a photographic emulsion as shown in Fig. 288.

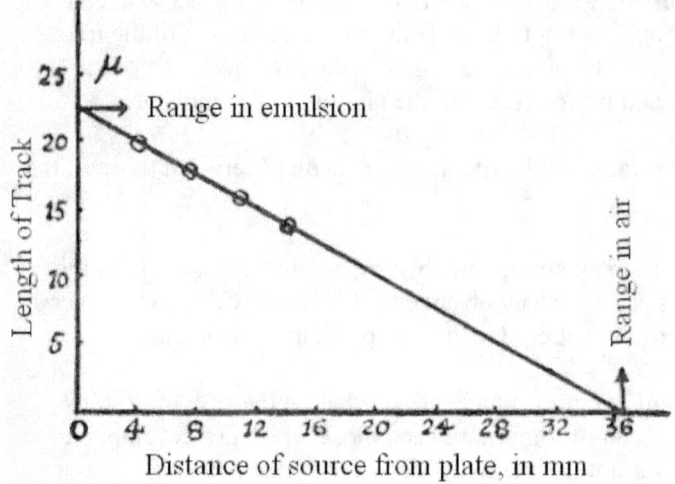

Fig. 288. Calibration of range in emulsions.

The range in air of an unknown α-particle can therefore be easily determined from the length of its track in the emulsion using the calibration curve and consequently its energy- calculated from the well established range-energy relation in air. The stopping power of the emulsion is, in general, about 1,700 times that of air, i.e., one micron of emulsion is equivalent to about 0.17 cm standard air. For precise work, each type of emulsion must be calibrated. For the Ilford B-2, C-1 and C-2 plates, it has been experimentally found that 1 MeV energy corresponds to a range of 14.5μ for protons and 3.52μ for α- particles, while 10 MeV energy to 564μ for protons and 58μ for α- particles.

Thirdly, as the emulsion is three-dimensional and the tracks recorded might-run in any direction, methods have to be devised for accurately determining the "angle of dip" of individual tracks in relation to the surface of the plate. If a particle enters the plate obliquely as shown in Fig. 289, the actual distance it travels (AC) is greater than the apparent distance traveled (BC). The true length of the track can be found as follows:

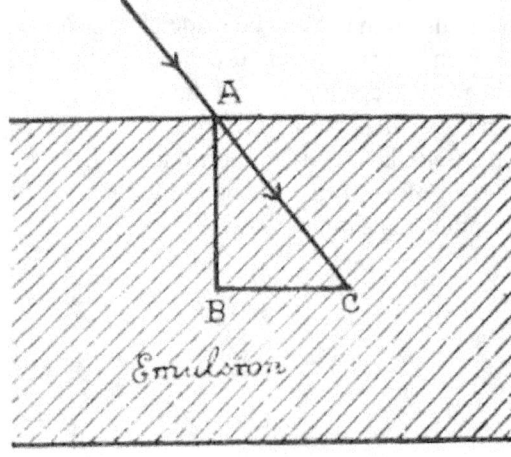

Fig. 299

With high power magnification two settings of the depth screw of the microscope having an eyepiece scale in microns are obtained, one on A, the first grain near the surface and a second on C, the last grain on the track. From the depth AB and the projection length BC, the actual range AC can be calculated.

The photographic emulsion method in use
Some of the applications already made will give us an idea of how this method has developed into a powerful tool.

(a) Decay of radioactive substances
The α-decays of radium and radiothorium have been studied by this method. The photographic plate is 'bathed' for a few minutes in a solution of a salt of the radioactive substance under test, then dried and kept in a light-tight box for three or four days. When the plate is then developed and examined under the telephanto, numerous *star-like* images are seen. Each arm of the star represents the track of an α- particle ejected from the radioactive substance, electrons and γ-rays leaving no recognizable tracks in the emulsion.

In the case of radiothorium the identification of the various arms as corresponding to a particular daughter element is done by measuring the length of the tracks. This method also permits the study of the decay of very weak radioactive substances, as was actually done with the element samarium by Lattes and Cuer who found long range α- particles emitted by this element.

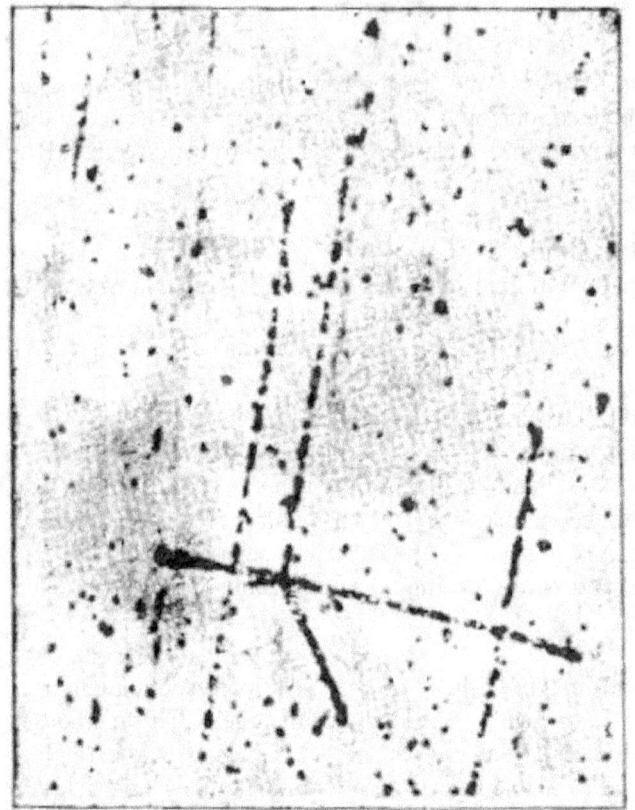

Disintegration of nitrogen into α-particles by high energy deuterons. (Powell)

(*b*) *Artificial transmutations*

Significant and convincing results have been obtained for nuclear disintegrations of elements by exposing the photographic plates to high energy particles according to the two methods already described. As an example, let us consider the *transmutation of nitrogen by high energy deuterons* studied by this method, as illustrated in the adjacent photo. The tracks of three deuterons of initial energy 8.9 MeV enter the emulsion from the top of the plate at a small glancing angle of incidence. One of them bombards a nucleus and leads to the ejection of four particles which from the grain spacing in the tracks can be identified as α- particles. From the observed ranges of the particles, their energy and momentum can be determined. The vector sum of the momenta is found to equal that of the incident deuteron and the total energy corresponds to that expected if the transmutation is correctly represented by the equation

$$_7N^{14} + {}_1D^2 \rightarrow 4 \, _2He^4$$

This result is produced evidently by the disintegration of a nucleus of nitrogen contained in the emulsion itself under deuteron bombardment.

The special type of disintegration, known as *nuclear fission,* ha s also been studied by this method. Very interesting and complicated fission tracks were obtained by Livesey

86

and Green at Cambridge and by Joliot at Paris. The recording of fission fragments present specific photographic problems; a special Kodak photographic emulsion for fission studies has recently been introduced.

(c) *Cosmic ray research*

The photographic method has proved to be of particular interest and importance in the field of cosmic rays, where the events are less frequent than in the case of artificial nuclear reactions. Powell and Occhialini, by merely exposing a set of special plates to cosmic radiation for a few weeks at high altitudes, have obtained a record a nearly 3,000 individual transmutations of complicated types. All of these have not yet been interpreted. A number of spectacular observations, however, have already been made.

In one type of disintegration, illustrated in the adjacent photo, a nucleus, probably of silver, has been disintegrated by a cosmic ray particle of very high energy of about 1000 MeV. The tracks of seven protons (very fine), five α-particles (thicker) and a number of fission fragments (dense and short, near the center) can be distinguished. Most of the particles pass out of the emulsion into the glass or out of the surface, so that their range and hence their energy cannot be determined accurately.

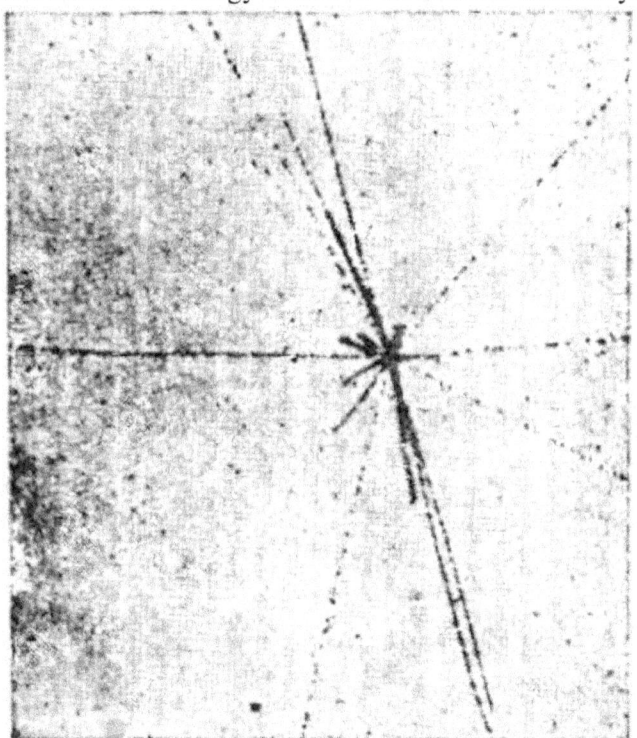

Explosive disintegration of a silver nucleus by cosmic rays (Powell and Occhialini)

Among other new types of disintegration, several examples have been found that could be attributed to the bombardment by the mesons of cosmic rays. It is chiefly in the exploitation of cosmic rays as a research weapon that the new technique of Powell is most clearly proving its worth. As an instrument to detect moving charged particles it is equivalent to the cloud chamber.

The special attractive features of the emulsion methods are:

- it is *continuously sensitive* during the whole of the time the plate is exposed, as contrasted with the cloud chamber which is sensitive only for a fraction of a second after an expansion and remains ineffective for several seconds between successive expansions;

- it is also *simple, inexpensive* and *light weight,* unlike the elaborate, costly and heavy cloud chamber; in special problems *it permits analysis from a single, direct and permanent record* while several thousands of stereoscopic cloud chamber photographs would be needed to obtain the same result.

The **two great limitations** of the method are:

- the *uncertainty in range measurements,* chiefly for steep angles of incidence, and

- the *impracticability of measurements of curvature in a magnetic field,* on account of the very short length of the tracks.

The cloud chamber proves itself superior to this method as regards these two points. It might therefore be said that at the actual hour *the photographic method is complementary to the Cloud chamber.*

RESULTS OBTAINED FROM THE EXPERIMENTAL STUDY OF ARTIFICIAL TRANSMUTATIONS

These results are so vast and varied that it is not possible to give here a detailed account of them; nor can this serve any useful purpose to the general student. It is more profitable to indicate first the *main guiding principles* used in the classification and interpretation of results, then study *certain prominent types of reactions* that throw light on the nuclear structure and finally give an account of some very *important discoveries,* such as the *neutron, artificial radioactivity* and *nuclear fission.*

GUIDING PRINCIPLES IN THE CLASSIFICATION AND INTERPRETATION OF EXPERIMENTAL RESULTS

(1) Bohr's theory of nuclear disintegration

In 1936, Bohr proposed a simple theory which is well adapted to a clear and easy understanding of the mechanism of disintegration. Considering the close packing of the constituent particles in the nucleus and the very strong forces that act at "short ranges" between them (these properties will be considered later), he assumed that *the nucleus is somewhat like a drop of liquid. When a projectile strikes a nucleus, it is first captured and an intermediate compound nucleus is formed.* Due to the special nuclear properties stated above, the energy initially concentrated in the projectile is very rapidly distributed among all the particles of the compound nucleus which therefore resembles a *"heated up" liquid drop* and may be considered to exist in a variety of *"quasi-stationary excitation states"*. For instance, in the case of the capture of an α-particle by a nucleus, once the compound nucleus is formed, the α-particle can no longer be considered as an independent entity, but as two protons and two neutrons among all the other protons and neutrons. For a similar reason, an α-particle ejected from the nucleus cannot be considered as "pre existing" in the nucleus as an individual unit.

Transitions take place between these excitation states in which one particle or the other of the compound nucleus is separated from the rest and the residual product nucleus is formed in the following manner. Each of the particles of the compound nucleus will have some energy, but none has sufficient energy to escape from the rest. Only after a comparatively long time, sufficient fraction of the excess energy is concentrated again on one of the particles to enable it to escape and we have a disintegration. This process may be compared to *the slow evaporation of particles from the surface of a liquid drop.*

It may be noted that the penetration of the projectile into the nucleus as well as the escape of a particle from the compound nucleus are governed by the same wave-mechanical principles, as already explained in connection with α-particle disintegration. Particles with much lower energies than are needed to overcome the potential barrier can penetrate into the nucleus, *i.e.,* can be "captured" by the nucleus. Likewise, with the compound nucleus in a high energy state caused by the capture of the projectile, there is a definite probability that one of the many energy changes will result in giving a single particle sufficient energy to leave the nucleus, although this energy is less than the height of the potential barrier.

(2) General scheme of nuclear reactions

According to Bohr's idea, the general scheme of nuclear reaction is

A + B → C → P + O

The projectile B strikes the nucleus A and combines with it to form the *compound* nucleus C. The latter then splits into an outgoing particle O and a residual product nucleus P.

Nuclear reactions may be classified into two main types according to the character of the outgoing particle O. It may be either a light quantum (emission of γ-rays) or a material particle. The first case is a *simple* or *radiative "capture"* process, while the second a *"particle' disintegration.* Both theory and practice show that she former process is ordinarily loss probable than the latter.

In any nuclear process, the residual nucleus P may be left in the ground state or in an excited state. To each of these excited states there corresponds a *"group"* of outgoing particles with a certain definite energy. The group of the highest kinetic energy will correspond to the nucleus P being left in the ground state, while the other groups of decreasing kinetic energy to excited states of increasing excitation energy. This will be true for simple capture as well as for particle disintegration. In the case of simple capture, the $h\nu$ of the γ-rays replaces the kinetic energy of the ejected particle. By measuring the kinetic energies of the various groups of the emitted particles or the spectrum of the γ-rays in the simple capture process, the excited energy levels of the product nucleus can be deduced.

The residual nucleus P left in an excited state may break up further, *provided the state is above its dissociation energy,* emitting material particles according to the scheme

*P $\rightarrow$ D + S,

where the asterisk denotes an excited state, D is the second residual nucleus and S the second emitted particle.

Experimentally, such a process will appear as a *three-particle disintegration* according to the scheme

A + B $\rightarrow$ C $\rightarrow$ *P + O$\rightarrow$ D + S + O

It seems that a breaking up into three particles in a single process is very improbable.

A particular case of three-particle disintegration is that in which one or both of the emitted particles S and O are of the same kind as the incident particle B, so that we have in effect the reaction

A + B $\rightarrow$ C $\rightarrow$ D + B + O
Or
A + B $\rightarrow$ C $\rightarrow$ D + S + B
Or
A + B $\rightarrow$ C $\rightarrow$ D + B + B

Such reactions, known as *"non-capture" disintegrations,* occur in several cases of fast neutrons.

If the residual nucleus P is left in an *excited state which is below its dissociation energy,* it will emit one or more γ-rays, until it finally returns to its ground state. These γ-rays from the residual nucleus must not be confused with the γ-rays emitted in the capture process. The latter are part of the main nuclear process itself while the former are emitted *after* the main process is finished. Further, the capture γ-rays have a continuous spectrum if the incident particles are not "mono-kinetic", while the γ-rays from the residual nucleus have a discrete spectrum depending only on the levels of the residual nucleus.

The nucleus which is finally formed, after all particles, γ-rays, etc., have been emitted, may be radioactive, usually β-active, though in some cases it may also be α-active. This is what is known as *artificial radioactivity*. It seems preferable not to include this β-decay in the nuclear reaction at all, because the time required for β-decay is of an entirely different order of magnitude from the time of nuclear reactions. Even the slowest nuclear reactions are completed in about 10^{-13} sec., while the shortest lifetime for a β-radioactive nucleus is 1/50 sec., with most of the lifetimes ranging from a few seconds to a few years. This is 10^{11}, or more, times longer than the duration of a nuclear reaction. The chief use of this β-activity of the final nucleus in the study of nuclear reactions is to provide a convenient means of detecting certain product nuclei. In fact, most of the capture processes with neutrons and many of the reactions produced by charged particles have been discovered through the radioactivity of the product nucleus.

Notations used for representing nuclear disintegrations

Nuclear reactions are often denoted by parentheses enclosing two symbols, the first of which represents the projectile, the second the emitted particle or ray; *a, p, d, n, γ* are used to denote, α- particle, proton, deuteron, neutron, γ-ray respectively. The best known types of nuclear reactions, using this symbolism, may be grouped as follows:

(α, *p*), (α, *n*), (*p*, α), (p, *d*), (*p*, *n*), (p, γ), (*d*, α), (*d*, *p*), (*d*, *n*), (*d*, 2*n*),. (*n*, α), (*n*, *p*), (*n*, 2*n*), (*n*, γ), (γ, *n*), (γ, *p*)

(*p*, γ) and (*n*, γ) represent simple "capture" processes, while the others are "particle" disintegrations.
2*n* denotes the simultaneous emission of 2 neutrons in a given *"non-capture"* disintegration.
(γ, *n*) is known as *photodisintegration*.
Reactions (α, *d*), (α, 2*n*), (α , γ), (*p*, 2*n*), (*a*, *d*), (γ, *d*) and (γ, 2*n*) have not been observed. Just a few (γ, *p*) reactions, perhaps observed, are not well established.

In *certain cases, all these different types of reactions result in the production of unstable residual nuclei* which can be readily detected by their radioactivity and identified by chemical processes.

(3) Complexity of nuclear reactions

The diversity of transmutations of one and the same element subjected to the same bombarding agent and consequently the multiple reactions obtained with different projectiles render the interpretation of data delicate and difficult. Thus, for example, Al may, be transmuted in the following different ways:

(*a*) Al *bombarded by α-particles of* Po exhibits a transmutation with production of *protons;*

Here the Al nucleus is supposed to capture the incident α-particle, the unstable intermediate compound nucleus formed expels a fast proton and the final stable product nucleus is silicon:

(α, p) $_{13}Al^{27} + {_2}He^4 \rightarrow {_{14}}Si^{30} + {_1}H^1$

With the same arrangement, Joliet and Curie discovered a different transmutation where the α-particle is captured, but the compound nucleus emits a *neutron* and the product nucleus is unstable radioactive *radio-phosphorous* which, in its turn, disintegrates with the emission of *positive electrons* and the final stable nucleus is silicon:

(α, n) $_{13}Al^{27} + {_2}He^4 \rightarrow {^*_{15}}P^{30} + {_0}n^1$

$\qquad {^*_{15}}P^{30} \rightarrow {_{14}}Si^{30} + e^+$ (T = 2.5 mins, 3.5 mins.)

A radioelement is marked ordinarily with an asterisk before the symbol.

(b) With protons as transmuting agent

Al transforms itself into stable Mg with the emission of *α- particle:*

(p, α) $_{13}Al^{27} + {_1}H^1 \rightarrow {_{12}}Mg^{24} + {_2}He^4$

(c) With deuterons as projectiles

A similar reaction takes place with the emission of *α- particle* and formation of stable Mg, although it is another isotope:

(d, α) $_{13}Al^{27} + {_1}D^2 \rightarrow {_{12}}Mg^{25} + {_2}He^4$

But deuteron produces also another reaction where *radioactive aluminum* is formed with the emission of *a proton.* The radioactive Al disintegrates emitting electrons and forming Si^{28} as the final stable nucleus.

(d, p) $_{13}Al^{27} + {_1}D^2 \rightarrow {^*_{13}}Al^{28} + {_1}H^1$

$\qquad {^*_{13}}Al^{28} \rightarrow {_{14}}Si^{29} + e^-$ (T = 2 mins, 33 secs)

(d) With neutrons as transmuting agent

Al gives *either radioactive* Na with the emission of *α- particle or radioactive Mg* with the emission of *proton.* Both the radioelements formed emit β-rays and the final stable product is Mg in the first case, while it is Al in the second, the same as at the beginning:

(n, α) $_{13}Al^{27} + {_0}n^1 \rightarrow {^*_{11}}Na^{24} + {_2}He^4$

$\qquad {^*_{11}}Na^{24} \rightarrow {_{12}}Mg^{24} + e^-, \gamma$ (T = 15 hrs)

(n, p) $_{13}Al^{27} + {_0}n^1 \rightarrow {^*_{12}}Mg^{27} + {_1}H^1$

$^*_{12}Mg^{27} \rightarrow {}_{13}Al^{27} + e^-, \gamma$ (T = 10 mins.)

Such multiple alternative reactions are frequently met with, and one must be careful in interpreting experimental data. The concurring evidence of different methods and the use of equations of transmutations are of great help in drawing correct conclusions. In spite of all these precautions, wrong inferences are still possible.

The fact that a given projectile is capable of producing quite a number of different reactions with a given nucleus can be explained by Bohr's idea of *effective "evaporation" of particles*. The transformation of the excited compound nucleus into a stable nucleus may take place by any one of several competitive processes, each with a quite different probability of occurrence. Thus the compound nucleus may emit a γ-ray, a neutron, a proton, or if sufficient energy is available, any combination of the particles. It may be noted that just as different new elements can be produced from a given element, so also one and the same element can be obtained from different elements by suitable choice of the projectiles: *e.g.* Al^{28} can be produced in five different ways.

(4) Equations of transmutations

Nuclear disintegrations are conveniently expressed in the form of equations, similar to the reaction equations used in chemistry. Such nuclear reaction equations are based on the universal law of conservation of energy, in which mass is regarded as one form of energy, according to the mass-energy equivalence of the theory of relativity. Incidentally every such equation must also balance when considered as an ordinary equation in either the mass numbers or the atomic numbers, since both masses and charges are conserved in a nuclear reaction.

A nuclear reaction equation, satisfying the universal principle of conservation of energy can be written in a general way as follows:

$$M_0 + (M_1 + E_1) = (M_2 + E_2) + (M_3 + E_3) \dots\dots\dots\dots\dots (1)$$

Where,
M_0 is the mass of the nucleus bombarded,
M_1 and E_1 the mass and kinetic energy respectively of the projectile,
M_2 and E_2 and M_3 and E_3 similar quantities of the two products or transmutation, *viz.* the outgoing particle and the product nucleus respectively.

We have not included the intermediate compound nucleus, as it is not required in the present consideration. It is to be noted that the two sides of the above equation can be expressed in terms of either energy or mass, thanks to the relativity principle of Einstein.

Nuclear reaction energy

In the nuclear reaction equation, an important factor, known as the nuclear reaction energy or the *disintegration energy,* usually denoted by the symbol Q, is involved. It represents the energy liberated or absorbed during the nuclear reaction and hence gives the energy balance in the reaction. It is equal to the difference between the sum of the masses of the particles that enter into the reaction and the sum of the masses of the particles that result from the reaction.

From equation (1)

$$Q = (M_o + M_1) - (M_2 + M_3) = E_2 + E_3 - E_1 \quad \text{......................} \quad (2)$$

The value of Q, may, therefore, be found in either of two alternative ways:

(a) If the masses of all the particles that are involved in the reaction are known, *i.e.,* M_o, M_1, M_2 and M_3, the value of Q is at once calculated by subtracting the sum of the masses of the final particles from the sum of the masses of the initial particles and then converting the difference into energy units by Einstein's mass-energy relation..

(b) If the three kinetic energies E_1, E_2 and E_3 are measured, the value of Q can be calculated from the second part of equation (2). The kinetic energy (E_1) of the projectile is readily known in most cases. For α- particles and γ-rays from natural radioactive substances, E_1 is the energy of the groups used as projectiles; for artificially accelerated protons, deuterons and helium nuclei, E_1 is given by the accelerating potential.

In the case of neutrons, *if fast,* the energy can be known from collision experiments; *if slow,* it is a negligibly small quantity; the kinetic energy E_2 of the emitted particle, which is always a proton or an α- particle or a neutron, is the quantity experimentally measured by the detecting devices; the kinetic energy E_2 of the residual nucleus can be calculated by a simple application of the law of conservation of momentum to experimental data.

The nuclear reaction equation can be written as

$$M_o + M_1 \rightarrow M_2 + M_3 + Q \quad \text{......................} \quad (3)$$

Q may be *positive* or *negative.* In the first case the sum of the masses of the final particles will be *less* than that of the initial particles and *energy* will be *evolved* in the reaction. In the second case the sum of the masses of the final particles will *exceed* that of the initial particles and *energy* will be *absorbed* in the reaction, being obtained from the kinetic energies of the particles. Accordingly, nuclear reactions can be classified as *exothermic* or *exoergic* (when Q is positive and energy is released) and *endothermic* or *endoergic* (when Q is negative and energy is absorbed).

Reactions equivalent to the simple addition of a deuteron, proton or neutron to a nucleus, such as (*d, n*), (*d, p*), (*p, γ*), (*n, γ*), as well as (*d, α*) reactions, are nearly always exoergic while reactions of the type (*p, n*), (*n, p*) are usually endoergic. To produce an endoergic reaction in a stable nucleus, the projectile must possess kinetic energy at least equal to Q, since energy equal to that amount disappears in the reaction and the final energy, in all its different forms, rest, kinetic and radiant, cannot be negative.

These considerations bring out *the great utility and importance of the equations of transmutation.* They enable us not only to *check up the actual disintegrations* represented by them, but also *to predict other possible reactions.* One can deduce from these equations whether a given nuclear reaction will be endothermic or exothermic, which is a. great asset to the experimental investigator of artificial transmutations.

Further, they offer a *means of accurate determination of atomic masses,* since in any one nuclear reaction expressed by a suitable equation, if three of the four masses are known, a measurement of the reaction energy is sufficient to deduce the mass of the fourth atom. *New isotopes, hitherto unknown, can be discovered by means of them.*

Also Einstein's mass-energy relation can be experimentally verified. For, if the reaction energy is measured in a nuclear reaction and the masses of the atoms involved are assumed from mass spectrograph data, the equation representing the nuclear reaction readily establishes the veracity of Einstein's law.

Examples
A few concrete cases of nuclear disintegrations may now be considered to illustrate the important results that could be obtained with the help of the equations of transmutation:

(*i*) *Nitrogen bombarded by α- particles*

It is the first observed case of artificial transmutation, where protons are ejected and the rare isotope of oxygen O^{17} is formed as the product stable nucleus. The best value of the reaction energy Q obtained from measurements of the kinetic energies of the particles involved in the process is — 1.26 MeV.

In interpreting these experimental data one meets, at the very outset, with the great difficulty common to experiments on transmutation with elements of two or more isotopes, viz., one does not know which of the isotopes is or are being disintegrated. This is, sometimes, welcome to the theorist who can ascribe the transmutation to whichever isotope happens best to fit his theory. Nitrogen has two stable isotopes of masses 14 and 15, the latter being only a very small proportion (0.38%) in ordinary nitrogen.

To, explain satisfactorily the observed phenomenon when α- particles bombard ordinary nitrogen gas, the equation of transmutation is written as

$$(\alpha, p) \quad _7N^{14} + _2He^4 \rightarrow _8O^{17} + _1H^1 + Q$$

Attached to the chemical symbol of each atom are its mass number as superscript and its atomic number as subscript. It is seen that the equation is balanced as regards both the mass numbers and the atomic numbers. Q is evidently the energy-balance in the reaction.

The atomic masses of all the particles figuring in the equation are accurately known to the fifth decimal place from mass spectrum analysis, *viz.,*
N^{14} = 14.00753,
He^4 = 4.00387,
O^{17} = 17.00450
and
H^1 = 1.00813.

Substituting these values in the equation,

14.00753 + 4.00386 = 17.00450 + 1.00813 + Q

Q = 18.01139 — 18.01263 = — 0.00124 (m.u.)

Converting mass into energy using Einstein's mass energy relation, according to which 1 m.u. = 931 MeV,

the calculated value of Q is

Q = —0.00124 x 931 = —1.16 MeV.

This agrees very well with the value obtained from the experimental measurement of the kinetic energies of the particles involved. The slight discrepancy is probably due to the poor geometry of the experiment, necessitated by the low intensities.

It is readily seen that Einstein's law of mass-energy equivalence has a quantitative verification here, since the agreement between the value of Q calculated from the equation of transmutation and that experimentally observed is possible only on the supposition of the validity of the law. There is also sufficient evidence to admit that the transmutation actually takes place in the manner represented by the equation.

(ii) Lithium bombarded by protons

This was the first successful disintegration performed by Cockcroft and Walton (1932) using protons artificially accelerated with their high tension multiplier to energies ranging from 100 to 700 KeV. Bombarding a lithium target with these fast protons, they observed the emission of α- particles which were detected first by the scintillation method and then with Geiger counters. They were found to have a range of slightly over 8 cms in air. The yield of α- particles was found to decrease rapidly as the energy of the bombarding protons was reduced, but it could still be traced for energies as low as 20 KeV.

The experiment was repeated in 1934, by Dee and Walton using protons of energy less than 200 KeV and a Wilson's cloud chamber for detecting α- particles. They found that pairs of α- particles shot out in almost opposite directions, as seen from one of their photos reproduced here.

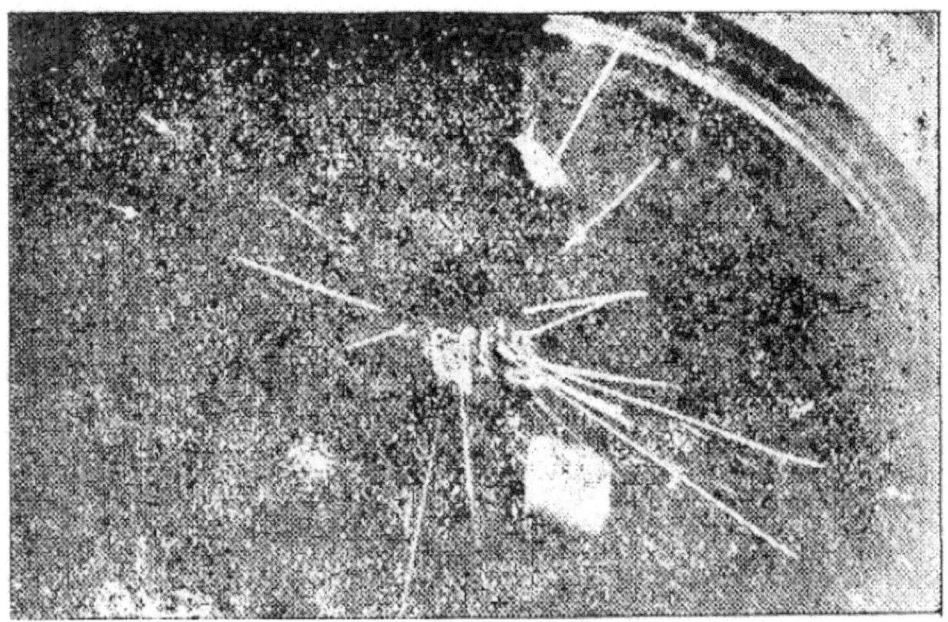

$_3\text{Li}^7 + _1\text{H}^1 \rightarrow 2 \ _2\text{He}^4$ (Dee and Walton)

Two pairs of a-tracks of equal length are indicated by the arrows; the tracks are not quite opposite due to the forward momentum given to the pair of α- particles by the impinging protons. Each α- particle has a range of 8.3 cms corresponding to an energy of about 8.6 MeV. Thus the products of disintegration have a total energy of over 16 MeV, while the bombarding proton provoking the disintegration has only an energy of about 0.2 MeV, which argues to a veritable explosion of the bombarded Li nucleus which supplies all that great excess of energy to the α- particles. The reaction energy Q, as determined from the momentum-energy consideration, is found to be 17.28 MeV.

In the interpretation of these experimental data by the use of transmutation equations, we again meet with the initial difficulty as to which of the two isotopes of Li of masses 6 and 7 is implicated. The relative abundances are 7.9% for Li^6 and 92.1% for Li^7.

Writing the equation of reaction as

$(p, α)$ $_3\text{Li}^7 + _1\text{H}^1 \rightarrow 2 \ _2\text{He}^4 + Q$

and substituting the values of the masses of the different nuclei from the mass spectrograph data

$7.01816 + 1.00813 = 2 \times 4.00386 + Q$
$Q = 8.02629 — 8.00772$
$\quad = + 0.01857$ (m.u.)
$\quad = + 0.01857 \times 931 = 17.29$ MeV.

This agrees very well with the experimentally measured value of Q, so that one might legitimately conclude that the reaction actually takes place in the fashion expressed by the equation and that Einstein's law is valid.

But, this is not the only way in which lithium is, transmuted by protons, for there are indications to believe that the less abundant isotope Li^6 is also disintegrated in the same proton bombardment, since Cockcroft and Walton observed also the presence of other short range α-particles. These were investigated in detail by Oliphant, Kinsey and Rutherford (1933) who showed that they formed two groups of 1.2 cms and 0.8 cm. respectively. They suggested that these α- particles are due to the transmutation of Li^6 according to the equation:

$(p, α)$ $_3Li^6 + {}_1H^1 \rightarrow {}_2He^4 + {}_2He^3 + Q$

Thus a *new rare but stable isotope of helium of mass number 3 was discovered.* Using the measured value of Q (3.94 MeV) and the known masses of Li^6, H^1 and He^4, the mass of the new isotope can be calculated and it turns out to be 3.01699.

Sir Marcus 'Mark' Laurence Elwin Oliphant (8 October 1901 – 14 July 2000) Australian

Gamma rays of very high energy (17 MeV) have also been observed as the result of the bombardment of lithium by proton. This has been interpreted as a simple capture process, represented by the equation

$(p, γ)$ $_3Li^7 + {}_1H^1 \rightarrow {}_4Be^8 + hv$

The nucleus of Li^7 captures a proton and the compound nucleus is formed in an excited state. This nucleus returns to its normal state with the emission of a γ-ray photon of high energy.

With fast protons of energy about 2 MeV, induced radioactivity (T = 43 days) can be observed, the product nucleus being a radioactive isotope of Be according to the equation

$$(p, n) \quad _3Li^7 + {_1}H^1 \rightarrow {^*_4}Be^7 + {_0}n^1 + Q$$

Thus we are made to realize the complexity of reactions arising even with a single target and a single projectile. Oliphant and Shire and Crowther, in 1934, by bombarding the pure isotopes of Li (separated by means of a mass spectrograph) with protons, were able to show which reactions were due to Li^6 and which to Li^7.

(iii) Deuterium bombarded by deuterons

This is another case of transmutation which is of great interest and importance. Oliphant, Hardtack and Rutherford were the first (1934) to study this disintegration. Deuterium targets were made either by freezing heavy water on a surface cooled by liquid air or out of some other solid compound containing heavy hydrogen. *Two different nuclear reactions* have been observed to occur with approximately equal probability and are detectable oven with-deuterons of energy as low as 100 KeV.

The first reaction is

$$(d, p) \quad _1D^2 + {_1}D^2 \rightarrow {_1}H^3 + {_1}H^1 + Q$$

It has been studied with the aid of a cloud chamber, which enabled the particles to be identified as isotopes of hydrogen of mass numbers 1 and 3. A photograph of this reaction taken by Dee is reproduced here.

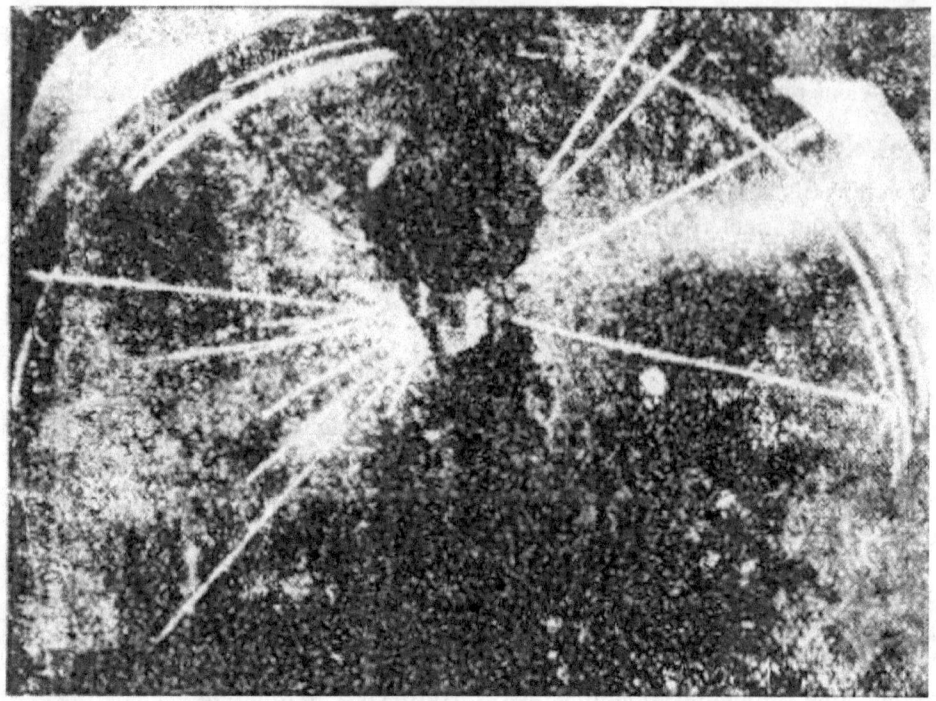

(d, p) $_1D^2 + {_1}D^2 \rightarrow {_1}H^3 + {_1}H^1$ (Dee)

The longer tracks are of ordinary protons H^1 and the shorter ones are due to the isotope H^3. The ranges of these particles have been found to be 14.7 cms and 1.6 cms respectively and the measured value of Q is 3.98 MeV. With the known masses of D^2 and H^1 and the value of Q, the mass of H^3 can be determined and is found to be 3.01700. This is at present the most accurate method of determining the mass of H^3. Thus a *new isotope of hydrogen, heavier than deuterium called "tritium", has been discovered,* although it is not found as one of constituents of ordinary hydrogen. It has been later shown by O'Neal and Goldhaber (1940) *that H^3 is radioactive* and disintegrates as follows:

$*{_1}H^3 \rightarrow {_2}He^3 + e^-$,

ejecting electrons with a half period T = (31 ± 8) years, and maximum energy of electrons of only 15 KeV. This explains its absence in the ordinary hydrogen.

The second reaction is

(d, n) $_1D^2 + {_1}D^2 \rightarrow {_2}He^3 + {_0}n^1 + Q$

The energy of the neutrons produced in this reaction has been investigated by observing the recoil tracks of protons projected by them, which are sometimes found in cloud chamber photographs apart from the two groups of particles due to the first reaction: From a. number of such photographs, Dee has shown that the neutrons constitute a homogeneous group. Recent measurements show that the energy of neutrons emitted at

$90°$ to the direction of the incident deuterons is 2.38 MeV + 1/4 of the deuteron energy. The value of Q has been found to be 3.16 MeV. With this value of Q and using the masses of D^2 and He^3, the mass of the neutron can be calculated; or assuming the mass of al, the mass of He^3 can be determined. This reaction provides a very convenient source of neutrons, chiefly when fast neutrons of definite energy are required.

(iv) Lithium bombarded by deuterons

The first experiments of this type were carried out, in 1933, by Prof. Lawrence who was able to obtain fast deuterons with his cyclotron. Using deuterons of 1.33 MeV energy he found that pairs of α- particles of equal range were shot out in opposite directions. The range of each α- particle was 12.7 cms corresponding to a reaction energy of 22.1 MeV, the largest measured for any nuclear reaction. Dee and Walton confirmed these results by the cloud chamber method.

Writing the equation for this reaction as

$$(d, \alpha) \quad _3Li^6 + _1D^2 \rightarrow 2\,_2He^4 + Q$$

and substituting the masses of the different nuclei involved from mass spectrograph data, the value of Q was found to be 22.17 MeV in excellent agreement with the experimental value.

Another transmutation in the same reaction is

$$(d, p) \quad _3Li^6 + _1D^2 \rightarrow _3Li^7 + _1H^1 + Q$$

which gives a value of 4.8 MeV for Q. There is evidence for this in the cloud chamber photographs obtained. Over and above the two heavy tracks in opposite directions due to the two α- particles formed in the previous reaction, there is a thin track in the foreground, most probably due to a proton of range 30.5 cms which can be accounted for by the present reaction. The range of the recoil Li^7 is not great enough to pass from the target into the chamber.

A third possible transmutation in the case of Li bombarded by deuterons is with the heavier isotope Li^7. Oliphant, Kinsey and Rutherford, using a pure deuteron beam (magnetically separated from proton impurity), first observed (1933) a continuous distribution of α- particles with ranges up to 6 MeV. The reaction most probably is

$$(d, \alpha)\ (d, n) \quad _3Li^7 + _1D^2 \rightarrow 2\,_2He^4 + _0n^1 + Q$$

Here we have a case of emission of three particles simultaneously. From studies of experimental data, it seems possible to classify the reaction as both (d, α) and (d, n) types. The available energy being distributed at random between the two α- particles and the neutron, the largest energy for an α- particle will be obtained, if this particle is emitted in a direction opposite to that of the other α- particle and the neutron, the latter escaping directions parallel to one another, according to the principle of conservation of momentum. In such a case the first α-particle takes nearly 5/9 of the total energy. Assuming that the observed upper limit for the energy of the α- particles can be explained in this way, the value of Q is 14.9 MeV approximately, in satisfactory agreement with the value of 15.04 MeV calculated from the masses of the nuclei taking part in the reaction.

According to Lauritsen and his co-workers, Li^7 can also be transmuted in the following way

$$_3Li^7 + {}_1D^2 \rightarrow * {}_3Li^8 + {}_1H^1 + Q$$

The isotope of Li^8 appears to be radioactive and disintegrates into Be^8 with the emission of an electron.

These illustrations show sufficiently clearly the importance of the study of transmutations with the aid of nuclear reaction equations which, on the one hand, testify to the validity of Einstein's mass-energy relation and, on the other, enable one to pick out reactions which actually occur among the many complicated and possible ones. Examples can be multiplied without restriction, thanks to the many experimental data already gathered in artificial disintegrations.

(5) Nuclear cross-section (σ)

In the study of nuclear reactions it is often instructive to evaluate a quantity known as the *cross-section,* usually represented by the symbol σ, which is a measure of the probability of the reaction under consideration. It is intimately related to another quantity known as the *yield* which is defined as the ratio of the number of emitted particles to the number of incident particles in a reaction.

The cross-section for any nuclear process, in general, may be visualized in a simple manner as follows:

Let I be the number of incident particles which penetrate completely a thin target of thickness t containing N particles (nuclei) per sq. cm and m be the resulting interactions of the type considered. Let σ be the effective cross-section.

The average volume per nucleus in the target = $(1 \times t)/N$ c.c.

Since each incident particle will sweep out a cylindrical volume of area σ and length equal to its depth of penetration, *i.e., t,* the effective volume swept out by the I incident particles in traversing the target $= I \sigma t$.

Hence the number of collisions

$m = I \sigma t / (t/N) = I N \sigma$
and

$$\sigma = m \, / \, I \, N \qquad cm^2$$

The quantity σ is not inherently determined by the dimensions of the bodies and for two given particles it may have different values 'depending on the particular process considered. In the various nuclear reactions it has values ranging from 10^{-20} to 10^{-32} cm^2. The Cross-section σ is always expressed in terms of a unit called the *"barn"* which is 10^{-24} cm^2. It is usual to speak of elastic collision, inelastic collision, capture and fission cross-sections.

The cross-sections for different nuclear processes have been worked out by different authors on the basis of wave-mechanical theory. It has been shown that *in the case of nuclear reactions involving charged particles, the cross-section, in general, increases with increasing energy of the incident particles.* But there are many instances in which the cross-section becomes very great for particular and even low values of the energy within narrow limits, which can be explained only on the principle of *resonance penetration* through virtual levels of the disintegrated nucleus. When the energy of the incident particle exceeds that corresponding to the potential barrier of the bombarded nucleus, the cross-section no longer increases with increasing energy of the incident particle.

If the incident particle is a neutron, the cross-section is comparatively greater than that in the case of a charged particle, evidently due to lack of a potential barrier for the uncharged neutron. Thus for instance, the cross-section for proton lies roughly between 10^{-29} and 10^{-27} cm^2, while that for neutron is between 10^{-26} and 10^{-24} cm^2 . Further, neutrons have the special property of very large cross-sections, when their energy is very low, of the order of an electron volt, and the process involved is a simple radiative capture.

This is evidently a resonance effect due to the existence of low energy neutron levels inside the bombarded nucleus.

The cross-section depends also upon the atomic numbers of the bombarded and the bombarding nuclei, decreasing as the atomic number increases. This explains the increasing difficulty of transmutations of heavy elements by protons, deuterons and α-particles. The same is true even in the case of neutrons when the reaction leads to the emission of charged particles as protons or α- particles, for although the incident neutron penetrates without difficulty into the nucleus, yet with increasing atomic number of the nucleus bombarded, the potential barrier preventing the charged particles from escaping becomes higher and higher and its transparency rapidly decreases.

CERTAIN PROMINENT TYPES OF NUCLEAR REACTIONS

1. Substitutional reactions

These refer to particle disintegrations in which one of the component particles of the bombarded nucleus may be considered to be *substituted* by the incident particle, so that the constituent particle thus replaced is ejected and the nucleus is transmuted. Many reactions of this type have been observed with the different kinds of projectiles, α-particles, protons, deuterons and neutrons.

A very important feature in substitutional reactions is the evidence they furnish for the existence of quantum states of excitation in nuclei by the readily observable "groups" of outgoing particles.

The incident particle, on entering the nucleus can be captured, not on the ground level, but on a level corresponding to an excited state of the nucleus.

In such a case, the emitted particle will have an energy less than normal by the amount of excitation and, if several excited states exist, one may observe several groups of emitted particles in complete analogy with the phenomenon of "fine structure" of α- particles met with in radioactive substances.

It is to be expected also that transformations which are characterized by several groups of emitted particles will be accompanied by γ-radiations corresponding to the transition of the product nucleus to the normal state after the reaction. The energies of the γ-rays should be equal to the differences in the energies of the various groups. It is generally assumed that the group of highest energy corresponds to the product nucleus left in the ground state; the others of lower energies correspond to higher excited states. The differences between the highest energy group and the others thus give directly the energies of the excited levels of the product nucleus. The value of the reaction energy Q usually given in connection with the reaction equations corresponds to the highest energy- in the groups of the emitted particles. These considerations may now be illustrated by a few examples.

(i) Al bombarded with α- particles

This is one of the early reactions which has been carefully studied by Pose, Chadwick and Constable and others. The reaction equation is

$$_{13}Al^{27} + _2He^4 \rightarrow {}_{14}Si^{30} + {}_1H^1 + Q$$

It is therefore a (α, p) reaction, in which protons are emitted and the product nucleus is silicon. The range and number of the emitted protons were observed for particular energies of the incident α- particles. When these data were plotted, the distribution- in-range curve of the protons presented the form shown in Fig. 290.

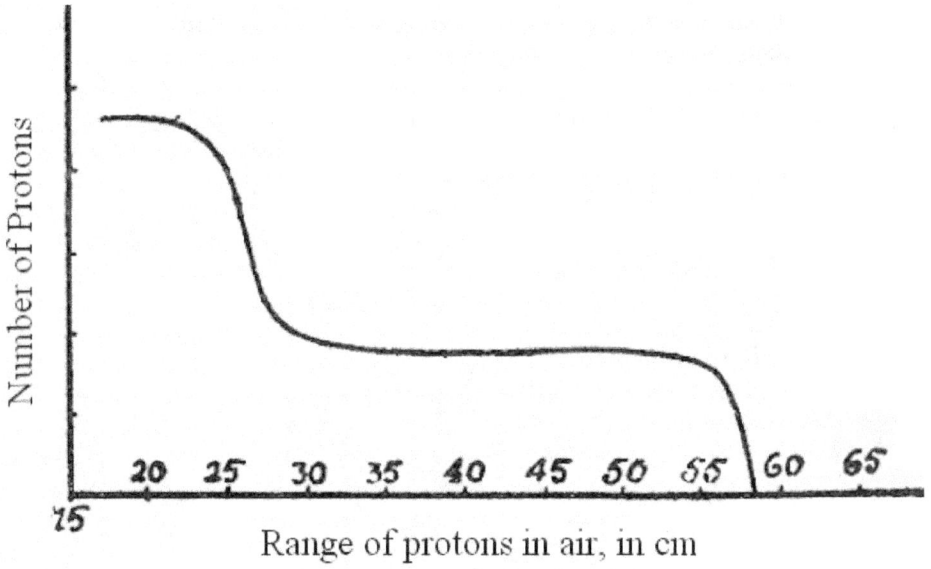

Fig. 290. Distribution-in-range curve of protons from (Al + α).

It is seen that the protons produced for a given two homogeneous α-particles form groups (represented by the flat parts), one of about 28 cms range and the other of about 58 cms range. Other homogeneous groups of protons were also observed, using α- particles of different energies, in all about 8 groups which require the intervention of resonance phenomenon for a complete interpretation as we shall see below. For the present, considering only the two groups of protons obtained for a particular energy of the incident α- particles, these groups indicate that the product nucleus silicon has been left in at least two excitation levels, the lower one corresponding to the ejection of protons of longer range. The expected γ-rays, as the excited product nucleus returns to the normal state, have actually been observed in the reaction, although the data are not accurate and extensive enough for a quantitative verification.

The proton groups emitted by boron, nitrogen, fluorine, sodium, magnesium, sulphur, etc., when bombarded by α- particles have been analyzed in a similar manner. In the case of nitrogen, only one group of protons has been found, evidently due to the fact of a very low reaction energy (-1.3 MeV) which makes excitation energetically impossible.

(ii) Be bombarded by deuterons

Bonner and Brubaker, (1930) have studied the neutron groups emitted in the reaction

$$_4Be^9 + {}_1D^2 \rightarrow {}_5B^{10} + {}_0n^1$$

Using 0.9 MeV deuterons, they determined the energies of the emitted neutrons from recoil proton tracks produced by them in a high-pressure cloud chamber containing methane. They limited themselves to those proton tracks which pointed almost straight away from the target and, in consequence, could be assumed as projected in the same direction as the incident neutrons.

Plotting the number of tracks against the measured energies the protons, they obtained a curve with four peaks as shown Fig. 291.

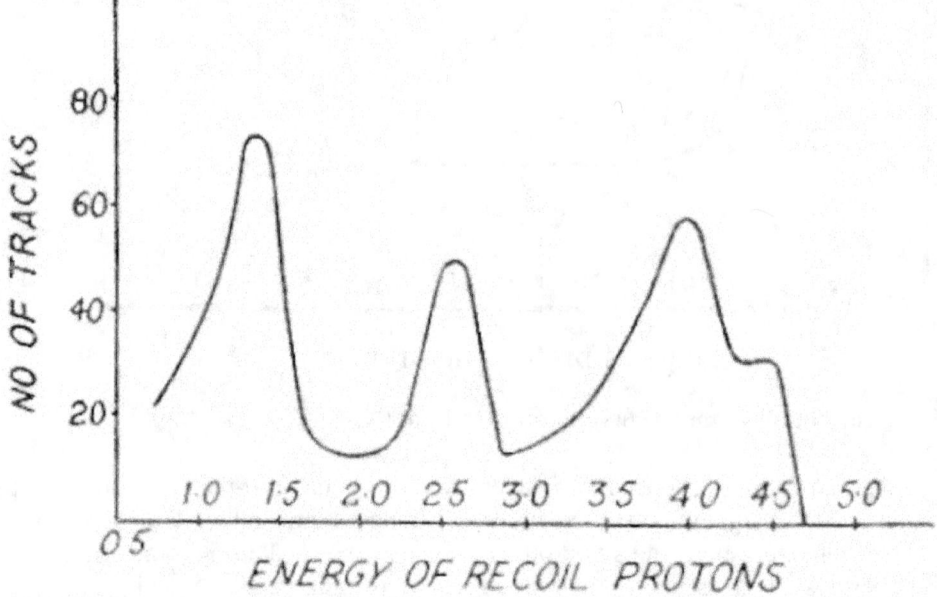

Fig. 291. Energy distribution of recoil protons due to neutrons from (Be + D).

These peaks indicate that the product nucleus B^{10} may be left in any one of four possible quantum states. For it is on account of this fact that the neutrons emitted in the bombardment of beryllium by deuterons of a given energy and moving in a direction at right angles to the deuteron beam have energies equal to one of the four sharply defined values and hence the protons that are projected straight ahead by the neutrons are limited to one of the four possible kinetic energies. Broad peaks instead of sharp lines are obtained due to the limitations of actual experimental conditions, such as the impact of the high-speed neutrons on the walls of the cloud chamber; lack of homogeneity in the incident deuteron beam, the slowing down of the deuterons in the target and the straggling of the proton tracks. The high energy edges of the peaks can, however, be taken as representing fairly well the energies of the four groups of neutrons that would be emitted under ideal conditions.

From the curve, the energies of these four groups are found to be 1.4, 2.6, 4.0 and 4.52 MeV, the last one corresponding to the lowest state of B^{10} nucleus, while the others to higher excited states. Applying the laws of impact, the values of the reaction energy Q in the four cases are found to be 0.82, 2.14, 3.68 and 4.25 MeV. Substituting the masses of the elements in the reaction equation, Q can be estimated and is found to be 4.39 MeV in good agreement with the value obtained for the group of the highest energy.

The energy levels of B^{10} can be readily found by subtracting from 4.25 the other three values of Q, viz., 0.82, 2.14 and 3.68. They are at 0.57, 2.11 and 3.43 MeV above the ground level, as shown in Fig. 292. Kruger and Green (1937) have been able to study the γ-rays from the same reaction and confirm the existence of the four energy levels. From the distribution of the Compton electron tracks ejected by the γ-rays from a thin mica sheet in a cloud chamber, they inferred that six γ-ray lines occur with energies 0.51, 1.07, 1.14, 1.96, 2.81 and 3.21 MeV.

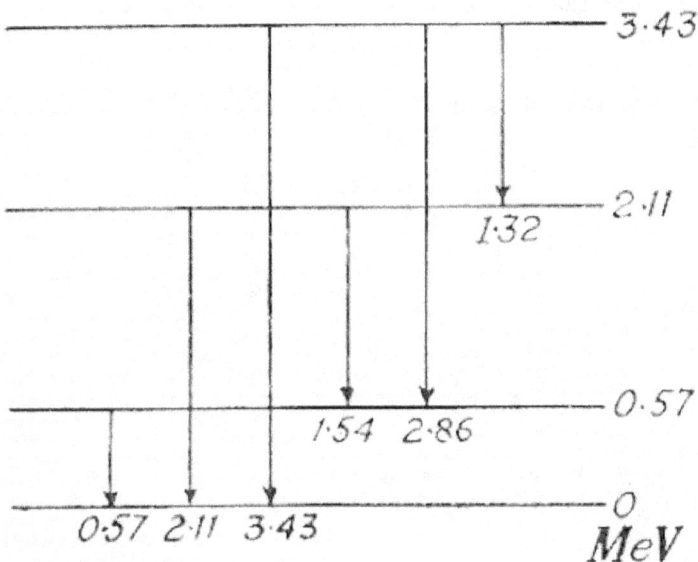

Fig. 292. Energy levels of B^{10}.

These γ-rays could be interpreted arising from different transitions between the four levels of the B^{10} nucleus, as indicated in the figure. The values calculated from the proton energies, shown in the figure, agree with those estimated from γ-ray energies within the limits of experimental error.

Bonner, in 1940, *bombarding fluorine with deuterons* and producing transmutation according to the equation

$$_9F^{19} + {}_1D^2 \rightarrow {}_{10}Ne^{20} + {}_0n^1 + Q$$

found seven neutron groups. Another case in which the group structure of neutrons has been carefully analyzed by Bernardini and Bocciarelli (1937) is the *reaction between Be and α- particles* according to the equations

$$_4Be^9 + {}_2He^4 \rightarrow {}_6C^{12} + {}_0n^1$$

And

$_4Be^9 + _2He^4 \rightarrow 3\ _2He^4 + _0n^1$

There is complete evidence for four neutron groups in these reactions.

2. Radiative capture reactions

When an incident particle penetrates into the nucleus it need not necessarily eject another nuclear particle (substitutional reaction) since it can be bound to the nucleus in a stable state, the excess of energy being got rid of by the emission of a γ-ray. Such a process is called a *radiative capture.*

In general, this simple capture reaction is less probable than particle disintegration, because the "radiation width" of nuclear levels is usually smaller than the "particle width". This type of reaction can, therefore, take place only when particle disintegration probability becomes extremely small for some special reasons such as:

(*a*) sufficient energy is not available to make up the masses of the possibly produced particles
or
(*b*) the potential barrier is too high for particles which, on energetic grounds, might be emitted,
or
(*c*) only enough energy is available for slow particles to be emitted so that the particle width is very much reduced
or
(*d*) the emission of particles is forbidden by selection rules.

Conditions (*b*) and (*c*) are fulfilled for the simple capture of slow neutrons by heavy nuclei, since the height of the potential barrier in such cases prevents the ejection of charged particles. Hence, the radiative capture process becomes very important in the case of slow neutrons reacting oil heavy nuclei.

The first and last conditions, (e) and (*d*), are realized in a few transmutations produced by bombarding light nuclei with protons.

The evidence for the radiative capture process cannot be had directly as there are no fast particles as products of disintegration, but only indirectly, either by detecting the γ-rays emitted in the process by their secondary effects, or by the product nucleus when it happens to be β-active, as in the case of many transmutations caused by the capture of neutrons.

Radiative capture of protons

Postponing the study of radiative capture of neutrons to the section on neutrons, we shall state here briefly the results obtained with the radiative capture of charged incident particles. The first experimental evidence of this process was furnished by Cockcroft

(1934) who noticed that a carbon target bombarded by an intense beam of protons becomes positron-active, the activity being evidently due to the formation of radioactive N^{13} according to the reaction

(p, γ) $_6C^{12} + {}_1H^1 \rightarrow {}^* {}_7N^{13} + h\nu$

$^* {}_7N^{13} \rightarrow {}_6C^{13} + e^+$ (T = 10 mins)

This process is evidently due to the fact that sufficient energy is not available to make up the masses of the possibly produced particles [condition (a)].

Other cases that have been discovered and studied extensively by Hafstad and Tuve, Crane, Delsasso, Lauritsen, Fowler and others are:

(p, γ) $_3Li^7 + {}_1H^1 \rightarrow {}^* {}_4Be^8 + h\nu$

(p, γ) $_9F^{19} + {}_1H^1 \rightarrow {}^* {}_{10}Ne^{20} + h\nu$

These reactions have been detected and analyzed either by the emitted γ-rays or by the radioactivity of the product nucleus or by both.

In the first case of the *γ-rays from the capture of protons by Li,* experiments conducted by Delsasso and his associates, in 1937, have shown that there is only one strong γ-ray line at 17.1 MeV and probably one or more weak ones between 10 and 17 MeV. The first line corresponds to the transition to the ground state of Be^8. The reaction energy Q calculated from the masses of the nuclei involved and the energy of the incident protons (440 KeV) is 17.4 MeV, in good agreement with the observed 17.1 MeV. A line at 14 MeV may be expected from a transition to the excited state of Be^8 of 2.8 MeV excitation energy known from other nuclear reactions.

The second case of *γ-rays from the capture of protons by fluorine* is interesting since it gives a good example of the working of selection rules [condition (*d*)]. There seems to be a single γ-ray line having an energy of 6 MeV about. The reaction energy Q deduced from nuclear masses arid the energy of incident protons (0.33 MeV) is 13.23 MeV. It appears, therefore, that the transition to the ground state is forbidden, evidently by selection rules.

3. Resonance disintegrations

It was early realized that in the (α, *p*) reactions certain groups of the emitted protons have a definite energy which does not vary with the incident α-particle energy. Furthermore, these groups are observed only for a definite energy of the α- particles. These facts indicate the existence of the *phenomenon, of resonance,* first suggested by Gurney on wave-mechanical principle and experimentally observed in the bombardment of Al with α- particles by Pose. It has been subsequently found by many other observers that the resonance disintegration is a pretty common occurrence, not only in the case of bombardment of light elements by α-particles, but also in proton and neutron produced processes.

The principle underlying this resonance process is as follows:

If the energy of the incident particle is such that the total energy of the system is close to one of the virtual levels of the compound nucleus, the probability of the formation of the compound nucleus and hence of the nuclear process will obviously be much greater than if the energy of the particle falls in the region between the levels. Therefore, one may expect to find characteristic fluctuations of the *yield* of a nuclear process with the energy, from high values at the resonance energies to low values between resonance levels.

These resonance phenomena are most pronounced with slow neutrons, but have also been observed in the radiative capture of protons and in transmutation caused by α- particles, where they were first discovered. The possibility of producing excited residual nucleus is also present in resonance disintegration and results in two or more groups of emitted particles of constant range{in particle disintegration) or of discrete γ-ray spectra (in capture process) Occurring simultaneously for a given resonance energy of the incident particle.

The study of these resonance phenomena is of great importance in nuclear physics. First of all, the *spacing between the neighboring levels* of the compound nucleus may be deduced from the resonances. Experimental data indicate that this spacing is of the order of 10 volts for *heavy* nuclei of atomic weight 100 or more and excitation energies of the order of 8 MeV, as obtained from slow neutron reactions. For *light* nuclei and excitation energies about 12-15 MeV, spacing of a few hundred thousands volts seems to prevail, as derived from proton capture, γ-ray spectra and resonance in α- particle disintegrations. The determination of the spacing as a function of the mass number and of the excitation energy enables theoretical ideas about nuclear structure to be checked.

Secondly, *the width of the resonance levels* is of great interest. Just as in the theory of atoms the width of an excited level is given by the probability of its emitting radiation, so also the width of a compound nucleus is given by the total probability of the emission of the particles, γ-rays, etc. Thus the width of the levels enables one to determine the probability of the concentration of energy on any one of the particles in the compound nucleus.

The width of the resonance levels may be obtained in various ways:

(1) If there are a few resonances of comparatively large width, the simplest procedure is to measure the total yield of the reaction from a thin target as a function of the incident particle energy.

(2) With a thick target, the intensity of a given group of emitted particles is studied as a function of the energy of the incident particle. From the incident particle energy at which

the group first appears and that at which it attains its full intensity, the width of the level may be deduced.

(3) The inhomogeneity in energy of the omitted particles of a given group will also give the width of the level. In general, the width tends to decrease with the increasing mass of the nucleus.

In the case of *light* nuclei, as the spacing of the resonance levels is wide, resonances will, in general be observable only if particles of suitable energy are available and if the width of the resonance levels is small compared to the spacing. For very high excitation energy, the ratio of the total width to the spacing can be shown to increase rapidly. Therefore, *it is improbable that the resonance phenomena will be observed with deuterons* as the incident particle, since the high internal energy of the deuterons makes the energy of any compound nucleus also very high.

In the case of *heavy nuclei,* since the spacing is extremely narrow; no available source can give fast charged particles homogeneous enough to observe such closely packed levels. Hence *one cannot hope to observe resonance levels in heavy nuclei with charged particles.*

One of the methods of experimentally studying the .phenomena of resonance is by means of the *excitation function curves* which are obtained by plotting the yield of the reaction against the continuously varied energy of the incident particle. Such a curve frequently shows sharp maxima indicating resonance.

To illustrate the main features of resonance disintegration, we shall now consider some experimentally observed and carefully studied reactions.

(*i*) *Aluminum bombarded by α- particles*

This reaction, which has already been cited in connection with the groups of emitted protons is also the one that gave the first experimental evidence for resonance phenomena. Pose, in his experiment on the transmutations of Al with α- particles from Po (1929), used thick layers of Al (≈ 0.04 mm) so that the incident α- particles were gradually reduced in energy and finally stopped in the target itself. In such a case, one should expect that the protons produced in the interaction of α- particles with Al nuclei at different depths should have different velocities, so that their observed energy distribution would be in the form of a continuous spectrum extending from an upper limit determined by the initial energy of the α- particles to much lower values. The measurements indicated, however, that the emitted protons belonged to several more or less discrete groups, as shown in Fig. 293, where the number of ejected protons is plotted against their range. This result can be explained only on the hypothesis that the observed groups of protons were produced at a number of particular depths in the Al target, at which the continuously slowing down α- particles have energies just corresponding to resonance penetration.

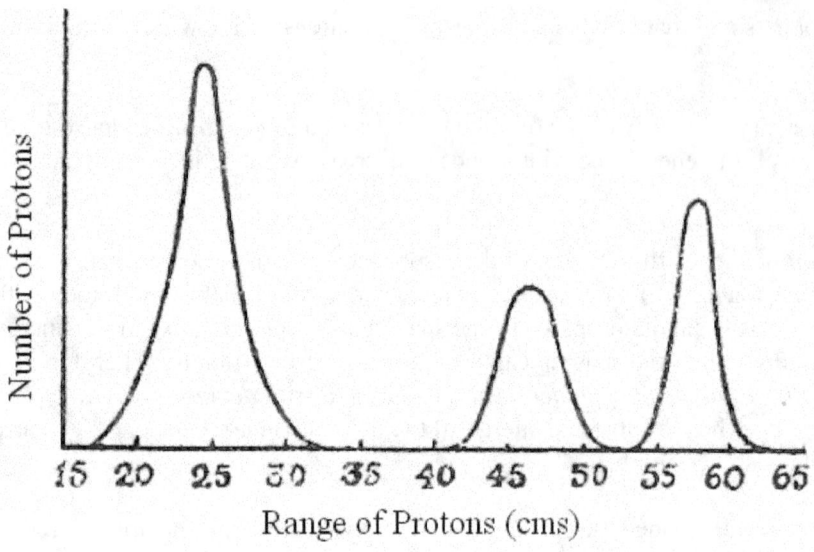

Fig. 293. Excitation function curve of protons emitted by (Al + α)

When the α- particles were independently reduced by suitable absorbers to {definite values of energy before bombarding the Al target and the experiment repeated, on plotting the number of ejected protons against their range for the different values of the energy of the incident α- particles, a series of curves were obtained with, the same "step" structure as in Fig. 288, thereby indicating the same groups of protons at the different resonance levels of the compound nucleus. This means that the α-particle entering the nucleus through any one of the different resonance levels might be captured in several distinct modes and consequently give rise to discrete groups of protons

. The resonance levels refer to the intermediate *compound* nucleus rather than to the initial Al nucleus, while the groups of protons at each resonance level give indications of the excitation levels of the product nucleus Si^{30}.

A more detailed study of the resonance phenomena in this reaction was carried out by Chadwick and Constable (1932) and Duncanson and Miller (1934). As a result of these researches, it is now admitted that

there are *six resonance levels* for 4.0, 4.44, 4.86, 5.25, 5.75 and 6.61 MeV α- particles, and *four proton groups* for each resonance level corresponding to the four Q values 2.07, — 0.16, — 1.53, — 2.67 MeV. The residual nucleus Si^{30} can therefore be left either in the ground state (Q = 2.07 MeV) or in one of the three excited levels of 2.23, 3.60, or 4.74 MeV above the ground state.

Duncanson and Miller also found that *non-resonance* protons were observed for α-particle energies greater than 6.7 MeV, which has been interpreted as penetration of the α- particle above the top of the potential barrier. Usually, on bombarding elements with α- particles of relatively low energy, only resonance penetration through virtual levels occurs and hence distinct proton groups are observed, whereas bombarding with α-particles of energy higher than the top of the potential barrier, disintegration can be produced by particles of any velocity and therefore no distinct proton groups are observed.

It may be noted that it is possible to deduce approximately the heights of the potential barriers in nuclei from the cessation of pronounced resonance maxima. Thus, in the case of Al bombarded by α- particles, this height is roughly 6.8 MeV. An estimate of the *width of the resonance levels* gives a value of about 0.1 to 0.3 MeV for the higher energy levels, while the lower levels would be expected to be narrower.

The resonance levels in the disintegrations of B^{10}, N^{14}, F^{19}, Mg^{21}, Be^9 and B^{11} produced by α- particles have also been studied by different workers. In the first four casks protons are emitted, while in the last two neutrons are ejected. One or more resonance levels have been observed and several groups of emitted particles from each resonance level: *e.g.,* with two resonance levels at 4.1 and 3.7, MeV, and 6 to 8 proton groups occurring in pairs; with F^{19} two resonance level at 4.2 MeV and five proton groups and so on.

(*ii*) *Fluorine bombarded by protons*

This may be taken as an example for the resonance phenomena met with in the *radiative capture of charged incident particle.* Hafstad and Tuve (1936) found resonance maxima in the excitation curve for γ-ray production at 0.328, 0.892 and 0.92 MeV. The width of the first resonance level at 0328 MeV was found to be very narrow, only 4 KeV. Herb, Kerst, Mekibben (1937) extended these data to higher energies and found a broad resonance at 0.6 – 0.7 MeV and a sharp one at 1.40 MeV and indications of another at 1.76 MeV. Delsasso, Fowler and Lauritsen (1937) found a single γ-ray line at 6 MeV by analysis of pair-production observed in a cloud chamber.

It may be noted that the observed resonance features, especially the small widths of resonance levels, have been used to specify the nature of the reaction as a capture process in the case of a few other elements bombarded by protons, such as Li^7, Be^9, B^{10}, B^{11}, C^{13}, Na^{23} and Al^{27}.

The resonance disintegration in the case of neutron bombardment will be dealt with in the section on neutrons.

4. Exchange reactions

Disintegrations by deuterons of the (*d, p*) type, *i.e.,* where protons are ejected, appear to follow a different mechanism from other reactions, which make them more probable and less dependent on energy than would be expected from the ordinary wave-mechanical theory of Gamow, Condon and Gurney (G—C—G). This new mechanism, called *exchange reaction*, first proposed, in 1935, by Oppenheimer and Phillips (0—P) May be stated as follows:

The deuteron, which is a complex particle made up of a proton and a neutron, in approaching close to a nucleus in the target might break up under the influence of the large electric field existing there into a neutron and a proton. The neutron, with no electric charge, has not to overcome the potential barrier and hence could easily enter the nucleus, while the proton, on account of its positive charge would be repelled or scattered away. Hence

the proton that is emitted in the reaction does riot really come from within the bombarded nucleus but only appears to. The neutron that has entered the nucleus could produce any of the characteristic neutron reactions ordinarily a capture.

The process may be considered as a *"partial entry"* of the incident particle, in order to distinguish it from other reactions, where the incident particle as a whole enters the nucleus to transform the latter.

Such an exchange process will give rise to two product nuclei which will then move apart with an energy determined by the energy balance appropriate to the reaction. From the point of view of the results obtained, it can be considered as an ordinary substitutional reaction in which a deuteron enters the nucleus and is captured and a nuclear proton is emitted. But there is an important difference in the process itself which leads to a different value of the disintegration probability.

The essential condition of the applicability of the O-P theory lies in the fact that there is a greater probability for the entry of a neutron than the ordinary wave-mechanical theory gives for the entry of the deuteron as a whole. A quantitative consideration of the problem shows that

(1) in the case of the deuteron, on account of its small dissociation energy (2.2 MeV) the process in general, possible;

(2) the O-P theory can, however, be used .only in the case of heavy nuclei (A > 100) for which, it will surely increase the disintegration probability considerably; for light nuclei the difference between the O—P theory and the G—C—G theory is unobservably small;

(3) the velocity of the incident deuteron must be small compared with the velocities of the particles inside the nucleus;

(4) in heavy nuclei the (d, p) process usually follows the O—P mechanism, while the entry of the deuteron as a whole is generally followed by emission of a neutron —(d, n) process according to the G—C—G mechanism; the probability of the two processes are comparable.

The O—P theory of exchange reactions was first applied with a certain amount of success to the (d, p) reactions with Al and Cu, studied by Lawrence and his associates (1935). The excitation curves in these cases revealed a less rapid increase of yield with increasing deuteron energy than for processes resulting in α-emission and they did not fit the theoretical curve based on the ordinary wave-mechanical concept. The shape of the curves could, however, be explained on the basis of the O—P theory.

A more convincing proof of the existence of the O—P exchange reactions has been realized in the case of certain heavy elements, where the probability for the two competing (d, p) and (d, n) processes can be determined. The results obtained with deuteron bombardment of bismuth by Cork, Halpern and Tatel (1940) may be cited as an example.

Bi bombarded by deuterons

Bismuth has a single stable isotope of mass number 209 and atomic number 83. When a bismuth target is bombarded with sufficiently high energy deuterons, the two competitive processes that can occur are:

(d, p) $\quad _{83}Bi^{209} + {}_1D^2 \rightarrow *_{83}Bi^{210} + {}_1H^1$

(d, n) $\quad _{83}Bi^{209} + {}_1D^2 \rightarrow *_{84}Po^{210} + {}_0n^1$

The deuteron-proton reaction yields a radioactive isotope-of bismuth (Bi^{210}) which is the same as RaE. This has a β-activity with a period of 5 clays and is transformed into RaF. Radium F emits only α- particles and has a period of 140 days.

The deuteron-neutron reaction produces Po^{210} which is identified with RaF, which will straightway decay emitting α- particles with a period also of 140 days.

It might be expected that the (d, p) reaction would have a low probability compared to that of the (d, n) reaction, since the uncharged neutron can pass through the potential barrier readily, while the charged proton not so easily. But experimental data prove the contrary. Using identical Bi targets and bombarding them with a beam of deuterons of the same intensity, the yields in the two reactions can be measured as follows:

The intensity of α-emission immediately after bombardment can be measured and it gives the yield of RaF produced in the (d, n) reaction. By recording the increase in α-activity on successive days, the initial yield of RaF produced in the (d, p) reaction can be obtained.

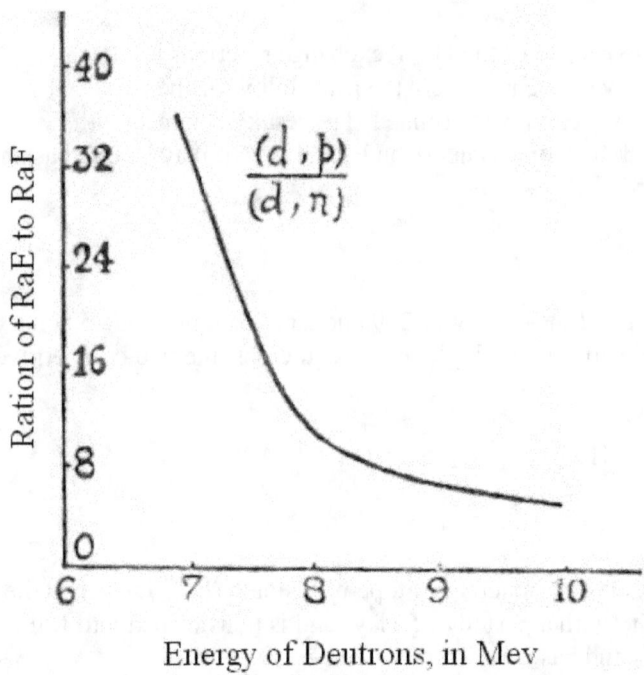

Fig. 294. Relative yields of radioactive atoms produced in the (*d, p*) and (*d, n*) reactions of Bi.

When the yields in the two cases are thus measured for different energies of the deuterons and the data plotted, the curve of relative yields is as shown in Fig. 294. It is readily seen that at low energies the (*d, p*) reaction is much more probable than the (*d, n*) reaction and at 9 MeV it is still five times more likely to occur than the other. This greater yield for the (*d, p*) process can be accounted for adequately only by the O—P theory. If the deuteron had entered the nucleus in every case the ratio of the activities would be represented by a curve different from that shown in the figure, since the greater yield would have been expected for the (*d, n*) reaction for reasons given above. But, even when the bombarding energy is increased up to 14.5 MeV, at which value the O—P process should become insignificant, the ratio (*d, p*) / (*d, n*) still seems to remain at a value of about 5, for energies from 9 to 14.5 MeV.

Thus the O—P theory, which considers the deuteron to be essentially "polarized", so that it splits up the intense field of a heavy nucleus into a neutron and proton, the former being absorbed by the nucleus and the latter repelled away; has been experimentally verified.

5. Photodisintegration

Disintegration of nuclei has been observed also under bombardment with high energy γ-ray photons. This nuclear analogue of the atomic photoelectric effect has been termed *photodisintegration* or *nuclear photo-effect*.

In most cases, the process results in the *emission of neutrons*. Consequently the product nucleus is always less in mass by unity than the bombarded nucleus. If the *product nucleus* does not exist in nature in a stable state, it will be *radioactive*, and in general, a *positron emitter*. The reaction can therefore be detected by either observing the neutrons emitted or more conveniently by the radioactivity of the residual nucleus.

Photodisintegration can occur only when the energy of the incident γ-*rays* exceeds the binding energy of the particle to be ejected. As the energy of the incident γ-rays steadily increased, the reaction begins at a certain "threshold" value of the photon energy, and. the yield then increases more and more rapidly as the energy is increased. Theoretical considerations by Bethe and Peierls show that the greatest probability of disintegration occurs when the γ-ray energy is just twice that of the binding energy of the ejected particle. The cross-section for the process is of the order of 10^{-27} cm^2. Although this is a small quantity, the great penetration of γ-rays exposes a large number of nuclei to bombardment, which results in an appreciable number of disintegrations.

In principle, all nuclei can be disintegrated by γ-rays of sufficiently high energy. It is to be expected also that particles other than neutrons, e.g., protons, α- particles or deuterons can be ejected in this process. The relative probability of these various types of photo-dissociation will be determined by the respective "particle widths" of the levels of the compound nucleus, *i.e.*, primarily by the penetrability of the potential barrier for the particle to be emitted; the probability will therefore be smaller for heavier nuclei.

Photodisintegration was first discovered in 1934 by Chadwick and Goldhaber who observed that deuterium gas, when irradiated by the γ-rays of ThC" of energy 2.62 MeV, was disintegrated, the deuterium nucleus splitting up into a proton and a neutron according to the equation

$$_1D^2 + h\nu \rightarrow {}_1H^1 + {}_0n^1 + Q$$

A calculation from the masses involved in this reaction shows a mass deficiency of 0.00234 m.u., which means that the reaction is *endoergic* (*i.e.*, Q is negative) and an energy Q equivalent to the mass defect or 2.178 MeV must be supplied. Hence to produce photodisintegration with deuterium, the incident γ-radiation must possess an energy of at least 2.178 MeV. Chadwick and Goldhaber proved this to be true from experimental data as follows.

The energy of the protons produced in the disintegration can be measured by means of a linear amplifier or a cloud chamber. Assuming that the neutron carries approximately the same amount of energy as the proton on account of their nearly equal masses, the total kinetic energy carried away by the two product particles can be taken as twice the measured value of the proton energy. The total energy thus estimated is 0.45 MeV. Subtracting this amount from the energy of the incident γ-rays, viz., 2.62 MeV the energy absorbed in the process is 2.17 MeV, which agrees well with the values calculated from the masses of the nuclei appearing in the equation. This result is further confirmed by the fact that photodisintegration of deuterium is produced also by the γ-rays of RaC' of energy, 2.198 MeV, although with much reduced intensity.

It is interesting to note that the *reverse of this reaction, viz.,*

$$_1H^1 + {}_0n^1 \rightarrow {}_1D^2 + h\nu$$

i.e., the formation of deuterium by the combination of a proton and a neutron, must be *exoergic,* liberating an energy of 2.17 MeV. It will be pointed out later that the experimentally measured γ-ray threshold for deuterium constitutes one of the most accurate methods of evaluating the mass of the neutron. Measurements of Wiedenbeck and Marhofer (1945) give the photo-disintegration threshold of deuterium at 2.185 ± 0.006 MeV.

Szilard and Chalmers in the same year (1934) discovered photo disintegration of beryllium, according to the equation

$$_4Be^9 + h\nu \rightarrow {}_4Be^8 + {}_0n^1$$

The threshold energy for this reaction is found to be at 1.69 MeV.

To produce *photodisintegration of heavier nuclei,* γ-rays more energetic than that of ThC" are required. The γ-rays of 17.5 MeV, obtained from lithium bombarded by protons, have been shown to be able to produce photodisintegrations in practically all the elements of the periodic table. With the advent of the betatron capable of producing X-ray photons of very high energy, as much as 100 MeV and more, the study of this reaction has been greatly extended. It has been found that several particles may be emitted from a nucleus bombarded with such high energy photons with a greater probability than the ejection of a single particle carrying a large energy. Thus, for instance, in the photodisintegration of O^{16} there appears to be sufficient evidence for the existence of radioactive carbon C^{11} of T = 20.7 mins along with other products, which would mean the ejection of 2 protons and 3 neutrons.

Hubber and his associates (1944) have found evidence for the following photodisintegration where a proton is emitted instead of a neutron:

$$_{12}Mg^{26} + h\nu \rightarrow *_{11}Na^{25} + {}_1H^1$$

$$*_{11}Na^{25} \rightarrow {}_{12}Mg^{25} + e^- \quad (T = 1 \text{ min.})$$

An accurate determination of the threshold value in photo-disintegration is of particular importance, as it gives the binding energy of the particle to be ejected. In the initial stages, where γ-rays of high and continuously variable energy were not available, the binding energy was estimated by the measurement of the proton energy, as we have seen above in the method of Chadwick and Goldhaber. More recently, with the X-ray photons from the betatron of high and readily variable energy, the photodisintegration thresholds in different nuclei have been measured by studying the radioactivity of the product nuclei. Thus, Baldwin and Koch, irradiating different elements from C to Ag with X-rays from the betatron, were able to determine the threshold values as follows:

For a given energy of the incident photons, the yield of the photonuclear effect is estimated by measuring the intensity of the β-rays emitted by the irradiated sample with a counter. The experiment is repeated by varying the energy of the photons. When the data thus obtained, i.e., the intensities of β-activity of the product nucleus for different incident energies are plotted, a curve is obtained which readily indicates the threshold for the element under test. The results obtained by these workers are given in the table below:

Elements	C^{12}	N^{14}	O^{16}	Fe^{54}	Cu^{62}	Zn^{64}	Mo	Ag
Threshold in MeV	19·0	11·1	16·3	14·2	10·0	11·6	13·5	9·5

In the case of C^{12}, N^{14} and O^{16}, it has been possible to calculate the reaction energies Q from the masses and they are 18:8, 10.6 and 16 MeV respectively, in good agreement with the observed values.

6. Disintegration by electrons

Early attempts to transmute elements using electrons as projectiles even up to energies of 1 MeV were unsuccessful. In 1939', however, Collins; Wildman and Guth considering the problem theoretically came to the conclusion that

it should be possible to produce disintegration with electrons if their energy exceeded the threshold value for photodisintegration.

They put this theoretical expectation to experimental test in the following manner. Using *beryllium,* whose photodisintegration threshold is only 1.63 MeV as *target,* they bombarded it with high energy electrons of 1.8 MeV from the Van de Graaff generator at Notre Dame University and observed the emission of neutrons for incident energies greater than 1.63 MeV. The reaction is readily represented by the equation

$$_4Be^9 + e^- \rightarrow {}_3Li^8 + {}_0n^1$$

The neutrons were detected by means of the radioactivity induced by them in foils of silver, rhodium and indium placed near the target. They established also that the disintegration was not due to γ- or X-rays arising from the ionization caused by the electrons in the target, since the same emission of neutrons was obtained from Be targets of very different thicknesses. The cross-section of the reaction at 1.73 MeV was estimated to be only of the order of 8×10^{-31} cm^2.

Following this first successful disintegration, many others and even heavy elements have been transmuted by bombardment with electrons of high energy produced in the betatron.

SOME IMPORTANT DISCOVERIES MADE IN THE STUDY OF ARTIFICIAL TRANSMUTATIONS

1. THE NEUTRON

Introduction

The neutron is an uncharged particle with very nearly the same mass as the proton. Its discovery may be considered to be one of the most important results of the researches in artificial disintegration, for it has lent itself to very many most interesting practical applications on account of its special property of readily inducing radioactivity in almost any element (*artificial radioactivity*) and of even breaking up some of the heavy unstable elements with liberation of enormous amount of energy (*fission*). It has also introduced a radical change in the theory of nuclear structure, since, according to actual conceptions, all nuclei are composed of protons and neutrons.

As early as 1920, when it was commonly assumed that the atomic nuclei were constituted with protons and electrons, Rutherford had predicted the existence of a nucleus of the simplest type, formed out of a proton and an electron more intimately united than in the hydrogen atom, so that it possessed a mass almost the same as that of the proton but without charge and hence called a neutron. But such a nuclear particle could not then be detected.

Sir James Chadwick (20 October 1891 – 24 July 1974) England

Discovery

The study of the disintegrations of light elements by α- particles led to the discovery of the neutron. Although three different groups of scientists, German, French and English, have contributed to this work, Prof. Chadwick of the English school justly deserves the merit of its final and definite accomplishment.

In 1930, Bothe and Becker, two German scientists, noticed that when beryllium was bombarded with α- particles from a Po source of energy 5 MeV, no protons were ejected as would be expected from the general type of reaction of α- particles with light elements. Instead, a *highly penetrating radiation* was emitted, which could be detected by a Geiger point counter, even when the source of α- particles and the target were completely surrounded by a screen of zinc or brass, thick enough to intercept all the α-particles as well as electrons and X-rays set free by the ionizing action of the α- particles. They, therefore, naturally assumed that the observed *penetrating radiation was of the nature of γ-rays.* Measurement of the absorption of these rays in lead indicated that the energy of the emitted photons should be about 7 MeV.

In 1932, Joliot and Curie of Paris, in their attempts to produce transmutations with these penetrating rays, discovered that they possessed the .very interesting property of expelling high speed protons from hydrogenous matter, such as paraffin, water, paper, cellophane, etc. The protons knocked out of paraffin by the rays had a range of 40 cms in air corresponding to an energy of about 5 MeV. They also found that a magnetic field of about 500 gauss established over 5 cms of the path of the rays did not produce any influence on them. Assuming, therefore, that the primary rays were γ-ray photons and that the protons were produced as the result of elastic impacts, as in Compton effect, calculations showed that the energy of the primary rays must be of the order of 50 MeV, which was entirely inconsistent with the value obtained from absorption measurements (about 7 MeV). Furthermore, since the supposed γ-rays arose from the capture of the α-particles by beryllium nuclei, writing down the equation proper to the assumed nuclear reaction, viz.

$$_4\text{Be}^9 + _2\text{He}^4 \rightarrow _6\text{C}^{13} + h\nu$$

the maximum energy available for the γ-ray emitted by the resulting carbon nucleus could be calculated from the known masses of the particles involved and it was found to be only 15 MeV about, far below the 50 MeV energy obtained from the recoil protons as stated above.

To overcome these difficulties, Chadwick of the Cavendish school in the same year (1932) reverted to the neutron hypothesis of Rutherford, as the result of a careful study of the projection of light nuclei such as He, Li, C, N, O, etc., by the rays coming from Be bombarded by α-particles. He found that if these rays were assumed to be γ-rays then the experimental data led to inconsistent values of the energy of these rays being dependent upon the nature of the projected nuclei.

For example, the protons ejected from paraffin had energies of about 5 MeV which gave a value of 55 MeV for the energy of the primary rays, while nitrogen nuclei projected by the same rays had energies of about 1.2 MeV, which led to a value of 90 MeV. In general, the energy of the same primary rays increased considerably with the increase in mass of the projected atom, which was evidently contrary to the principles of conservation of energy and momentum in elastic impacts.

Chadwick showed that all these difficulties disappeared completely, if it could be assumed that *the primary radiations arising from Be bombarded by α-particles were not γ-rays but neutrons which were uncharged particles with a mass approximately that of the proton* formed according to the equation.

$$_4\text{Be}^9 + {_2}\text{He}^4 \rightarrow {_6}\text{C}^{12} + {_0}n^1 + Q,$$

where ${_0}n^1$ represents the neutron indicating that its mass number is one, while the charge or atomic number zero.

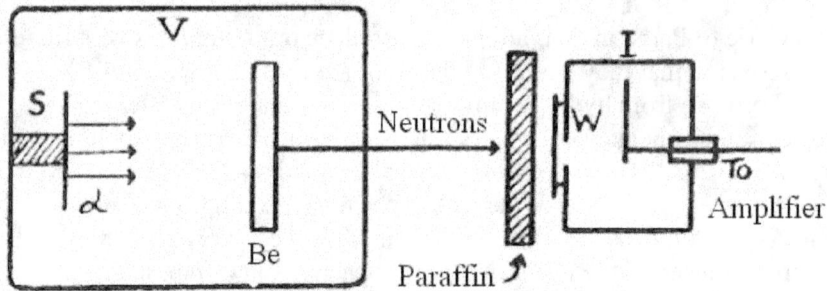

Fig. 205. Chadwick's apparatus for the detection of neutrons.

He next established the absolute correctness of this conclusion experimentally as follows:

The apparatus used for this purpose is shown in Fig. 295.
S is a polonium source S of α-particles;
Beryllium target Be was arranged inside the evacuated chamber V;
V id an evacuated chamber.
W is a thin aluminum window.
I is the ionization chamber of a linear amplifier.

The radiation emitted by Be when the α-particles fell on it passed through the thin Al window W of the ionization chamber I of a linear amplifier arrangement which registered the charged particles passing through I.

When the radiation from Be were allowed to pass directly into I, a few pulses per minute were registered. Interposing a lead sheet 2 cms thick in the path of the radiation before it reached .I, the number of pulses was not reduced appreciably, which showed the high penetrating power of the radiation.

If, however, a thin sheet of -paraffin was placed in front of the window W the number of pulses increased greatly. These results proved beyond doubt that the highly penetrating radiation, though carried no charge as it was not affected by a magnetic field, yet was not electromagnetic γ-radiation, since the introduction of lead absorbers did not decrease its intensity, but was made up of material particles which chased before them protons from the paraffin and caused the marked increase of pulses registered by the linear amplifier.

Using the cloud chamber and measuring the maximum velocities of different nuclei such as proton, nitrogen, etc., chased by the newly discovered particle radiation, Chadwick was also able to determine the mass and velocity of the new particle, the neutron, by the application of the laws of elastic impact. The mass was found to be approximately 1.15 times that of the proton and the velocity of the order of 3×10^9 cms/sec. Thus the neutron was definitely discovered.

Other elements which emit neutrons when bombarded by α-particles are: Li, B, F, Ne, Na, Mg and Al. In some of these disintegrations the product nuclei are stable, while in others radioactive with positron emission

In many cases, neutron groups with distinct energies are emitted, which are interpreted as corresponding to different excitation levels of the product nucleus. It is to be noted that in several of these reactions gamma rays are also emitted along with the neutrons, as shown by the appearance of tracks of the fast secondary electrons of energy ranging up to 5 or 6 MeV in the cloud chamber study, as well as by the asymmetry and complexity of the primary rays indicated clearly by their absorption curves.

Researches on the neutron

The discovery of the neutron in 1932 was followed at once by numerous other researches to study in detail the different properties of the neutron.

(i) Production of neutrons

Subsequent experiments have shown that

not only many light elements emit neutrons when bombarded with α-particles, but also almost any element bombarded with high energy particles, *e.g.*, protons, deuterons, neutrons and even photons, give neutrons.

They are classified as (α, n), (p, n), (n, 2n) and (γ, n) reactions. Although any one of these reactions might be used as a source of neutrons, in practice, some are chosen on account of their simple technique or great yield. Thus, for example,

(a) *Be, B or Li bombarded by α-particles from Po*

Though the yield of neutrons is comparatively low, this source of neutrons is often used on account of the absence of *γ-rays* and the pretty long period of 136 days of α-decay.

(b) *Be bombarded by α-particles of Ra or Rn*

The yield is quite good-20,000 to 27,000 neutrons per sec per millicurie, but the emission of γ-rays from Ra and Rn may be a source of inconvenience.

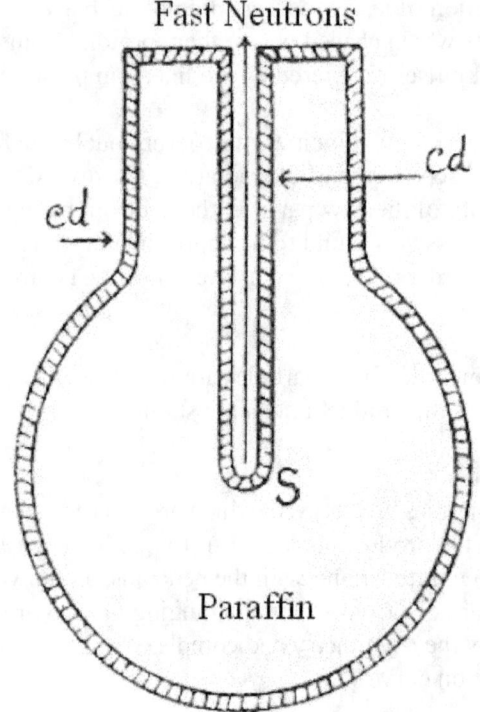

Fig. 296. Neutron howitzer.

All the same an ordinary standard source of neutrons is a small capsule containing a mixture of powdered beryllium and radon gas.

(c) *Bombarding light elements with deuterons*

This first attempted by Prof. Lawrence. Li, Be, or a substance containing deuterium such as "heavy" ice is used as target.

The respective reactions are:

$_3Li^7 + {}_1D^2 \rightarrow {}_4Be^8 + {}_0n^1$

$_4Be^9 + {}_1D^2 \rightarrow {}_5Be^{10} + {}_0n^1$

$_1D^2 + {}_1D^2 \rightarrow {}_2He^3 + {}_0n^1$

The last (*d, d*) reaction usually requires low voltages (about 200,000 volts), while the others much higher, several million volts. On account of the high yield of neutrons, of the order of 10^7 per sec, this method is of great industrial importance. Using *high-energy deuterons from a cyclotron* to bombard Be or Li, a very intense source of neutrons, equivalent to that from thousands of grams of Ra mixed with Be, is obtained.

Neutrons produced in these reactions possess very high energies ranging from 4 to 25 MeV, as can be shown from a calculation of the reaction energy Q involved. They are, in consequence, emitted with enormous velocities and are called *"fast"* neutrons. Their speed, however, can be rapidly *slowed* down by the use of hydrogenous materials, as we shall see presently.

The problem of collimating a beam of fast neutrons is difficult as the ordinary methods used for charged particles and γ-rays cannot be used. Dunning has devised (1938) a collimating arrangement known as the **"neutron howitzer",** shown in Fig. 296 for the neutrons from a (Ra + Be) capsule. The source S is placed at the bottom of a barrel inside a paraffin block shaped as in the figure. It sends neutrons with energies up to 10 MeV from the barrel outward. The thickness of the paraffin is sufficient to reduce the number and energy of the neutrons to a negligible quantity in every other direction except through the barrel. The outer surface of the paraffin as well as the walls of the barrel are coated with a layer of cadmium (Cd) which effectively absorbs the thermal neutrons resulting from the slowing down of the fast neutrons in the paraffin. A layer of lead over that of Cd is used to absorb any γ-rays arising from neutron capture.

In 1939, Aebersold has been able to obtain a well defined and intense beam of fast neutrons produced by bombarding Be with high-energy deuterons from a cyclotron, using a collimating arrangement as shown in Fig, 297.

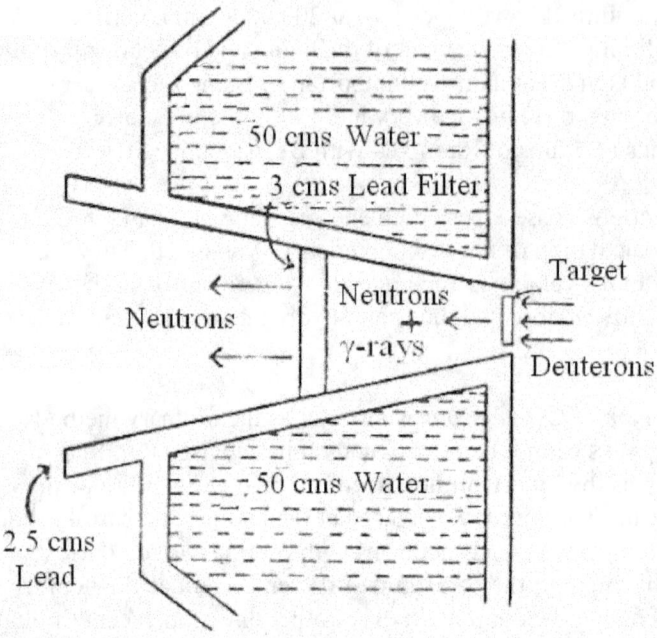

Fig. 297. Aebersold neutron collimator.

It consists of a slightly tapering channel through a 50 cm wall of water. The direct gamma rays from the bombarded beryllium are absorbed by a 3 cm lead filter. The large thickness of water required to produce a well-defined beam introduces considerable difficulty due to the scattering of radiation by the sides of the channel and the production of secondary radiation in the water column. This is minimized by lining the channel with a 2.5 cm lead wail.

Intense beams of monokinetic neutrons have been obtained by the use of suitable reflecting crystals and mirrors. Two types of reflection have been observed with strong neutron beams:

The first by crystals according to the Bragg law for the de Broglie wavelengths corresponding to their energy. Crystals of calcite and lithium fluoride are found to be satisfactory. The intense reflected beam consists of an ideal monokinetic group of neutrons.

The second by total reflection with highly polished surfaces of selected materials and angles smaller than their critical angle. Mirrors of graphite, aluminum, beryllium, copper, nickel, etc., produce intense reflected beams at glancing angles up to 10 mins of arc. It may be noted that these reflections of neutrons are similar to the well-known corresponding phenomena with X-rays.

(*ii*) Detection of neutrons

Since the neutrons carry no charge they have no ionizing power and hence can be detected only *indirectly* by the secondary effects they produce. There are several methods in use.

(a) Elastic collision method
When neutrons collide with atomic nuclei, chiefly with light ones, the latter are chased. These recoil nuclei may be made perceptible by the ionization chamber with a linear amplifier or by the cloud chamber and thus the invisible neutrons which projected them can be detected. Further; from measurements made on the recoil nuclei the intensity and energy of the neutrons can be deduced with a fair amount of accuracy. This method is well suited for the defection of *fast* neutrons.

(b) Transmutation methods
Neutrons are capable of transmuting a great number of elements. This property is utilized in the detection and study of the neutrons in a variety of ways.

Supposing a transmutation by neutron occurs in a *cloud chamber* and results in the emission of a charged particle from the transmuted nucleus, a V-shaped double track will be seen, caused by the residual nucleus and the ejected particle, the neutron itself producing, of course, no tracks.

The products of transmutation caused by neutrons can be measured also with ionization chambers. Such chambers coated with Li or B or filled with BF_3 gas are frequently used, especially for *slow* neutrons, the ionization being due to the α-particles emitted in one of the two reactions:

$$_3Li^6 + {_0}n^1 \rightarrow {_1}H^3 + {_2}He^4$$
Or
$$_5B^{10} + {_0}n^1 \rightarrow {_3}Li^7 + {_2}He^4$$

A G-M counter lined with boron (B^{10}) or filled with BF_3 gas, known as a *neutron counter,* is used for a quantitative counting of neutrons, on account of the large cross-section of boron for transmutation by slow neutrons.

Fast neutrons may be counted also with such a counter by the use of a moderating shield surrounding the counter.

The residual nuclei in the transmutations caused chiefly by *slow neutrons* are found *radioactive* in many cases. This radioactivity has been utilized also to detect the presence of neutrons. Thus, for instance, a counter whose outer cylindrical electrode is made of *silver,* a metal greatly affected by neutrons, forms a good neutron detector. Such a counter operates under the action of electrons produced in the decay of the radioactive element formed when neutrons bombard the silver electrode.

(*iii*) Distribution in speed of neutrons—fast and slow neutrons

The velocity of the neutrons can be determined by measuring the ranges of the nuclei projected by them with an ionization chamber or a cloud chamber. The neutrons emitted in any given reaction have different velocities according to the direction in which they leave the target, maximum in the same direction as the bombarding particles and minimum in the opposite direction. For instance, the velocity of neutrons from (Po + Be) varies from 1.6 to 3.5×10^9 cms/sec.

The fast neutrons obtained from a disintegration can be slowed down by causing them to make elastic collisions with light nuclei. Hydrogen nuclei are found to be the best for this purpose, since they have nearly the same mass as the neutrons. It can be shown that neutrons colliding at random with protons whose velocities are much smaller, will have their energy reduced in the ratio 1/e at each collision, on an average, (e = Naperian base 2.718). Calculation indicates that neutrons with an initial energy of 5 MeV can be slowed down to energies of thermal equilibrium after some 20 collisions with protons.

Water and paraffin, which contain an abundance of hydrogen nuclei, are the two substances most widely used for slowing down neutrons. Thus, for instance, placing a source of fast neutrons (Rn + Be) at the center of a paraffin sphere of about 10 cms, radius, slow neutrons with thermal energies emerge from the surface of the sphere.

Initial researches made it possible to distinguish roughly between fast and slow neutrons, the former having energies varying from 100 KeV to 15 MeV, while the latter from a few thousand volts down to thermal energies. Further work in selective absorption of slow neutrons, chiefly by Fermi and his collaborators, has shown that what is ordinarily called slow neutrons consists of several groups.

(iv) Transmutations produced by neutrons

Since neutrons carry no electric charge, they experience no potential barrier difficulty in penetrating atomic nuclei, even if these are heavy and highly charged. They ought to be, therefore, *very effective transmuting agents*. In fact, many more transmutations have been produced with neutrons than with other projectiles. Not only *fast* neutrons are capable of disintegrating nuclei, but also and especially *slow* neutrons have been found to be extremely effective. A great deal of research has been carried out with slow neutrons and the data obtained have formed the basis of Bohr's liquid drop model of the nucleus.

Feather was the first, in 1933, to observe transmutation by neutrons. Using a cloud chamber filled with nitrogen gas at low pressure and allowing neutrons to pass through the chamber, he found in the photographs obtained, besides single tracks of recoil nitrogen nuclei produced in elastic impacts with neutrons, a number of forked tracks, one branch of which was short and thick, while the other longer and finer. He interpreted them as disintegrations according to the equation:

(n, α) $_7N^{14} + {}_0n^1 \rightarrow {}_5B^{11} + {}_2He^4 + Q$ .... (1)

Some others, like Meitner and Phillipp, Harkins, Bonner and Brubaker, followed Feather in studying the disintegrations produced by neutrons in several elements, *e.g.,* oxygen, fluorine, neon, argon, aluminum, carbon, copper, etc., with the same cloud chamber technique.

The results obtained by Bonner and Brubaker (1936) in the case of *nitrogen transmuted by neutrons* are instructive as they indicate the complexity of nuclear reactions. The cloud chamber photographs showing three different reactions between nitrogen and neutron are reproduced here.

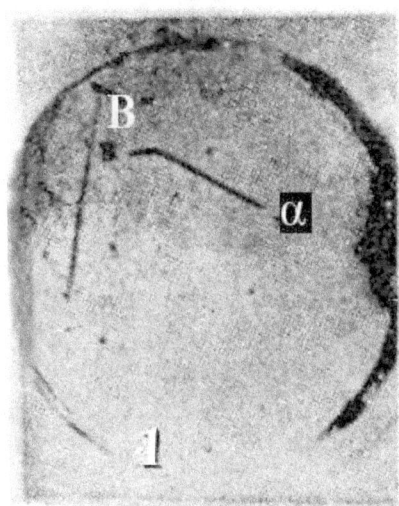 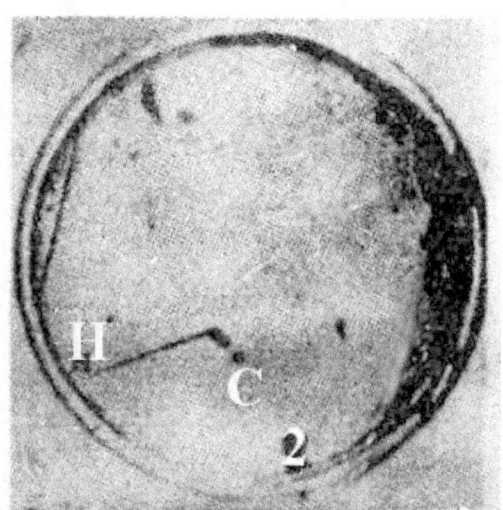

One of them marked (1) represents a very frequent transmutation of nitrogen into boron (short thick branch to the left marked B) with the emission of longer branch to the right marked α), according to the equation given above, detected by Feather.
The one marked (2) is a less frequent reaction in which a radioactive carbon atom marked C is formed with the emission of a proton marked H, according to the equation:

$$(n, p) \quad {}_7N^{14} + {}_0n^1 \rightarrow *_6C^{14} + {}_1H^1 + Q \quad \dots\dots\dots\dots (2)$$

The stereoscopic double photo marked (3) is also a rare transmutation in which *three particles* are formed, two of them α-particles and the third lithium, according to the equation

$$(n, \alpha) \quad {}_7N^{14} + {}_0n^1 \rightarrow {}_3Li^7 + 2\,{}_2He^4 + Q \quad \dots\dots\dots\dots (3)$$

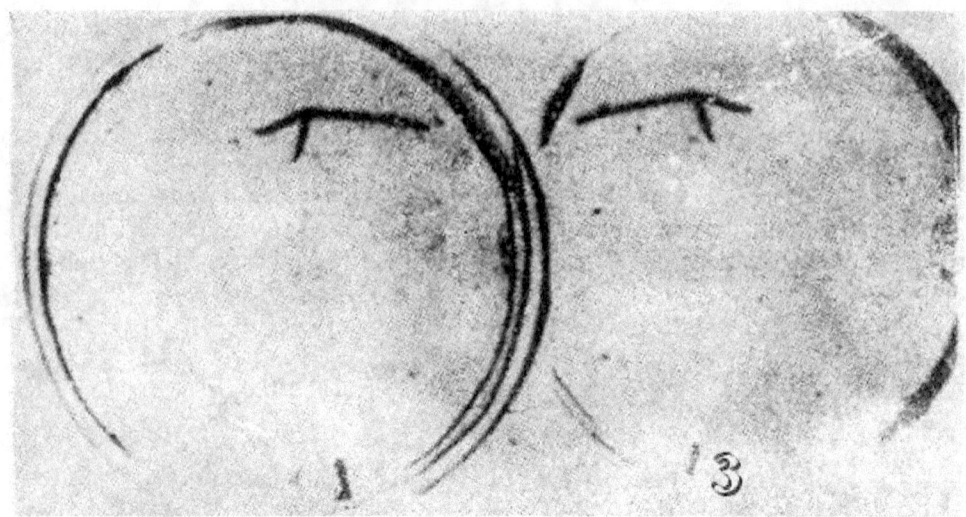

(The arrows in the figures indicate the directions of the incident neutrons).

We shall now briefly state some special interesting features of the different reactions induced by neutron bombardment.

Almost all elements can be disintegrated by neutrons. But the type of reaction depends largely on the energy of the incident neutrons.

While some disintegrations require fast neutrons, many others are produced by slow neutrons. The different reactions give rise to any one of a number of particles, *e.g.*, an α-particle, a proton, more than one neutron, or a photon. In most cases, radioactive elements are formed whose atomic numbers are either the same as or smaller by one or two units than that of the original nucleus.

(*n*, *α*) reactions are often *exothermic;* hence in the case of *light elements* they are readily produced by *slow neutrons*. Two cases, other than nitrogen considered above, have been studied extensively. They are:

(1) $_3Li^6 + _0n^1 \rightarrow _1H^3 + _2He^4 + Q$

(2) $_5B^{10} + _0n^1 \rightarrow _3Li^7 + _2He^4 + Q$

These two reactions have been observed directly in the cloud chamber. On account of their very large cross-sections (60 barns for Li and 500 barns for B) they are *used for detecting slow neutrons,* which would not give observable recoil protons.

In the case of *heavier elements, fast neutrons* are required to produce this type of reaction, since the α-particle cannot escape through the high potential barrier involved unless it possesses additional energy due to the impinging neutron. The *product nucleus is usually radioactive: e.g.*

(3) $_{11}Na^{23} + _0n^1 \rightarrow *_9F^{20} + _2He^4 + Q$

$\quad *_9F^{20} \rightarrow _{10}Ne^{20} + e^- \ (T = 12 \ secs)$

(4) $_{13}Al^{27} + {}_0n^1 \rightarrow * {}_{11}Na^{24} + {}_2He^4 + Q$

　　$* {}_{11}Na^{24} \rightarrow {}_{12}Mg^{24} + e^-$ (T = 14.8 hrs.)

These reactions may be represented in a general form as

$_ZX^A + {}_0n^1 \rightarrow * {}_{Z-2}Y^{A-3} + {}_2He^4 + Q$

(n, p) reactions are usually *endothermic with the formation of a radioactive element,* which decaying with *electron activity* of high energy (1 to 5 MeV), gives as the final stable nucleus the original nucleus. Now, since the masses of neutron and proton are nearly equal, the large amount of energy released must be supplied by the bombarding neutrons and hence *reactions of this type are produced only by fast neutrons.* Some typical reactions are:

(1) $_7N^{14} + {}_0n^1 \rightarrow * {}_6C^{14} + {}_1H^1 + Q$

　　$* {}_6C^{14} \rightarrow {}_7N^{14} + e^-$ (T > 10^3 yrs)

(2) $_{12}Mg^{24} + {}_0n^1 \rightarrow * {}_{11}Na^{24} + {}_1H^1 + Q$

　　$* {}_{11}Na^{24} \rightarrow {}_{12}Mg^{24} + e^-$ (T = 14.8 hrs.)

(3) $_{13}Al^{27} + {}_0n^1 \rightarrow * {}_{12}Mg^{27} + {}_1H^1 + Q$

　　$* {}_{12}Mg^{27} \rightarrow {}_{13}Al^{27} + e^-$ (T = 10.2 min)

It may be noted that the *same materials can be produced in different ways,* e.g., $* {}_{11}Na^{24}$ and $_{12}Mg^{24}$ are produced both in (n, α) and (n, p) reactions, the target in the first case being Al, while in the second case Mg. This fact also indicates the possibility of producing artificial radio-elements in practical quantities and on an economical basis.

The general scheme of this type of reaction is

$_ZX^A + {}_0n^1 \rightarrow * {}_{Z-1}Y^A + {}_1H^1 + Q$

(n, 2n) reactions were first observed by Bothe and Gentner (1937) with *fast neutrons.* Pool, Cork and Thornton have made a systematic study of this type of reaction with many elements, using high energy neutrons (20 MeV) produced by the bombardment of Li with 10 MeV deuterons. *In most cases, the product nucleus is unstable and an isotope of the target of mass one unit lower.* Many of them decay with a *positron activity.* The general scheme of these reactions is

$_ZX^A + {}_0n^1 \rightarrow * {}_ZY^{A-1} + 2 {}_0n^1 + Q$

As concrete examples, we may give:

(1) $_{29}Cu^{63} + {}_0n^1 \rightarrow * {}_{29}Cu^{62} + 2 {}_0n^1 + Q$

　　$* {}_{29}Cu^{62} \rightarrow {}_{28}Ni^{62} + e^+$ (T = 1.5 mins)

(2) $_{19}K^{39} + {_0}n^1 \rightarrow *_{219}K^{38} + 2 {_0}n^1 + Q$
  $*_{19}K^{38} \rightarrow {_{18}}A^{38} + e^+$ (T = 7.7 mins)

More recently (1940) evidence has been reported of a *"three particle" disintegration,* either two neutrons and a proton, or three neutrons, being ejected when high energy neutrons are used as projectiles.

(n, γ) reactions, known as the *"simple* or *radiative capture"* process, were first definitely discovered in 1934 by Fermi and his associates who made a rather exhaustive survey of this type of reaction for the available elements of the periodic table. They used the neutrons from a strong. (Ra+Be) source of high energy ranging up to 10.9 MeV. The radioactivity persisting in the bombarded element in many cases was taken as an indication of the capture reaction. In some cases, however, no radioactivity was observed, but γ-rays were found to be emitted during the bombardment, which was interpreted as the capture of neutrons with the formation of the heavier stable isotope of the target. It was later found that the efficiency of these reactions could be increased considerably by using neutrons, slowed down by passage through a hydrogenous material, such as paraffin or water. Since the reactions with slow neutrons yielding particle products are in general not probable in heavy elements, the observed "water sensitive" activity is taken as a strong evidence for this type of reaction.

The compound nucleus formed by the capture of a neutron will evidently be an *isotope of the target, of mass one unit higher.* If this isotope exists, the reaction must be *exoergic* by an amount approximately equal to the binding energy of the neutron, which energy is given off in the form of γ-radiation. Further if the isotope does not normally exist in nature, it is bound to be *radioactive.* Hence: the reaction is represented, in general, as

$$_ZX^A + {_0}n^1 \rightarrow * {_Z}Y^{A+1} + h\nu$$

Since there is no emission of charged particles, the *capture reaction occurs for heavy elements with even higher probability than for light ones* as heavy nuclei possess many closely packed resonance levels, Moreover, as these levels are of low energy, the *probability of this reaction becomes great for slow neutrons.* With fast neutrons the probability is small but observable.

We may now give a few concrete examples of this type of reaction:

(1) $_1H^1 + {_0}n^1 \rightarrow {_1}D^2 + h\nu$

The probability of this reaction, in which deuterium is formed by the combination of a proton and a neutron, is very small for fast neutrons and fairly large for slow neutrons.

This reaction has been indirectly established by a comparison with photodisintegration, of which it is the reverse.

(2) $_{29}Cu^{65} + _{0}n^{1} \rightarrow *_{29}Cu^{66} + hv$
$*_{29}Cu^{66} \rightarrow _{30}Zn^{66} + e^{-}$ (T = 5 mins)

(3) $_{79}Au^{197} + _{0}n^{1} \rightarrow *_{79}Au^{198} + hv$
$*_{79}Au^{198} \rightarrow _{80}Hg^{198} + e^{-}$ (T = 2.7 days)

It is to be noted that in these two reactions *copper is transformed into zinc and gold into mercury ultimately.*

Fission reaction

Another very interesting reaction, recently discovered, is what is known as the *nuclear fission* where heavy nuclei, rendered unstable by the capture of neutrons, instead of emitting small particles such as α-particles, protons, etc., break up into two more or less equal parts. On account of the great importance of this process, we shall deal with it in detail in a separate section.

(v) The mass of the neutron

One of the assumptions; on which the discovery of the neutron was based, was that the masses of the neutron and the proton were nearly the same. Chadwick proved the correctness of this assumption by evaluating the mass of the neutron from experimental data obtained from the projection of light nuclei and showing it to be only 1.15 times the mass of the proton. But in view of the probable errors involved in the measurement of the velocities of the recoil nuclei which amount to about 10%, this result could not be taken to give the mass of the neutron to a high degree of accuracy. On the other hand, it was both interesting and important to determine as accurately as possible the mass of the neutron, as it was expected to throw light on the precise nature of the neutron in relation to the proton.

Since the neutron, carrying no charge, cannot be deflected in either a magnetic or electric field, the conventional methods used for charged particles cannot be applied to determine its mass. Only the indirect methods of nuclear disintegrations are available for this purpose and have been used by many workers. Any reaction involving the neutron is suitable, provided the reaction is established beyond doubt, the atomic masses of the nuclei appearing in the reaction are known with sufficient accuracy and the original and final kinetic energies of the particles can be determined accurately.

Chadwick was the first to evaluate the mass of the neutron by this method using the reaction,

$$_{5}B^{11} + _{2}He^{4} \rightarrow _{7}N^{14} + _{0}n^{1}$$

The masses of $_{5}B^{11}$, $_{2}He^{4}$ and $_{7}N^{14}$ were known to a high degree of accuracy from mass spectrograph measurements. The kinetic energy of the incident α-particles from Po used was also known. The kinetic energy of the neutrons was determined by measuring the maximum energy of the hydrogen nuclei projected by them. Substituting these known quantities in the equation, the mass of the neutron could he evaluated and it was found to be 1.0067 ± 0.0005, making allowance for the probable errors involved in the atomic masses and in the energy measurements. *This value is less than the proton mass, 7.00813.*

Several investigators such as Joliot and Curie, Lauritsen and Crane, Chadwick and Goldhaber, Feather and Brestcher, etc., have used different well-ascertained nuclear reactions to estimate the mass of the neutron and have consistently obtained *a value slightly greater than that for proton.*

One of the best reactions used is the photodisintegration of deuterium by the γ-rays from ThC" which produces protons and neutrons according to the equation

$$_1D^2 + h\nu \rightarrow {}_1H^1 + {}_0n^1$$

The energy of γ-rays used is known to be 2.62 MeV. Since the proton and neutron have nearly equal masses, these two particles will be ejected with approximately equal energies. The energy of the proton can be measured readily. By doubling the proton energy thus obtained, it is found that the two particles together have a kinetic energy of 0.45 MeV. The masses of D^2 and H^1 are known accurately from mass spectrograph data. Using these values, the mass of the neutron, the only unknown quantity in the equation, can he calculated. *The* value *thus determined is 1.00893, definitely greater than that* of *the proton.*

The slightly larger value of the mass of the neutron as compared with that of the proton has led to a number of interesting speculations about the precise relation between the proton and the neutron which, according to actual conceptions, are the fundamental constituent particles of all atomic nuclei. We shall consider this point when we study the structure of the nucleus.

(*vi*) Absorption of neutrons in matter

On account of the uncharged nature of the neutrons, their absorption in matter cannot take place in the same manner as in the case of the charged particles, viz. by interactions with the peripheral electrons resulting in the ionization of the atoms and with the positively charged atomic nuclei according to the law of inverse squares. The neutrons can, however, experience a force when they come within *extremely close range of nuclei.* Hence *their absorption. in matter is due exclusively to their 'short range interactions' with nuclei.* This interaction may be regarded as a collision which may be either *elastic* or *inelastic.*

Elastic collision

This will give rise to a mere *"scattering of neutrons",* according to the laws of conservation of energy and momentum. A portion of the energy of the striking neutron will be transferred to the struck nucleus. If this latter is very heavy, the energy loss will be very small. Light nuclei, on the other hand, will at each collision reduce appreciably the energy of fast neutrons. This explains the slowing down action of hydrogenous materials. A beam of fast neutrons is reduced to half its intensity by 5 gm / cm^2 of water or paraffin, while 50 gm./cm^2 of lead is required to produce the same result.

> The *scattering cross-section for fast neutrons is found to increase with the atomic number of the nucleus,* in the same manner as the geometrical cross-section of the nucleus.

But its value is of the order of 10^{-24} cm^2 for all nuclei, varying from 1.4×10^{-24} to 5.5×10^{-24}. These results were obtained by the researches made by Dunning with neutrons from (Ra + Be) source.

Inelastic collisions

This may be subdivided into two categories:

(i) partial, where the neutron is *scattered anomalously* and the struck nucleus undergoes internal changes and is raised to an excited state of higher energy;

(ii) total, where the neutron is *captured* by the nucleus leading to a real transformation.

The first kind called *'inelastic scattering'* may be detected either by measuring the energy of the scattered neutrons or by the γ-rays that are emitted by the nucleus when it returns to a state of lower energy. This type of scattering has been observed by Little, Long and Mandeville (1946) in magnesium with *fast* neutrons. They found that the cross-section for inelastic scattering was roughly one-third that for elastic scattering. Similar anomalous, scattering of *thermal* neutrons in iron has been noted by Block, Condit and Staub (1943), who found that

> magnetization of iron almost to saturation decreased the neutron scattering by about 10%. This result has been interpreted as an evidence for the existence of the *magnetic moment of the neutron,* in spite of the fact that it carries no charge.

The second kind of inelastic collision, known as the *capture process,* plays a prominent role in transmutations produced by neutrons.

Fermi effect

The study of the *capture cross-sections* was initiated by the important observation made by Fermi and his associates (1934) that the activity induced by neutrons in silver, rhodium and many other elements was enormously increased when the neutron source and the target were surrounded by hydrogenous material, such as paraffin and water or when such material was interposed between the source and the target. This is known as the *Fermi effect,* which has been explained by assuming that the

> *neutrons slowed down* by elastic impacts with protons in the hydrogenous material, *have much larger capture cross-sections than the fast neutrons.*

Fermi also found that:

(1) only those reactions which gave rise to an isotope of the target and hence were caused by the radiative capture of a neutron were sensitive to the effect

and

(2) the slow neutrons responsible for the effect were strongly absorbed in the same elements which were activated and also in many others which did not become radioactive, like boron, cadmium, etc.

Capture cross-sections for slow neutrons

A systematic study of the capture cross-sections for slow neutrons in different elements was undertaken, in 1935, by Dunning and Pegram who used for the detection of the slow neutron not the induced radioactivity but the lithium or boron reaction. The results obtained by them may be summarized as follows: *The values of the cross-section vary greatly and irregularly, perhaps with a certain periodicity, from element to element.* In some elements such as Cd and the rare earths, σ is as large as 3000 x 10^{-24} and more, whereas for some other elements it is even less than 0.1 x 10^{-24}.

The existence of such large cross-sections (many times greater than the nuclear size) in several cases can be explained only on a wave-mechanical basis, as has been done by Fermi, Bethe and others. The wave-mechanical theory indicates that *at sufficiently low speeds, the capture cross-section should vary inversely as the velocity v of the neutron, i.e.,*

$$\sigma \propto 1/v.$$

This means that probability of capture of slow neutrons is directly proportional to the *time of interaction with the nucleus* which is greater the slower the speed of the neutron. This law has been experimentally verified in the case of silver and boron but not for cadmium. The large value of cross-section in Cd has in consequence been accounted for as a *resonance effect* with a selective absorption of thermal neutrons.

A layer of Cd 0.5 mm. thick effectively absorbs all thermal neutrons out of a beam. For this reason Cd screens are often used to filter out thermal neutrons and obtain faster ones.

Selective absorption and groups of slow neutrons

Experiments on selective absorption of slow neutrons, carried out chiefly by Fermi and his collaborators, showed that what is generally termed as "slow" consists of several groups. The main classification is *thermal* or *"C"* neutrons and *resonance neutrons.*

The thermal neutrons which are present in large quantities in the interior of paraffin are
- completely absorbed in 0.5 mm of Cd; they have, in paraffin,
- a Maxwellian distribution of velocities and
- a scattering cross-section for proton of 43x 10^{-24} and
- a capture cross-section of 0.31 x 10^{-24},

which means that they will have about 150 elastic collisions before being captured by a proton to form deuterium. Their mean life in paraffin is about 10.6 x $10^{-4 \ sec}$, which explains why neutrons are not found in a free.state far a long while.

The resonance neutrons, which are readily obtained by the use of a Cd filter, have been further divided into several subgroups, based on their selective absorption with extremely large cross-sections in certain elements; *e.g.,*
- the *"D" group* strongly absorbed in rhodium,
- the *"A" group,* selectively absorbed in silver, and
- the *"I" group* strongly absorbed in indium.

These various groups are differentiated solely by the difference in speeds or energies:

The "C" neutrons have thermal energies about 0.4 eV,
The "D" group about 1 eV,
The "I" group about 2 eV and
The "A" group about 3 to 4 eV, as experimentally estimated.

The main features of the experimental results in the absorption of slow neutrons by the radiative capture process can be readily understood with Bohr's idea of nuclear process. In the highly excited *compound* nucleus, formed by the capture of a neutron by the initial nucleus, the strong interaction among the constituent particles produces the effect that the excitation energy is immediately distributed among a large number of them and is transferred frequently from one to another. Then we may speak of the excitation energy of the nucleus as a whole, but not that of any particle. Further, the number of excitation levels which depends both on the excitation energy and the number of constituent particles will be great; chiefly in the case of heavy nuclei, while the distance between neighboring levels will be only of the order of a few electron volts. In the case of light nuclei the spacing of the levels will be much larger, of the order of several ten thousands of volts.

Under these conditions,

it seldom happens that enough energy is concentrated on one neutron to enable it to escape from the nucleus although there is just sufficient energy in the nucleus for the process to occur. The more probable process will be a transition to a stable state with emission of radiation. This explains why *radiative capture occurs more frequently than scattering.*

The large number of closely packed sharp excitation levels in the compound nucleus explains the frequent occurrence of resonance of slow neutrons.

The selective absorption of given "groups" of slow neutrons by given nuclei is also understood as a resonance effect of the neutrons with a virtual energy level of the compound nucleus. As the resonance levels will be different from nucleus to nucleus, each nucleus will, in general, have its own characteristic group or groups of neutrons which are easily captured.

Theory shows that the 1/v law holds if the neutron energy is small compared to that of the first resonance level or is small compared with the width of this resonance level. Hence, in heavy nuclei whose levels are dense and narrow, 1/v law holds only for a very small energy region, while in light nuclei it holds up to rather high energies corresponding to the large spacing and width of the levels.

These are experimentally verified in the case of Ag for heavy nuclei and boron for light nuclei. In Cd the non-validity of 1/v law in the thermal region appears to indicate that the resonance level is very close to zero energy.

(*vii*) Diffraction of neutrons

We have already remarked that intense beams of monokinetic neutrons can be produced by two methods, viz., by reflection from crystals according to Bragg's law, and by total reflection from highly polished surfaces, in a manner analogous to X-rays. The first technique sorts out the higher energy neutrons whose de Broglie wavelengths are within the Bragg reflecting limit, while the second the low energy neutrons of longer wavelengths that are totally reflected.

The fact that neutrons can he scattered like X-rays leads to the idea of refractive index of various materials for a neutron beam. Experiments on the regular reflection of neutrons from various crystal planes indicate that $(\mu - 1)$ is of the order of 10^{-6} and that it *is* negative in sign (i.e., $\mu < 1$) for most elements, which is a condition for total reflection, and positive (i.e., $\mu > 1$) for some, *e.g.,* manganese. For the commonly used wavelengths of one to ten Angstroms, in the case of neutrons, the critical angles of reflection are in the whereabouts of 10 to 30 minutes of arc.

These findings, involving wave properties of a neutron beam, have logically led to the study of diffraction of neutrons by crystals.

 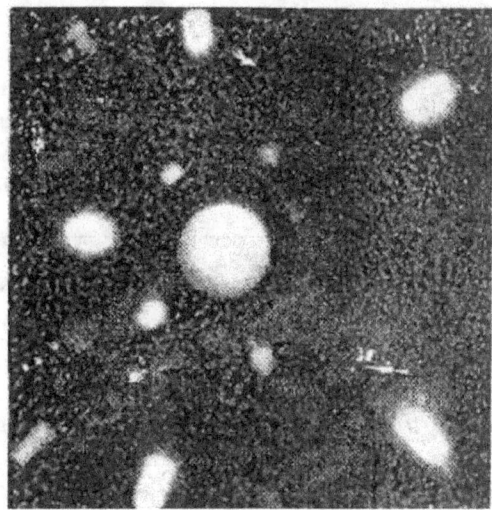

Rocksalt   Lead

Laue patterns obtained with neutron beams.

If a beam of thermal neutrons from a chain-reacting pile, possessing all velocities or wavelengths in the thermal spectrum, is collimated and made to fall upon a single crystal,

138

a series of diffracted beams are observed emanating from the crystal. If the diffracted beams are photographed, a Laue pattern is obtained. The photo above gives two such neutron Laue patterns, one obtained with a crystal of rocksalt and the other with a lead sheet 5/8 inch thick. The pattern obtained with lead clearly brings out the much greater transparency of matter to neutrons than to X-rays which would be confined to a very thin surface layer alone.

The technique of the diffraction of neutrons by crystalline materials has other advantages also that are not to be had with X-ray and electron diffraction methods. The following cases that have been studied may be cited as illustrations:

(a) Crystal structure
The relative scattering power of the scattering centers making up the crystal is of great importance in the study of crystal structure. Neutrons, having no charge, are unaffected by the peripheral electrons of atoms and hence can be scattered only by the atomic nuclei. This nuclear scattering is effected by the nuclear force field, about which very little is known and hence theoretical considerations are not useful in describing relative scattering amplitudes. But many nuclear scattering cross-sections have been measured experimentally. The result of these measurements indicates pronounced differences between neutron and X-ray scattering. While the X-ray scattering cross-sections vary regularly with increasing electron content of heavier atoms, there is no such variation for the neutron scattering cross-section. This finding has been utilized in the study of the crystal structure of substances inaccessible with X-rays.

For example, in the study of the *crystal of ice,* the effects with X-rays are almost entirely produced by the oxygen atoms in the crystal since the X-ray scattering cross-section of hydrogen is extremely small compared with that of oxygen, and, in consequence, the scattering effects due to hydrogen are not detectable. But with neutrons, if the deuterated form ("heavy ice") is used, the scattering cross-sections of the deuterium atoms and the oxygen atoms are very nearly equal and a direct analysis of the positional arrangement of these atoms in the crystal can be and has been made.

(b) Proton-neutron interaction
Neutron scattering by materials containing hydrogen has a very important significance in nuclear physics, since it involves the interaction of the two constituent particles of the nucleus, viz., the proton and the neutron. Measurements of the scattering amplitude in crystal diffraction and in total reflection with hydrogenous compounds have given not only the magnitude of the proton-neutron interaction, but also the nature of the interaction, viz., that it is *spin dependent, i.e.,* depends upon the relative alignment of the nuclear spins. For example,

in the case of deuterium formed with one proton and one neutron, the spins of the two particles are parallel.

(c) Structure of magnetic materials—Anti-ferromagnetism
Since the neutrons have a spin and associated magnetic moment, magnetic interaction between neutrons and atomic electrons responsible for magnetic properties of materials

may be expected. We also know that in ferromagnetic and paramagnetic materials the unfilled third shell in metallic atoms or ions contains electrons with unpaired spins which are responsible for the magnetic behavior of these materials. In ferromagnetic materials the spins are ordered, while in paramagnetic materials they are oriented essentially at random.

Many years ago, X-ray diffraction studies were attempted on magnetic materials in order to analyze their internal structure, but without success; because the necessary magnetic interaction was wanting. Neutron diffraction technique, on the other hand, provides the required magnetic interaction and since the scattering of neutrons will depend upon the orientation of the magnetic moments in the crystal lattice, the neutron diffraction pattern is bound to reveal the internal magnetic structure.

The study of the scattering and diffraction of neutrons by paramagnetic and anti-ferromagnetic crystals has been initiated and some interesting results have already been obtained. For the case of completely uncoupled moments, the magnetic scattering effects show up as a diffuse scattering in the pattern, with a form factor decrease in intensity with angle of scattering. This form factor is interesting in itself, because it permits a determination of the radial distribution function for the specific electrons in the atom, which are responsible for the atomic moment. When the atomic moments in a crystal are coupled together, coherence in the magnetic scattering is introduced and the scattering is concentrated in the Bragg reflections from the crystals, which manifests itself in a sharp and well-defined diffraction pattern. In this way, it has been shown that in ferromagnetic and anti-ferromagnetic substances all the moments are rigidly aligned.

In paramagnetic substances, some like manganese in the manganese sulphate, have their magnetic axes oriented at random with no coupling forces existing between the individual moments, while others have their moments coupled together presumably by the exchange forces which act between the electron spins of neighboring atoms.

The case of MnO, which was formerly thought to be paramagnetic, is interesting. In neutron diffraction experiments with MnO, it was found that

when the crystal is well above the Curie point there are no magnetic interference effects, but below the Curie point additional diffraction peaks appear corresponding to double the spacing of the unit cell as determined by X-rays.

At temperatures moderately above the Curie point, the extra peaks do not disappear completely but persist in a diffuse form. These observations have established that MnO is anti-ferromagnetic.

In a class of crystalline substances, such as FeO, FeCl$_2$, CoO, NiO, Mn, MnO, Cr, etc., the magnetic vectors of adjacent atoms are oppositely directed throughout the lattice, which cause the interaction of next-to-nearest spins to become more important than that between those which are immediate neighbors. For example, in MnO, the coupling between spins of Mn ions is through excited states, where the O ion plays the role of an intermediary. When the atoms Mn—O—Mn are collinear, the coupling between two Mn ions (next-to-nearest neighbors) is very strong, so that in such structures the magnetic vectors of adjacent Mn ions, when separated by intervening ions, point in opposite directions, resulting in anti-ferromagnetism. Anti-ferromagnetic substances have two characteristic temperatures, the Curie point θ and a characteristic temperature θ_s for each substance. At the Curie point the susceptibility is maximum and below this temperature the anti-ferromagnetic order establishes itself. As the temperature is lowered below θ, the lattice vibrations become smaller and the opposed alignments of neighboring spins become more perfect, so that in this range the susceptibility χ varies directly with temperature.

Above the Curie point, thermal oscillations of the Lattice become larger and the spin alignments become increasingly random. In this region, paramagnetic behavior is established and the susceptibility obeys the law:

$$\chi = C / (\theta + \theta_s).$$

From neutron diffraction experiments θ and θ_s can be estimated. For instance, in the case of MnO, $\theta_s = 122°$ K and $\theta_s = 610°$ K. From the diffraction patterns it is found that in MnO the spins are oriented along (100) crystal axes, whereas in FeO they are aligned along the (111) axes. It has also been found possible to alter the neutron diffraction patterns by the influence of a powerful constant magnetic field on the spin orientations. The appearance of diffraction peaks at double the spacing of the unit cell is due to the fact that in anti-ferromagnetic crystals the parallel spins have twice the period of the lattice spacing.

(viii) Mean life of free neutrons
Neutrons in the free state exist for only a very short time. A neutron freed from a nucleus continues in its motion until it encounters another nucleus and is absorbed by the latter. Even if the neutron is slowed down to thermal velocities, the duration of its free state is only 10^{-4} sec, during which alone its behavior can be studied and its radioactivity, if any, detected.

The neutron should decay according to the scheme

$n \rightarrow p^+ + e^- + v.$

Robson working in Canada (1950) has counted the number of decay electrons in an intense beam of neutrons from a nuclear reactor pile. From a study of the variation of the number of electrons with the intensity of the neutron beam, he has found the half life of neutrons to be about 13 mins This time is so large that neutrons are usually absorbed in matter before disintegration occurs.

(ix) Some practical applications of neutrons
The biological and medical sciences are being greatly benefited by the discovery of the neutrons. It has been found that neutrons produce intense biological effects, even greater than X-rays or γ-rays on normal and tumor tissues. Further, while X-rays and γ-rays are

absorbed much more rapidly in the bony rather than the fleshy tissues, neutrons are absorbed more rapidly in the fleshy tissues, very probably due to the greater concentration of hydrogen nuclei there.

Experiments on animals have indicated that neutrons are more destructive to neoplastic than normal tissue. Hence they are being applied to cancer therapy and patients suffering from cancer. Earlier medical applications were conducted regularly with neutrons from the 60-inch and 220 tons cyclotron of the medical biological laboratory in the University of California. Recent experiments performed in that laboratory have demonstrated another possible application of neutrons in cancer therapy. Cancers from mice placed in non-toxic concentrations of boric acid and irradiated with slow neutrons have been found to be destroyed with doses which are harmless to tissues not in contract with boron. The mechanism involved is probably as follows:

The slow neutron is captured by the boron nucleus, resulting-in the emission of two heavy ionizing particles in opposite directions, viz., an α-particle and a lithium nucleus, which traverse a distance of about 7μ in the tissue and thereby produce a kind of explosion within the cell.

2. ARTIFICIAL OR INDUCED RADIOACTIVITY

Introduction

In the different artificial transmutations described in the previous sections, we have frequently met with cases where the product nucleus is unstable and undergoes further disintegration emitting electrons or positrons until a final stable element is formed. Since this β-disintegration, which follows the first instantaneous nuclear reaction, is characterized by the same properties of *spontaneity, delayed action, exponential decay* and a *continuous energy distribution* as the natural β-radioactive elements, it is to be considered as true radioactivity. But as it is produced by *artificial means,* it is rightly called *artificial* or *induced radioactivity.* The great and welcome surprise caused by this phenomenon is easily realized, if it be remembered that before its discovery it was taken for granted that artificial transmutations could be only instantaneous processes, while radioactivity was entirely beyond control and restricted to some forty naturally occurring heavy elements. Little did scientists suspect then that today one could actually produce by artificial means some 500 more radioactive elements which do not occur normally in nature.

Irène Joliot-Curie (12 September 1897 – 17 March 1956)

Jean Frédéric Joliot-Curie (19 March 1900 – 14 August 1958)

Discovery

Artificial radioactivity was first discovered by Mme. Curie Joliot and her husband M. Frederick Joliot, in 1934, in the course of their researches on neutrons emitted when light elements as B, Al, Mg, etc., were bombarded by α-particles. They found that the bombarded substance continued to emit radiations even after the source of α-particles had been removed.

This delayed action, similar to that found in natural radioactivity, disproved that it could be a *direct* effect of the bombardment. Ionization measurements and magnetic deflection experiments showed that the radiations were positrons. They next discovered that this positron activity of the bombarded element behaved very much like the ordinary β-activity of the natural radio-elements. For, when the target, after a few minutes exposure to the bombarding beam of α-rays from Po, was removed and placed by the side of a thin-walled Geiger-Muller counter, hundreds of counts per minute disclosed the emergence of fast flying particles from the target. And the activity was found to decrease *exponentially* with time, just as in the case of natural radio-elements.

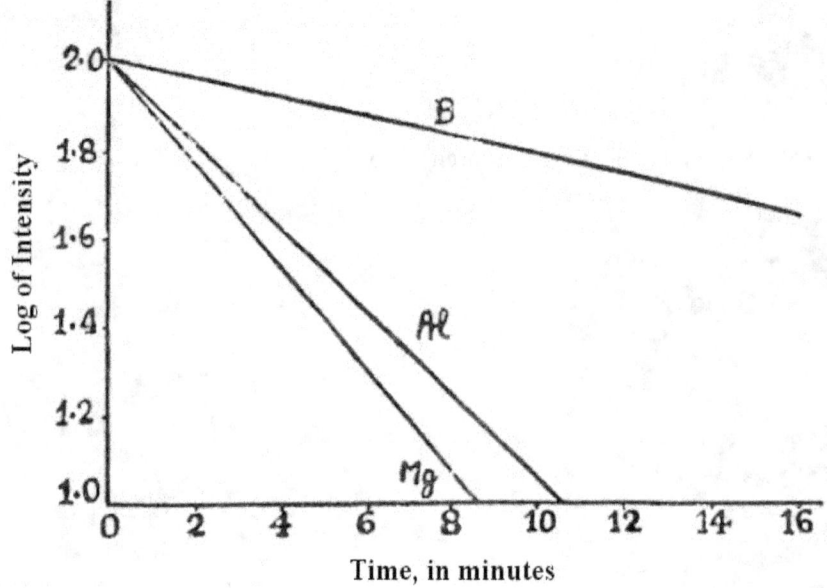

Fig. 298. Exponential decay of the radioactivity induced by the bombardment of B, Al, Mg with α-particles. (I. Curie & F. Joliot)

Plotting the experimental data, i.e., the logarithm of intensity against time, a straight line graph was obtained as shown in Fig. 298, from which the half-period T of the activity could be readily measured. The values of T thus obtained were different for different targets as with natural β-emitters, *e.g.,* for boron 14 mins, (more accurate value later obtained being 10 mins) aluminum 3 mins 15 secs, magnesium 2 mins 30 secs These observations made the Curie-Joliot draw the very important conclusion that the product nucleus formed in the first (α, n) reaction in the cases studied was unstable and hence disintegrated with the emission of a positron according to the reaction equations:

$$_5B^{10} + {}_2He^4 \rightarrow *_7N^{13} + {}_0n^1 + Q$$

$$*_7N^{13} \rightarrow {}_6C^{13} + e^+ \ (T = 10 \text{ mins})$$

$$_{13}Al^{27} + _2He^4 \rightarrow *_{15}P^{30} + _0n^1 + Q$$

$$*_{15}P^{30} \rightarrow _{14}Si^{30} + e^+ \ (T = 3 \text{ mins and } 15 \text{ secs})$$

The product nuclei $*N^{13}$ and $*P^{30}$ formed in the first part of the reactions were not found among the known stable isotopes, while the final residual nuclei C^{13} and Si^{30} that resulted after the emission of positrons were known stable isotopes. To decide definitely whether the reactions actually took place as represented by these equations chemical tests were applied. The general method of procedure was to dissolve the irradiated substance and then add to the solution small quantities of inactive neighboring elements. The different elements were then separated by chemical methods, generally, the precipitation of an insoluble salt and sometimes the formation of a gaseous compound. The identification of the radio-element could then be readily made.

For example, in *the boron reaction,* Curie and Joliot irradiated boron nitride (BN) with α-particles for a few minutes and then heated it with caustic soda. One of the products of this chemical reaction was gaseous ammonia (NH_3). When the various products were tested for positron activity, it was found that only NH_3 had it, which thereby indicated that nitrogen was the radio-element produced in the experiment. Its half period was found to be the same as that produced in other irradiated boron targets, while no such radioactivity was observed when ordinary nitrogen was used as target.

In *the aluminum reaction,* when the irradiated aluminum was dissolved in HCl, the activity was carried away by the hydrogen, in the gaseous state, which could be collected in a tube. The chemical reaction must be the formation of phosphine PH_3, if phosphorus had been produced in the α-bombardment of the aluminum target. To make sure of this point the irradiated Al was dissolved in aqua regia, without evolution of gas, and a compound of zirconium was added which would precipitate phosphorus as zirconium phosphate, the aluminum remaining in .solution. The positron activity was found only in the precipitate and not in the solution. Hence the substance emitting the positrons was definitely not aluminum and was almost certainly an isotope of phosphorus.

Such tests established without ambiguity that the observed positron activity was really due to the new radioactive elements, *radio-nitrogen* and *radio-phosphorus,* produced according to the equations given above. These are isotopic with the ordinary nitrogen and phosphorus, but do not normally exist in nature. Being unstable, they break down into stable isotopes C^{13} and Si^{30} respectively by the spontaneous emission of positive electrons.

As soon as Curie and Joliot announced their discovery, the study of artificial radioactivity was taken up in many other laboratories as Cavendish in Cambridge and Pasadena in California, and new cases of it were announced at a rapid rate. It was soon found that the phenomenon occurred not only under α-particle bombardment, but also with other projectiles, as protons, deuterons and neutrons and that the artificially produced radio-elements not only emitted positrons but also electrons in several cases.

Wilson's cloud chamber method was applied in the analysis of the activity of the artificial radio-elements by Ellis and Henderson at the Cavendish Laboratory and by the workers at Pasadena and beautiful photographs of the tracks of the emitted particles curved by magnetic fields were obtained in confirmatory evidence of the discovery. One such photograph is reproduced here. It was obtained by Anderson in California, when a carbon target, after having been bombarded for several minutes with protons of 0.9 MeV energy, was placed right into the chamber. The tracks obtained have the specific aspect of the electron tracks and in the applied magnetic field of 800 gauss they are curved in a sense which proves the particles to be positive.

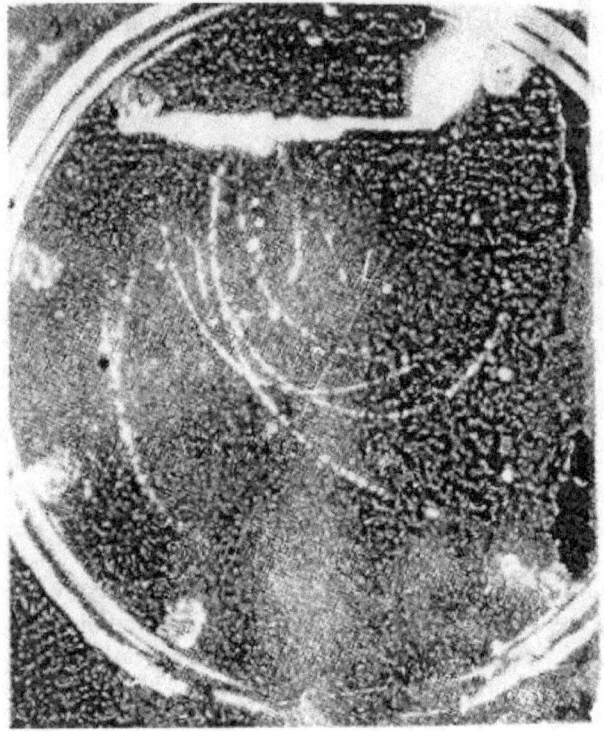

Induced positron activity from carbon bombarded by 0-9 MeV protons. {Anderson)

The positron activity is evidently due to the formation of radio-nitrogen N^{13} in the "capture" (*p, γ*) reaction of proton with carbon, already described.

The *nature of the dependence of the induced radioactivity on the kinetic energy of the bombarding particles* was also investigated. The Joliots were able to establish that the half period was independent of the kinetic energy of the projectile when the energy of the α-particles of Po was varied from 5.3 to 1 MeV. This proved that a *single kind of unstable nucleus alone resulted from the reaction.* But the intensity of the emitted positron was found to diminish as expected when the energy of the α-particles was decreased becoming almost imperceptible at about 3 MeV for boron, and at 4 to 4.5 MeV for Al and Mg. Ellis and Henderson at the Cavendish Laboratory working with 8.3 MeV α-particles found the same result, *viz.,* the number of positrons steadily increased as the kinetic energy of the α-particles was raised, in the ratio of 15: 1 for a change the energy from 5.5 to 7 MeV.

Energy spectrum

The complete similarity of induced radioactivity with natural radioactivity was finally proved from nature of the energy distribution of the particles emitted by the bombarded targets. Their energies and number were estimated by the magnetic spectrograph or the cloud chamber, as described in connection with pray spectra. When the data were plotted the distribution-in-energy curve presented a continuous form with a definite upper limit as shown in Fig. 299, drawn from data obtained by Anderson and Neddermeyer by the cloud chamber method.

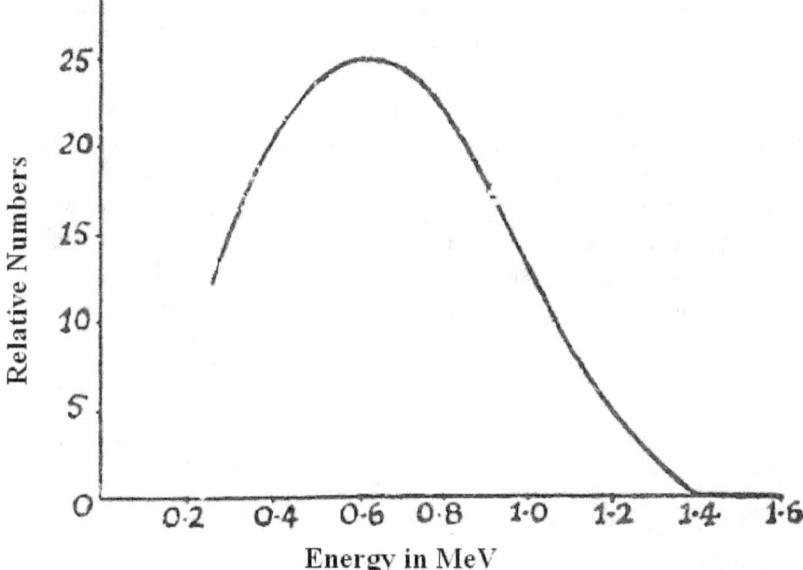

Fig. 299. Distribution-in-energy of positrons of the induced radioactivity resulting from bombardment of carbon with protons.(Anderson)

Ellis and Henderson also were able to observe a continuous distribution of energies of positrons ranging from 1 to 2.5 MeV in one of the cases of radioactivity induced by α-particle impact. Subsequently, this was found to be true in every case analyzed, regardless of whether positrons or electrons were ejected. This general result, in its turn, showed that the newly discovered artificial radioactivity followed the same mechanism involving a simultaneous ejection of a neutrino as in the case of β-activity of natural radioactive material. Thus, the new phenomenon frequently met with in artificial transmutations was definitely established as genuine radioactivity, though induced by artificial means.

Researches on induced radioactivity

As the main problem of identification of induced radioactivity as true radioactivity was satisfactorily solved within less than a year after it was first suggested, thanks to the spirit of collaboration and enthusiasm of several investigators, the researches that followed the discovery chiefly concerned themselves with the accumulation of more cases of the phenomenon, their classification and interpretation, the study of some special types of radioactive decay such as *nuclear isomerism* and *K electron capture,* and especially with the practical application of induced radioactivity, extending beyond the limits of artificial transmutations to the realm of medicine and biology, where they have now opened out a very important and wide field of investigation. A short account of these developments will now be given.

(1) Production of artificial radioactivity

Since the first discovery in 1934, **more than 500 artificial radioactive isotopes** have been found, provoked with all the five commonly used projectiles, α-particles, protons, deuterons, neutrons and γ-rays. A great majority of them have been identified as regards their mass and atomic numbers, which indicate that one or more radioactive isotopes exist for practically all the elements in the periodic table.

Many of the light elements between boron and calcium give rise to artificial radio-elements when *bombarded with the α-particles* from natural radioactive sources. *Many* of these radio-elements *emit positrons* as Be^{10}; N^{14}, Na^{23}, Al^{27}, etc., but *some* as Si^{29}, Ca^{40}, P^{31}, etc., *electrons.* Mg yields positrons as well as electrons, but this has been shown to be due to different primary reactions with the two isotopes Mg^{24} and Mg^{25}, the former producing radioactive $*Si^{27}$ with the emission of a neutron, and the latter forming radioactive $*Al^{28}$ with the emission of a proton, according to the equations:

$$_{12}Mg^{24} + {}_2He^4 \rightarrow *{}_{14}Si^{27} + {}_0n^1$$

$$*{}_{14}Si^{27} \rightarrow {}_{13}Al^{27} + e^+ \ (T = 6.7 \ mins)$$

$$_{12}Mg^{25} + {}_2He^4 \rightarrow *{}_{13}Al^{28} + {}_1H^1$$

$$*{}_{13}Al^{28} \rightarrow {}_{14}Si^{28} + e^+ \ (T = 2.3 \ mins)$$

With protons as projectiles, (p, n) and (p, γ) reactions are known to give rise to artificial radioactivity, emitting positrons or electrons.

With deuterons, chiefly of high energy produced by the cyclotron, many elements, even heavy ones as Bi and Pt, acquire radioactivity, emitting electrons or positrons.

With heavy elements like Bi, α-radioactivity is also produced.

With neutron bombardment induced radioactivity has been extended to most elements. Chiefly with the discovery of the Fermi "water effect", where slow neutrons of thermal energies become very effective transmuting agents, a great number of radioactive isotopes, even with many heavy elements, has been formed. For the three most important processes which are known to produce induced radioactivity, *viz.,* (n, p), (n, α) and (n, γ) *emission of electrons is common,* though in a few cases positrons are also observed. In *photodisintegration processes,* which ordinarily cause the emission of neutrons, the radio-elements formed are *positron* emitters.

Yield

In the artificial radioactivity produced by means of *α-particle bombardment the yield is very low:* e.g. for every 10^7 α-particles of high energy (8.3 MeV) incident upon an Al target only six are effective in the production of radio-phosphorus. Further, even this small yield falls rapidly as the energy of the α-particles is diminished

In the radioactivity provoked *by neutron, bombardment the yield is much greater,* which is still further increased when the neutrons are slowed to thermal energies, in some cases as much as 40 to 100 times.

High energy deuterons produce also large yields. Chiefly the *(d, p)* reaction results in intensities comparable with the natural radioactive sources.

Period and energy ranges

The *half periods* of the different artificially produced radio-elements, measured with Geiger counters or ionization chambers equipped with sensitive electrometers, range from 0.02 secs (B^{12}) to more than 10 years (Be^{10}). These limiting values have both been obtained from deuteron induced processes, in which cases the high energy of the projectile and the consequent high yield have made these extreme cases just observable. The shortest half period of B^{12} was estimated by counting the number of electron tracks in a cloud chamber arranged to be automatically expanded at short known intervals following the cutting of the deuteron beam irradiating a boron target placed in the chamber. The longest period of Be^{10} was found by studying the activity in a beryllium target which had been bombarded by deuterons for a period of six months as a neutron source. It may be noted that both the artificial radio-elements considered here are produced by *(d, p)* and not by *(d, n)* reactions:

$$_5B^{11} + {}_1D^2 \rightarrow *{}_5B^{12} + {}_1H^1$$

$$*{}_5B^{12} \rightarrow {}_6C^{12} + e^- \ (T = 0.02 \text{ secs})$$

$$_4Be^9 + {}_1D^2 \rightarrow *{}_4Be^{10} + {}_1H^1$$

$$*{}_4Be^{10} \rightarrow {}_5B^{10} + e^- \ (T \approx 10 \text{ yrs})$$

The *maximum energies* of the electron and positron emissions from artificial radio-elements, as measured by absorption and deflection techniques, are found to vary from below 0.3 MeV to 13 MeV, although in many cases the upper limits are in the whereabouts of 2 to 3 MeV. The maximum energies and the half periods of several of the artificial radio-elements appear to satisfy approximately the Sargent relation, as in the case of the β-rays from natural radioactive elements.

(2) Classification and interpretation of artificial radioactivity

On the basis of the proton-neutron model of the nucleus and the tendency for the ratio N/P, (*i.e.,* of the number of neutrons to that of protons in nuclei) to change towards a stable value, a general classification and interpretation of radioactivity can be made. The number of excess neutrons in the stable nuclei is found to increase more or less regularly with increasing charge. The unstable radioactive nuclei above or below this region of stability will tend to return to the stable region with the emission of those particles which best accomplish this result. *Hence, .radioactive nuclei formed with an increase in (N - P) and in consequence heavier than the known stable ones tend to exhibit electron activity,* resulting in a product nucleus of the same mass but higher charge. On the other hand, *radioactive nuclei formed with a decrease in (N — P) and consequently too light to be stable will emit positrons,* giving rise to a product nucleus of the same mass but lower charge. Emission of an electron increases P at the expense of N, while emission of a positron has the opposite effect. In either case, a nucleus having a more stable ratio of N/P is produced. The following table based on the above-stated general law gives the classification of the principal types of reactions with respect to the kind of β-activity, to which they give rise. Many exceptions, however, occur.

Activity	e^-			e^+	
Change in N—P	2	1	0	— 1	— 2
Reactions	(n, p)	$(a, p), (d, p)$ $(n, \alpha), (n, \gamma)$	(d, α)	$(\alpha, n), (d, n), (p, \gamma)$ $(n, 2n), (\gamma, n)$	(p, n)

The additional important points to be noted in this connection are:

(*a*) *Radioactive isotopes ordinarily lie close, on the mass scale, to stable isotopes of the same element,* lighter ones tending to be transformed with a decrease of charge (positron activity) and heavier ones with an increase of charge (electron activity): *e.g.,* Na^{22} (positron active), Na^{23} (stable), Na^{24} (electron active).

In most cases the radioactive isotope is found to have a mass number differing by one unit from the extremes of the stable isotopes of that element: *e.g.,* N^{14} and N^{15} (stable) are flanked by N^{13} (positron active) and N^{16} (electron active).

In many instances, the radioactive isotope fills the gap between two known stable ones; *e.g.,* Ag^{108} (electron active) lies between the stable Ag^{107} and Ag^{109}.

In a few cases, the unstable isotope differs by two or more mass units from the stable ones *e.g.,* Al^{29} (electron active), while Al^{27} (stable); $Sb^{116,118,120}$ (positron active), $Sb^{121,123}$ (stable), $Sb^{124,127,129,131}$ (electron active).

In some instances, *branching processes* occur, yielding *positrons and electrons from the same radioactive isotope.*

(b) *The same radioactive isotope can usually be produced in different ways:* e.g.,$*_{13}Al^{28}$ which decays to stable $_{14}Si^{28}$ with the emission of electron can be produced in the following *five* ways:

$$_{12}Mg^{25} + _2He^4 \rightarrow *_{13}Al^{28} + _1H^1$$

$$_{13}Al^{27} + _1D^2 \rightarrow *_{13}Al^{28} + _1H^1$$

$$_{13}Al^{27} + _0n^1 \rightarrow *_{13}Al^{28} + h\nu$$

$$_{14}Si^{28} + _0n^1 \rightarrow *_{13}Al^{28} + _1H^1$$

$$_{15}P^{31} + _0n^1 \rightarrow *_{13}Al^{28} + _2He^4$$

$$*_{13}Al^{28} \rightarrow _{14}Si^{28} + e^- \ (T = 2.4 \text{ min})$$

Similarly Ag^{106} which decays to stable Pd^{106} with positron emission (T =25.5 mins) can be produced in at least *eight* ways.

(c) *Radioactive isotopes may be formed of elements hitherto unknown, which means production of new elements.* In the periodic table of elements, there were three places vacant for atomic numbers Z = 43, 85 and 87. With the different types of projectiles available and the variety of nuclear reaction known, it has been possible to produce these elements artificially as radioactive isotopes by bombarding nuclei of neighboring atoms with suitable particles.

At least eight reactions result in the formation or radioactive products with Z = 43, now known as *technetium* whose chemical properties will be given by those of the above-mentioned radioactive isotopes.

The element of Z = 85 has been produced by bombarding bismuth with 32 MeV α-particles, It is called *astatine* and occurs in the same vertical column of the periodic table as iodine; the resemblance between the two has been experimentally established.

Element 87, called *francium,* has been found as one of the products in the natural radioactive disintegration of actinium. It decays emitting electrons with a half-period of 21 mins Radioactive forms of *transuranic elements* with Z = 93 (neptunium), Z = 94 (plutonium), Z = 95 (americium), Z = 96 (curium), Z =97 (berkelium), Z = 98 (californium), Z = 99 (einsteinium), Z = 100 (fermium), Z = 101 (mendelevium) and Z = 102 (nobelium) have also been produced recently.

(d) *Many artificial radio-elements emit also γ-rays along with the β-rays.* In order to distinguish these γ-rays emitted in the decay of the artificial radio-element from those which might be emitted in the primary reaction in which the radio-element itself is formed, it is usual to use the symbol hv in the latter case, while the symbol γ is reserved for the former. The γ-rays emitted along with the betas in the radioactive decay are in general monoenergetic and are much too strong in intensity to be due to conversion of the betas themselves. Hence they are considered as coincident with the β-emission and as *representing a transition of the final residual nucleus from an excited state* in which it is left after the β-emission.

In some cases, for example, $*Li^8$, $*Bi^{12}$, $*N^{16}$, $*F^{20}$, the β-emission leads practically always to an excited state of the residual nucleus. This will result when the β-transition to the round state is highly forbidden. In other cases, for example, $*Cl^{38}$, $*A^{41}$, $*K^{42}$, $*Mn^{56}$ and $*As^{76}$, the β-decay leads sometimes to the ground state and sometimes to an excited state. There are also instances where the β-transition takes place to several alternative excited states: *e.g.,* $*Na^{24}$. In the first process where the transition is limited to a single excited state, the γ-ray energy is additive to the β-maximum, and experimentally a *simple β-ray distribution* would be observed and the number of γ-quanta would be equal to the number of radioactive electrons. In the other two transformations *a complex β-spectrum* consisting of several groups," analogous to the heavy particle groups in ordinary nuclear reactions, would be observed. The residual nucleus left in excited states will, in its turn, go down through other excited states to the ground state with emission of more γ-rays, which will, in consequence, increase the complexity of the phenomenon.

The *transformation of radioactive* Na^{24} affords an instructive example of the complex process stated above. According to Kruger and Ogle (1945) who have studied this case very carefully, β-rays emitted fall into three groups of energies, 1.84, 1.63 and 1.07 MeV corresponding to three alternative levels in which the final Mg^{24} nucleus may be formed. There are also at least seven different γ-rays emitted by the Mg^{24} nucleus, as indicated by the analysis of pair particles generated in a cloud chamber by the γ-rays. Their measured energies are 2.56, 2.76, 3.24, 2.68, 2.89, 1.26 and 1.38 MeV. These facts argue to the existence of at least five excited levels above the ground state in the Mg^{24} nucleus.

(3) Two interesting types of decay of artificial radioactivity
Disintegrations of the artificially produced radio-elements are known to occur also by processes other than the simple ejection of positrons or electrons. Two such novel processes, which are of special interest, are known as *K electron capture* and *nuclear isomerism.*

K ELECTRON CAPTURE

According to the theory of β-disintegration, a proton by emitting a positron becomes a neutron. But a proton might as well transform itself into a neutron by absorbing an electron. This cannot happen, however, in stable nuclei of low energies, since the mass of the electron (0.00055 m.u.) is insufficient to supply the increase of mass in the formation of the neutron, which is equal to $1.00893 — 1.00813 = 0.00080$ m.u. This explains also why under normal conditions nuclei containing protons exist without any spontaneous absorption of the peripheral electrons.

In an excited nucleus, on the other hand, where, an additional store of energy is available, there is a possibility of a proton turning into a neutron by absorbing an electron, instead of emitting a positron.

In 1935, Yukawa, from more rigorous theoretical considerations, put forward the hypothesis that in a positron emitting nucleus there exists a certain probability of its decay by capturing an electron. Since its own K electrons are the most readily available, it seems reasonable that the nucleus might absorb one of these, thereby leading to *disintegration by K electron capture.*

When an atom decays by K electron capture, the vacancy created in the K shell will soon be filled by one of the outer orbital electrons, giving rise to a characteristic X-ray line of the K-series. Further, as the nucleus has actually changed by the absorption of the K electron, before the X-ray is emitted, the X-ray will be characteristic of the product atom and not of the original atom. It was observation of such characteristic X-rays by Alvarez, in 1937, that gave the first evidence of this type of disintegration.

A good number of the artificially radioactive nuclei were soon found to disintegrate in this way, decreasing their atomic number by one unit just as they would have by emission of a positron. The case of vanadium ($_{23}V^{49}$) disintegrating into titanium ($_{22}Ti^{49}$) offers a straightforward example of the K electron capture process, since Walke, Williams, and Evans who very carefully studied this reaction (1939) found that no radiations of any other kind, except the characteristic X-rays, were emitted to complicate interpretation of data. When titanium was bombarded with deuterons, radioactive vanadium was presumably formed according to the reaction:

$$_{22}Ti^{48} + {_1}D^2 \rightarrow {^*_{23}}V^{49} + {_0}n^1$$

since the active product could be identified as an isotope of vanadium of mass 49 by chemical separation methods. But it emitted no electrons, positrons or γ-rays but only X-rays which were found to be the K_α radiation of titanium. The intensity of these X-rays diminished exponentially with time and at the end of 600 hours the intensity was reduced to half its initial value, which showed that the period of the activity was equal to 600 hours. The only conclusion that could be drawn from these data was that vanadium (Z = 23) captured one of its K electrons and became titanium (Z = 22) according to the equation:

$$^*_{23}V^{49} + e^- \rightarrow {_{22}}Ti^{49} \quad (\text{T} = 600 \text{ hrs})$$

Titanium thus formed had one electron missing from the K-shell, which caused the subsequent emission of X-rays characteristic of titanium. The fact that no radiation other than X-rays was emitted proved

(i) that $_{23}V^{49}$ changed into $_{22}Ti^{49}$ by no other process except a capture of an electron and (*ii*) that the product nucleus $_{22}Ti^{49}$ would be left in the ground state.

It is to be noted that the case of $_{23}V^{49}$ illustrating the process in a clear-cut manner due to the absence of all other radiations except the characteristic X-rays of $_{22}Ti^{49}$ is rather an exception, the general rule being that radiations such as γ-rays and positrons accompany the process.

Many cases of disintegration by K electron capture *are known to emit γ-rays,* which is bound to complicate matters, since these nuclear γ-rays also give rise to characteristic X-rays of the product atom through the intermediary of internal conversion phenomenon. One must, therefore, be careful to distinguish between these X-rays and those which are considered to give evidence of K electron capture. This is ordinarily achieved by using counters and observing coincidences. Examples of such analysis are: $_4Be^7$ obtained by bombarding $_3Li^6$ with deuterons disintegrates into $_3L^7$ by K electron capture of period 43 days, emitting a γ-ray of energy 0.485 .MeV; $_{25}Mn^{54}$ produced by bombarding iron with deuterons, decays by K electron capture of period 500 days to form $_{24}Cr^{54}$ accompanied by a γ-ray of 0.835 MeV.

There are also several cases in which, the disintegration takes place either by positron emission or by K electron capture so that both the processes are observed in the same radioactive element, e.g., the isotope of vanadium of mass 48 disintegrates by this branching process. Walke, Williams and Evans found that in about 30% of the cases K electron capture occurs, while in the rest positron emission; likewise, it has been found that in the case of $_{19}K^{40}$, while the predominant mode of decay is by emission of a β-particle forming $_{20}Ca^{40}$, about 10% of the transformations are by K electron capture (or emission of positron) forming $_{18}A^{40}$. Theory shows the branching ratio in such cases to depend upon the energy available, the density of the electrons and the spin change associated with the positron emission.

It may be remarked that since the capture of an electron by the nucleus results in the change of a proton into a neutron, this process must be accompanied by the emission of a neutrino in order to conserve the angular momentum of the nucleus. Allen, in 1942, observing the recoil energy of Be^7 nucleus involved in its decay by K capture to form Li^7, found it to be about 45 eV in good agreement with the theoretical value of 58 eV and thus obtained an indirect proof for the existence of the neutrino, as already stated.

NUCLEAR ISOMERISM

It has been found that in several cases of the artificial radio-elements, *one and the same nucleus, having the same mass number A and the same atomic number Z, disintegrates in different ways with different decay periods.* This manifestation of different radioactive properties argues to certain differences in internal structure of nuclei which are otherwise identical in all respects. They are called **nuclear isomers** by analogy with chemical isomers, frequently found in organic compounds, in which a given group of atoms forming a molecule may have different spatial arrangements giving rise to different chemical and physical properties.

Soddy was the first, in 1917, to suggest that such isomers might exist among the natural radioactive elements, and Hahn was able, in 1921, to establish experimentally the isomeric property of UX. and UZ which were the only case of nuclear isomerism known for a long time. In recent years, however, several isomeric pairs have been discovered among the artificial radioactive nuclei. The study of this phenomenon is of special interest to nuclear physics, as it argues to *the existence of low lying metastable states in nuclei*. A brief summary of the experimental study and theoretical interpretation of nuclear isomerism will now be given.

Experimental study
Nuclear isomerism has been experimentally investigated in the following three ways, *viz.,*

(*i*) *a first identification of isomeric cases,*
(*ii*) *a closer examination of the internal mechanism* of *isomers* and
(*iii*) *separation of isomers.*

Identification of isomers consists chiefly in deciding whether the two different period β-activities experimentally observed in certain cases are isomeric or not. This is done by a simple "cross-checking" with different types of nuclear reactions, in which the element in question and its immediate neighbors are implicated, along with the knowledge of the masses and relative abundances of the corresponding stable isotopes. It must be emphasized that the simultaneous existence of two decay periods is, by itself, no sufficient criterion for the identification of isomers, but that "cross-checking" must necessarily be used to arrive at a safe conclusion. Let us illustrate by the isomerism of $_{35}Be^{80}$, which was the first case to be discovered among artificial radio-elements.

The isomers of Be^{80}
There are only two known stable isotopes of bromine of mass numbers 79 and 81 and relative abundances 50.7% and 49.3% respectively. When bromine is bombarded by *slow neutrons*, radioactive elements are formed which disintegrate by the emission of electrons with three different periods, 18 mins, 4.2 hrs and 36 hrs. Chemical tests show that the radioactive products are isotopes of bromine. Hence the reaction must be a *simple capture (n, γ) process*, represented by

For mass 79:

$$_{35}Br^{79} + {_0}n^1 \rightarrow {^*_{35}}Br^{80} + h\nu$$
$$^*_{35}Br^{80} \rightarrow {_{36}}Kr^{80} + e^-$$

For mass 81:

$$_{35}Br^{81} + {_0}n^1 \rightarrow {^*_{35}}Br^{82} + h\nu$$
$$^*_{35}Br^{82} \rightarrow {_{36}}Kr^{82} + e^-$$

Br^{80} and Br^{82} are therefore the only two radioactive isotopes responsible for the three observed periods, so that to one of these must be attributed a double period.

When bromine is bombarded with *γ-rays of* very high energy (17 MeV), radioactive elements chemically identified as isotopes of bromine, are formed, which decay emitting again with three different periods, 6.3 mins, 18 mins, and 4.2 hrs. The reaction is evidently a *photonuclear (γ, α) effect* represented by
For mass 79:

$$_{35}Br^{79} + hv \rightarrow *_{35}Br^{78} + _{0}n^{1}$$
$$*_{35}Br^{78} \rightarrow _{34}Se^{78} + e^{+}$$

For mass 81:

$$_{35}Br^{81} + hv \rightarrow *_{35}Br^{80} + _{0}n^{1}$$
$$*_{35}Br^{80} \rightarrow _{36}Kr^{80} + e^{-}$$

Here also there are only two radioactive isotopes for bromine formed, of masses 78 and 80, which are, however, responsible for the three observed periods, so that again to one of these must be attributed a double period.

Cross-checking these two different reactions, it is seen that two of the observed periods, viz., 18 mins and 4.2 hrs are common to both which must therefore be assigned to that radioactive isotope which is also common to both reactions and hence to bromine of mass number 80. Thus Br^{80} *consists of isomeric nuclei emitting electrons of two different periods.* The remaining two of the observed periods, viz., 6.3 mins, and 36 hrs, must be attributed to Br^{78}, a positron emitter, and Br^{82}, an electron emitter, respectively. Confirmatory evidence is had from the fact that the isomeric pair of Br^{80} can be produced by bombarding bromine with deuterons and fast neutrons, as was shown by Such (1937) or by proton bombardment of selenium, analyzed by Du Bridge and others (1938).

Study of internal mechanism of isomers
Attention has been directed towards two points chiefly:
(a) In the cases where the isomers are formed by, the "capture" process, nature of the capture level or levels responsible for the phenomenon is investigated. This is usually done either by measuring the "branching ratio" of the two isomeric activities produced under varying conditions of activation or by determining the resonance energies involved in the process by the "boron absorption" method. Thus Amaldi and Fermi by the `branching ratio" experiments (1936) and Von Halben by the "boron absorption" experiments (1937) were able to demonstrate a single capture level in the case of rhodium (Rh^{104}) which exists in two isomeric forms of periods 45 sees. and 4.3 mins

(*b*) In the second place, the possible relations that could exist between the two isomers of a radioactive isotope are analyzed by measuring the maximum energy limits of the two isomeric activities and constructing energy level diagrams that would represent the modes of disintegration. This second method of investigation has led to more positive results than the first. It is found that *some isomeric pairs have more or less the same maximum energy* (*e.g.,* Br^{80}: 18 mins, 2 MeV and 4.2 hrs, 2.05 MeV; Rh^{104}: 45 sec, 2.74 MeV and 4.3 mins, 2.76 MeV) *while others different energies* (*e.g.,* Ag^{106}: 24.5 mins, 1.9 MeV and 8.2 days, 1.3 MeV; In^{116}: 13 secs, 3.1 MeV and 51 mins, 1.4 MeV).

Separation of isomers

This has been achieved in several cases using the Szilard and Chalmer method. The γ-rays ordinarily emitted along with the β-activities is utilized for this purpose. The γ-rays may be internally converted and the conversion electrons thus produced may have enough energy to break the chemical bond. Or, if the γ-rays have a high energy, their emission will be accompanied by the recoil of the nucleus, which may ho sufficient to liberate the atom from the compound.

In 1939, Segrè and his associates have thus been able to separate the two isomers of bromine. Likewise, Seaborg and his associates have chemically separated the three isomeric pairs of tellurium (127, 129, 131). The results are of great importance as they indicate (i) that the isomers can be considered as different atoms, though possessing the same mass and the same chemical and physical properties and (*ii*) that the isomeric pairs may be genetically related, one growing out of the other, since it has been found that a in a number of cases the short period isomers grow from the long period ones, e.g., in the case of bromine the 18 mins period isomer grows from the 4.2 hrs period isomer; in rhodium the 45 secs isomer from the 4.3 mins one.

Results

At least *seventeen pairs of isomer have* so far been detected. *Some of them disintegrate by positron emission, while the majority by the ejection of electrons.* For example, scandium ($_{21}Sc^{44}$) having a positron activity of maximum energy of 1.33 MeV has two isomers of periods 3.92 hrs and 2.44 days; silver ($_{47}Ag^{106}$) has two. isomeric forms, one disintegrating with the emission of positrons of period 24.5 minutes, while the other by K electron capture with a period of 8.2 days; manganese ($_{25}Mn^{52}$) likewise has two isomers, one with positron activity of 21 mins, and the other either positron emission or K electron capture of 6.5 days. On the other hand, with $_{27}Co^{60}$, $_{30}Zn^{69}$, $_{35}Br^{80}$, $_{45}Rh^{104}$, $_{52}Te^{127,129,131}$ and UX_2 -UZ, both the isomers decay with the emission of electrons.

Theoretical interpretation

Wiezsäcker (1930) has proposed a very satisfactory theory of nuclear isomerism based on the existence of *low excited states whose angular momentum* (*i.e., nuclear spin) is considerable,* differing by several units from that of the ground state. Such low energy excited states with great spin would be *metastable* and the nucleus may remain in this state for a sufficient length of time for it to be observed as a different nucleus.

As a general rule, a nucleus formed and left excited in any transformation emits γ-radiations until it arrives at its ground state and the time required for this is very short, of the order of 10^{-12} sec But if there exist selection rules that make the radiative transition from a certain (in general, the first) excited state much *less* probable than in normal cases, then that excited state becomes metastable, which would have its own decay period and may decay in the following different ways:

Isomeric Nuclei

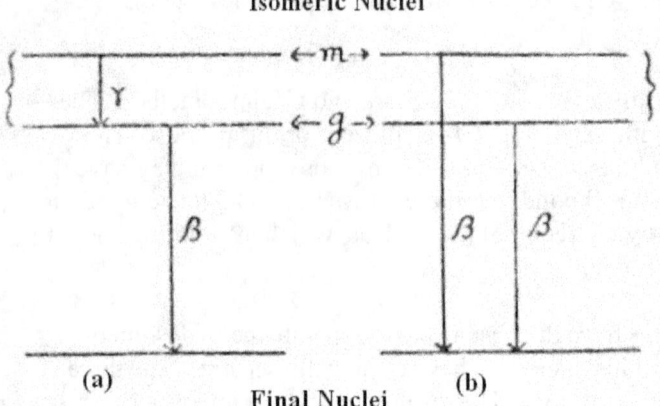

Fig. 300. Energy level scheme for isomeric nuclei *(a)* isomers genetically related; *(b)* isomers not genetically related; m = metasable state, and *g* = ground state of the isomeric nuclei.

(a) *The metastable state may go over to the ground state with the emission of γ-radiation whereupon the ground state will decay further with β-emission [Fig. 300 (a)].* The period of the activity due to the metastable state is then determined by the probability of the delayed γ-transition, and can be observed as a separate period only if it is longer than the natural period of the ground state. In such a case, the energy spectrum of the β-rays is expected to be the same for the two isomeric periods and the isomers are genetically related, one form growing out of the other. Further as regards the γ-transition from the metastable to the ground state, it can be shown that when the spin difference between the two states is large and the energy difference small, the probability of γ-absorption by the peripheral electrons of the disintegrating atom itself (internal conversion) becomes very great, which, in consequence, would affect the transition in the following two was:

(*i*) the life-time of the higher isomer against transition to the lower state is considerably reduced compared to the long life-time of a year or more against direct γ-radiation;

(*ii*) the transition will usually be accompanied not by γ-radiation but by the emission of a conversion electron.

All these theoretical conclusions are amply verified in the case of several nuclear isomers such as Sc^{44}, Co^{60}, Zn^{69}, Brʋ, Rh^{104}, Te^{127}, , Te^{129}, UX_2 —UZ. In all of them, maximum energies of the two isomers are nearly the same. The higher metastable isomer has a longer period than the lower one, though it is limited from minutes to days, instead of being long, a year or more. The energy of the γ-rays, as estimated from the observed conversion electrons is low. The two isomers are also genetically related.

Considering, for instance, the **concrete case of Rh^{104}**, the upper energy limit is the same for the two observed periods, about 2.75 MeV. If the 4.3 mins is referred to the transition of the metastable state to the ground state with γ-emission, the decay of the metastable state will have a proper period, experimentally observable since it is greater than 45 secs period of the ground state. The comparatively short and measurable life-time of the γ-transition verifies, in its turn, the very important condition of the theoretical explanation, viz., the considerable spin difference involved between the metastable and ground states. The low γ-ray energy of about 80 KeV, as measured by means of conversion electrons or 30 to 60 KeV accompanying the 4.3 mins, satisfies the other theoretical prediction of small difference of energy between the two isomeric states. The 45 sec β-emitter grows out of the 4.3 mins γ-transition and hence they are genetically related. Because of the sufficiently large difference between the periods of the two isomers it is possible to make a separation of the two isomers and show that the excited metastable state decays with 4.3 mins period, while the ground state itself decays with a period of 45 secs The chemical separation of the two isomers, in its turn, indicates that the two can be considered as different nuclei of a special type.

(b) *The metastable state may emit a β-particle directly,* [Fig. 300 (b)]. This will happen if the probability of γ-emission is smaller than that of β-emission for the given metastable state. Ordinarily the β-emission whose decay period is several seconds or more cannot compete with the extremely short-time of normal γ-emission, so that β-particles cannot, in general, be emitted from excited states of nuclei. But in certain cases, on account of the special nature of the excited state, its very low energy with considerable spin, may cause selection rules to intervene and ender the quick γ-process less probable and the slower β-process more probable so that the excited state is effectively destroyed not by the former but by the latter. *The life-time of the metastable state in such cases may be longer or shorter than that of the ground state.* The metastable and the ground states undergo independent radioactive transitions to the final nucleus, thus accounting for the *existence of isomers which are not genetically related.* Further, *the upper beta energy limit will not be the same for the two isomeric periods.* The transitions from the metastable and the ground states will, in general, lead to different states of the final nucleus, because the angular momenta are very different. Thus the two independent transitions may be followed by γ-rays from the residual nucleus.

These theoretical predictions appear to be experimentally verified in the case of the isomers of Ag^{106}, Sr^{89}, Mn^{52}, etc., which have very different upper energy limits for the two periods: *e.g.,* the positron emission of Ag^{106} with T = 24.5 mins has a maximum energy of 1.9 MeV, while the disintegration by K electron capture with T = 8.2 days has the upper energy limit of 1.3 MeV. There is also evidence of emission of γ-rays of energy 1 MeV.

Theoretical considerations further show that *sufficiently metastable states will, in general, not exist for light nuclei,* firstly because of the small density of levels for such nuclei and secondly because of the relatively small angular momenta. For heavy nuclei, on the contrary, one may expect metastable states as a pretty common feature.

Metastable states are not to be restricted to radioactive nuclei alone they will exist also in stable nuclei. But they are detected with great difficulty, since the transition to the ground state will be accompanied by emission of very soft γ-rays only. There is a chance (very small) that some stable nucleus may possess a metastable state with a life-time of the order of 10^{10} years. Such a nucleus might then appear under normal conditions in two modifications distinguishable by their spins, which would, in consequence, argue to isomerism in stable nuclei.

Importance of induced radioactivity

The study of the artificially produced radio-elements is of very great *theoretical importance,* since it gives new arguments in favor of the *neutron-proton constitution* of the nucleus. The $p \rightleftharpoons n$ transformations alone are able to explain adequately the special features of the phenomenon such as electron emission, positron emission and decay by K electron capture. The isomeric property of several of the induced radioactive elements reveals, in its turn, the existence of low-lying metastable excited levels in nuclei, chiefly in the heavy ones.

The importance of induced radioactivity is still greater in its many *practical applications.* In the field of **pure nuclear research,** it provides a very sensitive method for the study of the processes of nuclear disintegrations. Instruments for the detection of the artificially provoked activity can be made compact and extremely sensitive and can be isolated from the source of the particles used for producing the original disintegration. Observation of such an activity is just a complete proof of the validity of the primary reaction as the observation of the product particles, and sometimes, more definite in interpretation, especially if supplemented by chemical separations.

In other **applied branches of science,** such as medicine and biology, the importance of artificially produced radio-elements seems not only to be definitely assured but also to be ever on the increase. Their use as "tracers" has already been mentioned. *The potential value of radioactive isotopes of elements which are important in physiological processes, such as phosphorus, sodium, iron, calcium and iodine, is quite apparent.* By choosing a radioactive isotope of a material which the body localizes in some definite place, it is possible to apply the desired treatment in that part of the body needing it. For example, it is known that a good part of the iodine taken internally goes to the thyroid gland. Hence for a beneficial treatment of this gland, a radioactive isotope of iodine can be administered, which will for the greater part get localized in the thyroid and thus produce the desired effect without affecting the other parts of the body.

In general, the same radioactive isotope serves both the purposes of medical discipline, viz., diagnosis and treatment, being used in small quantities as 'tracers' for diagnosis and in larger doses for treatment Evidently right choice of isotopes, as regards life, nature of radiation, α, β or γ-rays, etc. is important in the matter of treatment. For example, *radio-phosphorus* (P^{32}), period = 14.3 days, β-emitter is found good for surface applications (skin diseases); *radio-cobalt* (Co^{60}), period = 5.3 year, γ-emitter, has been utilized in the treatment of deep cancers, in which it is implanted in the form of needles; *radio-gold* (Au^{198}), period 3 days, γ-emitter, is fired right into the cancer in the form of small bullets from a 'gun'; these bullets can be allowed to remain permanently in the system, since gold, after it has lost its activity, is inert and harmless to the body tissues. It may be noted that the early hopes that these radioisotopes would prove the 'magic bullets' in the fight of forms of cancer have not been realized; but in some special forms of cancer they have given relief and even complete cure.

Radio-carbon is another interesting radioactive isotope, which is obtained by slow neutron bombardment of nitrogen according to the reaction

$$_7N^{14} + _0n^1 \rightarrow *_6C^{14} + _1H^1 \ldots\ldots\ldots\ldots\ldots\ldots\ldots (1)$$

The radioactive carbon decays by emitting an electron,

$$*_6C^{14} \rightarrow _7N^{14} + e^- \ldots\ldots\ldots\ldots\ldots\ldots\ldots\ldots\ldots (2)$$

with a period of about 5,700 years. It has many uses as a "tracer" in biological research and for studies of petroleum products and organic chemicals where radio-carbon is substituted for natural carbon. But an important application known as *radio-carbon dating* is the determination of archaeological and short geological times that are of the order of magnitude of the mean life of radio-carbon. Cosmic rays contain an appreciable amount of secondary thermal neutrons, which on reacting with atmospheric nitrogen produce radio-carbon by reaction (1). The production of $_6C^{14}$ and its subsequent decay are in equilibrium in the earth's atmosphere. Hence $_6C^{14}$ is present in equilibrium. concentration in all living plants. The carbon is absorbed in the form of CO_2, so that dead plants that no longer take in CO_2 from the atmosphere lose their C^{14} by radioactive decay. Consequently, the ages of pieces of wood, grain, peat, charcoal, etc. can be pretty accurately determined by measuring the C^{14} they contain relative to the equilibrium amount present in living plants. The method assumes that the average cosmic ray intensity has remained constant for at least 20,000 years and that atmospheric mixing is rapid compared to the life-time of C^{14}. The archaeologist has only to find some wooden object in the excavations of an ancient settlement and determine its radio-carbon content. Then by a simple calculation he is able to estimate the time of time death of the plant and hence the time of existence of the ancient settlement. Libby using this method and working for a number of years (1949-1951) has examined many samples for their C^{14} content and has established dates when life ceased in the samples, which range from 1000 to 15,000 years.

It may be noted that *triton* (H^3), which is also formed by the fast neutrons of cosmic rays according to the reaction,

$$_7N^{14} + _0n^1 \rightarrow _6C^{12} + *_1H^3 \ldots\ldots\ldots\ldots\ldots\ldots\ldots\ldots (3)$$

has been used for radioactive dating. As its period is only 12.5 years, it is limited to estimating times of this order of magnitude. Its most important applications will probably be for meteorology and hydrology. For example, it is already established that the triton content in rain and snow increases almost proportionally to the distance from an ocean.

This is because the air currents carrying moisture to such regions must travel farther and at greater heights, so that the total exposure to cosmic rays is correspondingly increased.

In biology, a number of interesting applications have been made

(*a*) *Metabolism and biosynthesis*
It is the study of how a living organism uses its foodstuffs to provide itself with energy and prepare its own peculiar complex compounds from simpler ones. Schoenheimer and Rittenberg, using fatty acids containing deuterium (rare though not radioactive) as food for animals, have proved that the concept of a chemically *static* body is wrong, by finding that the injected deuterium, instead of being excreted quickly in the waste products of fat metabolism, remains deposited in the fat throughout the animal's body; hence the living body is a *dynamic* chemical system in a constant state of flux and interaction with its environment.

(*b*) *Photosynthesis in plants*
This is the chain of chemical reactions in which green plants use the sun's energy to build up carbohydrates, proteins and fats from inert carbon dioxide and water, releasing oxygen in the process. Exposing plants to $C^{14}O_2$ (*i.e., carbon* dioxide containing radio-carbon, C^{14}) in the presence of sunlight for different intervals of time, then separating the compounds in the plant and chocking them for C^{14}, biochemists are able to determine what compounds the plant is making from carbon dioxide. After one minute's exposure it is found that at least 50 compounds are tagged with radio-carbon and within two minutes, the carbon isotope is present in proteins and fat also.

(*c*) *Fertilizers*
Radio-phosphorus has been used to study how plants assimilate phosphate fertilizers. The *great merit* of the artificial radio-elements lies chiefly in the fact that *they form* a *relatively cheap and easily controlled "substitute" for the natural radioactive materials* in the treatment of diseases like cancer, etc. Natural radioactive substances can be had only with great difficulty from mineral ores in small quantities and at high costs, while the artificial radio-elements can be produced at will and to any desired amount, thanks chiefly to the modern nuclear reactors; there are actually 500 such radioactive isotopes available for use. Furthermore, among these artificial radioactive materials, there are many which have relatively short periods, say of the order of a few hours. Use of such short-lived materials eliminates the need of recovering them after treatment to prevent an excessive dose and thereby allow them to be administered even internally. This has proved of very great importance in the treatment of internal diseases.

3. NUCLEAR FISSION

Introduction
The continuous development of artificial disintegrations for over a period of two decades and especially the study of induced radioactivity finally culminated in perhaps the most portentous discovery of our times, the so-called *nuclear fission* which showed the way of tapping enormous amount of energy from nuclear explosions.

In ordinary nuclear transmutations, both natural and artificial, the nucleus is simply chipped off rather than broken and accordingly the amount of energy set free, though far superior to the actual sources of energy, is still small, 10 to 30 MeV.

In 1938, it was discovered that heavy unstable nuclei, such as uranium, when bombarded with neutrons, explode more radically into two more or less equal fragments which fly apart with prodigious energy of about 200 MeV.

Such a profound nuclear disruption has been named "fission", a term borrowed from biology, where it denotes the division of a cell into two cells of roughly the same size.

Discovery

The starting point in the discovery of nuclear fission is to be traced to the attempt of Fermi, in 1934, to produce "transuranic elements" of atomic number greater than 92, by bombarding uranium with neutrons. Since typical neutron reactions often resulted in the formation of a radioactive isotope of the bombarded element which frequently emitted electrons and thus became an element of atomic number one unit higher than the original element. Fermi conceived the idea that the heaviest known element uranium ($Z = 92$) bombarded by neutrons might follow the same process and thereby give rise to transuranic elements. Putting this surmise to experimental test, he and his associates found four new radio-activities emitting β-rays with periods 10 secs, 40 secs, 13 mins and 90 mins Since uranium is an α-emitter with a long period, these β-activities indicated that some new process was taking place. Further, since each successive β-disintegration would increase the atomic number by one, it was naturally assumed that the observed four different periods might have originated from a succession of β-disintegrations of elements having atomic number higher than 92. Hence, the observed phenomenon was interpreted as the formation of transuranic elements, chiefly, as no one could suspect at the time that the new substances formed in the reaction could differ greatly in atomic number from uranium.

Several chemical tests were next applied to substantiate this conjecture. Two of the "activities", which could be chemically separated from uranium, were found to be neither the isotopes of uranium nor those of any known elements in the range of atomic numbers 86 to 92.

On the other hand, the element 93, if it was formed, should have chemical properties similar to manganese, its homologue.

On making a chemical separation of manganese from the bombarded uranium, the precipitate of MnO_2 showed β-activity. Similarly, separations from the irradiated uranium of iron, rhodium and palladium which would be the chemical homologues of elements of atomic numbers 94, 95 and 96 respectively, were all found radioactive. Thus the conclusion of Fermi that hitherto unknown transuranic elements were being formed in the process seemed to be established.

During the years 1936 to 1938, the experiments of Fermi were repeated in other laboratories, particularly in Germany by Hahn, Meitner and Strassmann and in France by Curie-Joliot and Savitch. Although Fermi's results were confirmed and extended, it became necessary to assume that the supposed "transuranic" elements in decaying, gave rise to other radioactive elements, which would be isomeric with other heavy atoms.

Further, the number of different substances reported became so numerous that it was difficult to fit them into any plausible scheme. It must be said that chemical tests in this region of the periodic table was extremely difficult as the elements were new and could be obtained only in very small quantities.

The French scientists in one of their chemical separation tests found an "activity" of 3.5 hrs period, which could be precipitated by lanthanum (Z = 57). This result might well have been the key to the right solution of the complicated phenomena observed, thereby allowing Curie and Savitch to be the discoverers of fission; but they missed their chance by sticking to the then commonly accepted view that reactions of the type analyzed could not give rise to elements of such low atomic number as lanthanum, although they fully realized the difficulty of finding a place for an element having chemical properties similar to lanthanum among the transuranic elements.

It was left to the German scientists, Hahn and Strassmann, to clarify the issue in the following manner:

In the course of their investigations, they discovered in uranium bombarded with neutrons, a radioactivity of 8.6 mins period, which could be precipitated by barium (Z = 56). They also at first thought, in consonance with the opinion then prevalent, that the observed activity might be a new form of radium, since the chemistry of radium is very much like barium. But radium (Z = 88) could only be formed from uranium by the emission of two α-particles and such a process had never been known to follow neutron capture.

Moreover, on further careful tests, such as fractional precipitations and crystallizations of the kind used for separating radium from barium, they found that the activity definitely followed barium, not radium. Hence they proposed, not without hesitation, however, because of the strange results, a radically new hypothesis that

uranium nucleus after capturing a neutron might, instead of being chipped off by a small fragment as had always been observed previously in all particle disintegrations, break up into two large fragments, each of the size of a moderately heavy atom.

They justified their opinion by showing that among the radioactivities observed was also one with a period the same as that of a known isotope of krypton (Z = 36). On combining the atomic numbers of barium and krypton, (56 + 36) the number 92 could be obtained, which was the atomic number of uranium, so that the process might well be represented by

$$_{92}U^{238} + {}_0n^1 \rightarrow {}_{56}Ba + {}_{36}K + {}^*{}_0n^1$$

Hence, they concluded that the β-activities previously ascribed to transuranic elements were probably produced by radioactive isotopes of elements of lower atomic number, as barium and krypton. In this way, for instance, the activity found in lanthanum (Z = 57) by the French scientists could be formed by β-decay of barium (Z =56). Thus a new type of disintegration, in which a heavy nucleus splits into two other nuclei of comparable size, hence called "nuclear fission" was discovered.

As soon as these results were announced early in 1939, physicists in other laboratories immediately repeated and confirmed these experiments. In the light of the "fission" idea all the facts, which were found disconnected and difficult of synthesis, could be readily coordinated. Many of the fission products could be identified with substances of much lower atomic number than uranium, already known in the study of artificial radioactivity. Thorium (Th) and protoactinium (Pa) wore also shown to undergo fission when bombarded with neutrons. Thus the newly discovered process of nuclear fission was definitely established though a theoretical justification was deemed necessary for its formal and final acceptance. But a satisfactory explanation was proposed, almost immediately after the discovery, by Meitner and Frisch, who wore also the first to suggest the name of "fission" for the phenomenon.

Otto Robert Frisch (1 October 1904 – 22 September 1979), Austria

Theoretical interpretation
The problem concerning "nuclear fission" may be stated as follows.

How can the fairly moderate activation of a nucleus resulting from the capture of a neutron lead to such a cataclysmic disruption ?

Why does it occur only in a few heavy elements such as U Th and Pa ?

The answer to these questions was first given by Meitner and Frisch on the basis of the liquid drop model of the nucleus. Bohr and Wheeler developed it further with definite and quantitative results that could be checked and verified. A nucleus is analogous to a drop of liquid in many ways, such as the close packing of particles, constant density, short-range forces, etc. The resultant effect of these nuclear characteristics is to endow the nucleus with a property analogous to the surface tension of liquids. Considering a liquid drop, we readily see that the dissipative forces acting between the constituent molecules are counteracted by the surface tension forces. Further, any liquid drop, not subject to outside forces, tends to assume a spherical shape under the influence of surface tension, since a sphere presents the smallest surface for any given volume. Very large drops are impossible, because the surface tension forces arc feeble and there is a maximum size of the drop which cannot exist permanently.

As this size is approached the drop becomes less and less stable, gets deformed. At this stage, it begins to execute the so-called surface *tension oscillations* with the slightest provocation from outside, which lead to the disruption of the drop into two or more smaller droplets.

In atomic nuclei, the short range attractive forces, keeping the constituent particles, *viz.,* the charged protons and the uncharged neutrons, stay together in a stable state, play the role of surface tension, while the electrostatic repulsive forces between the protons that of disruptive forces tending to destroy the stability. If the surface tension forces prevail, the nucleus will never break up by itself, and two nuclei coming into contact with each other, will have a tendency to *fuse j*ust as two ordinary liquid droplets do.

If, on the contrary, the electric forces of repulsion have the upper hand, the nucleus will show a tendency to break spontaneously into two or more parts, which will fly apart at high speed; this is the phenomenon of *fission.*

Calculations made by Bohr and Wheeler concerning the balance between the surface tension and electrostatic forces in the nuclei of different elements led to the very important conclusion that while the surface tension forces predominate in the nuclei of all elements in the first half of the periodic table (approximately up to mass 110), the electrostatic repulsive forces prevail for all heavier nuclei. Hence all nuclei of mass greater than 110 are *potentially unstable,* and under the action of **a** sufficiently strong disturbance from outside would break up into two or more parts with the liberation of **a** considerable amount of internal nuclear energy,

On the other hand, a spontaneous fusion process should take place whenever two light nuclei, whose total mass is less than 110, come close together.

It is to be remarked; however, that neither the fusion of two light nuclei, nor the fission of a heavy nucleus would normally take place, unless by some means the light nuclei are brought close together against the repulsive forces acting between their charges, or the heavy nucleus is made to vibrate with a sufficiently large amplitude.

Thus the reason why fission can take place only with heavy nuclei under proper initial excitation is obtained.

Considering now heavy nuclei of mass greater than 110, which are therefore potentially unstable, and starting from a spherical shape for the nucleus, with increasing charge and size of the nucleus, actual stability conditions would require an increasing deformation from the spherical shape, since the electrostatic repulsive force is on the increase, and for a given volume the spherical shape is the most unstable if only the electrostatic forces were important. The nucleus would therefore become "flattened" in order to be stable. Beyond a certain stage, presumably for a value of nuclear charge Z close to 100, the deformation would be so far advanced as to split the nucleus into two. Now since the nuclei of uranium, thorium and protoactinium having values for Z in the whereabouts of 90 lie close to the limit of complete deformation, it appears reasonable that they should have only a narrow margin of stability against change of .shape and hence be easily excited upon receiving a moderate energy of excitation to oscillations like those produced in a liquid drop by surface tension, which lead to their "rupture" into two lighter nuclei,

Thus the fact why fission takes place with only a few of the heaviest elements is explained.

Bohr and Wheeler, by applying Bohr's theory of nuclear processes to the phenomenon of fission, were able to obtain a quantitative relation between the critical deformation leading to fission and the energy required to reach it. According to the general theory, nuclear fission takes place in two steps:

(i) the formation of the intermediate compound nucleus, in which the energy is temporarily distributed among all the nuclear particles in a manner similar to that of thermal agitation of a liquid
and
(ii) the transformation of a sufficient portion of this energy into potential energy of deformation of the compound nucleus which will lead to its fission. On the analogy of the liquid drop, the compound nucleus becomes 'very hot' and its oscillations result in conditions which in a very short time lead to its violent breaking up into two parts.

The possibility of the occurrence of fission by a heavy nucleus capturing a neutron depends, therefore, on threshold energy, *i.e.*, the difference between the critical energies of such unstable deformation of the nucleus and the excitation energy which is given by the binding energy of the added neutron. Making estimates of these quantities in the case of the three isotopes of uranium of masses 238, 235, 234 (relative abundances 99.28%, 0.71%, 0.006%) they showed that the rare isotope 235 was responsible for the fission produced by slow neutrons, since the binding energy of a neutron is much larger in nuclei of even mass than of odd mass, so that U^{236} will be excited to a higher degree than U^{239}, which means that the threshold energy is much lower for U^{236} than for U^{239} and in consequence the former will break up even under the influence of a slow neutron.

This theoretical prediction was completely confirmed by the experiment conducted by Nier, Booth, Dunning and Grosse who, by extracting a small amount of U^{235} by means of a mass spectrometer, showed that a large number of fissions occurred when it was bombarded with slow neutrons, while practically no fission with U^{238} was observed. It was further shown that U^{235} could also undergo fission but only when bombarded by high energy neutrons.

Research on nuclear fission
Within the next few years after the discovery, the essential features of the "fission" process were analyzed methodically by different workers and placed in clear evidence. A summary of these investigations will now be given.

(1) Means of inducing fission
In general, it may be said that any element with Z > 90 can succumb to fission under bombardment with proper energy. *The first cases of fission in U, Th and Pa were produced with neutrons.*

The number of fissions produced in uranium was found to vary markedly with the speed of the incident neutrons. With natural uranium, fission can be produced with fast neutrons of energy 1 MeV and more or by thermal neutrons, but not by neutrons of intermediate speed. The fission cross-section for fast neutrons is about 0.1×10^{-24} cm^2 , while for thermal neutrons about 2×10^{-24} cm^2. This selectivity is evidently due to the fact that the two isotopes of uranium of masses 235 and 238 undergo fission under different conditions.

The lighter isotope, whose relative abundance is only 1/140 of the total, undergoes fission with thermal neutrons with a large cross-section of 400×10^{-24} cm^2 , while to produce fission in the heavier isotope forming the greater portion, the neutron must have an energy of about 1.5 MeV (Nier and Dunning-1940). The small cross-section in natural uranium for thermal neutrons is evidently due to the very small amount of lighter isotope present and the intervention of other processes which impede the action.

The natural isotope of thorium of mass 232 has been shown to undergo fission with fast neutrons of 1.1 MeV (Nishina and his co-workers 1939), while Pa of mass 231 with neutrons of energy more than 1 MeV (Grosse, Booth and Dunning-1939).

It has also been found that *.fission can be produced by bombardment with particles other than neutrons.* Thus fission has been induced in both U and Th by *high speed protons* of energy 6.9 MeV (Dessauer and Hafner-1941), by *deuterons* of energy greater than 8 MeV (Gant-1939 and Jacobson and Lassen-1940) and by *α-particles* of 32 MeV energy (Fermi and Segré-1941). High energy *photons* such as the γ-rays obtained from nuclear reactions (Haxby, Shoupp, etc. —1941) or *X-rays* from a betatron (Baldwin and Koch-1945) have been found effective in producing fission. The cross section in these cases is of the order of 1 to 7 X 10^{-27} cms^2. These results bring out the fact that

> *almost any means, which will make an already unstable uranium or thorium nucleus more unstable beyond a critical stage can be used to produce fission.*

(2) Detection of fission

The fission process may be detected by the following means

(*a*) *Chemical identification* of the fission products as in the original observation of Hahn and Strassmann.

(*b*) The fission fragments may be caught upon a receiving surface placed close to the irradiated fissionable material and detected subsequently by means of their *radioactivity,* as was employed by Joliot.

(*c*) *Ionization chamber method.* Coating the walls of an ionization chamber with the fissionable substance and allowing the projectile to bombard the walls, huge pulses of ionization produced by the fission fragments can be observed.

(*d*) *Cloud chamber method.* The heavy tracks made by the fission fragments in a cloud chamber may be photographed.

(*e*) *Photographic plate method.* This technique has been applied with success to the detection of fission. The fissionable material, say uranium, is introduced into the plate by a "bathing" process using a solution of uranium citrate. When the plate is exposed to slow neutron bombardment, heavy tracks of the fission fragments are recorded. A photo of one such record with somewhat unusual characteristics obtained by D. L. Livesey is reproduced below.

Fission tracks with secondary α-particles. (Livesey)

The nucleus splits into three fragments. The track of longer range is much thinner than those of the fission particles and is presumed to be due to an α-particle. There is also a second fission track, unrelated to the first, which passes under the long track.

(3) Fission products

The fission fragments have been studied by several experimenters. The methods used in their identification are:

(*i*) chemical tests,
(*ii*) X-rays emitted by the excited atoms produced during fission and
(*iii*) comparison of the periods of the fission products with those produced in other nuclear reactions.

It is found that many different elements are produced by the fission of uranium and thorium bombarded with neutrons. Radioactive isotopes of more than 20 elements have been identified among the fission products, which fact indicates that the *fission process may take place in many different ways.* The products have been found, however, to fall into two groups, one of atomic number in the range 35 to 43 and atomic mass in the range 80 to 110, the other of atomic number from 51 to 57 and atomic mass from 125 to 160.

These two groups in the fission products indicate that two among them, formed first, are unstable and disintegrate further either by the ejection of neutrons or by β-emissions producing step by step other members of their respective groups until a final stable nucleus is reached. Both the neutron and β-processes have been observed. Further, some of the β-processes have been followed from the first fission product to the last stable element. Thus, for instance, Segré and Wu have observed the following series of β-disintegrations in one of the fission products of uranium

$$_{51}Sb^{133} \text{ (10 mins)} \rightarrow {}_{52}Te^{133} \text{ (60 mins)} \rightarrow {}_{53}I^{133} \text{ (22 hrs)} \rightarrow {}_{54}Xe^{133} \text{ (5 days)} \rightarrow {}_{55}Cs^{133}$$

While antimony has only two stable isotopes of masses 121 and 123, the isotope that is formed in fission has a mass 133. This means that the fission product has at least ten neutrons in excess, which makes it unstable. The excess of neutrons can be removed by β-process, where a neutron is transformed into a proton. On the other hand, α-ray emission, positron decay and K electron capture are not possible as they tend to increase the neutron-proton ratio.

(4) Energy liberated in fission

The far-reaching importance attached to the discovery of nuclear fission is partly due to the enormous amount of energy released in the process. As in all nuclear disintegrations, here also the energy liberated results from the small loss in mass involved, which according to Einstein's mass-energy relation ($W = mc^2$) corresponds to a very large amount of energy. In the present case, the loss of mass is relatively much larger than in other reactions.

A general idea of the energy that can be liberated in nuclear fission may be obtained by converting the difference between the mass defects of the fission products into energy by the use of the above relation. It is found that approximately 200 MeV per disintegration should be released in the fission of uranium. Evidently the amount of energy liberated will depend upon the particular process involved and this in turn will depend upon the primary fission products.

Further, the energy will be divided between the two fission products in such a way as to conserve momentum and energy in the process. Thus, for instance, if two products of equal mass are formed, they will travel in opposite directions with equal energies of about 100 MeV each. If, on the other hand, the products are unequal in masses, such as barium and krypton, the energies carried by them will be different.

Meitner and Frisch (1939) found that by placing uranium over water, the fission particles of uranium were driven into the water, from which they could be precipitated. This indicated that the fission products must be exceedingly energetic to be forced out of the uranium and driven into the water. McMillan (1939) by covering a sample of bombarded uranium with a number of thin foils and noting to what foil the radioactivity reached was able to make an estimate of energy per particle.

More direct and precise measurements of the energies of the fission products were made by the ionization chamber, cloud chamber and calorimetric methods:

Ionization chamber method

Hafstad and Dunning (1939), using an ionization chamber in conjunction with a linear amplifier and a cathode ray oscillograph, obtained neat records of huge pulses produced by fission particles from uranium bombarded, with neutrons, like the one shown in the adjacent photo.

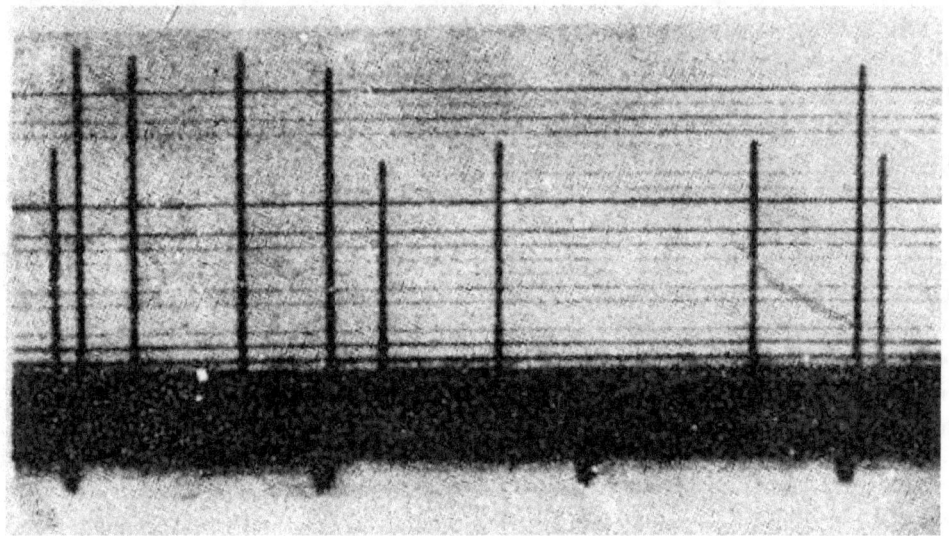

Linear amplifier oscillograph record of fission particles iron, uranium. (Dunning)

The varying heights of the peaks give the range of energies with which the fission particles are given off. The α-particles emitted by non-fissioned uranium are in the dark band at the bottom. It is easily realized that the much higher peaks of the far heavier fission particles as compared with those of the α-particles must be due to the enormous energy realized in fission.

Kanner and Barschall (1940), by observing a sufficiently great number of fission peaks and noting the number in each energy interval, were able to draw the distribution-in-energy curve, as the one shown in Fig. 301.

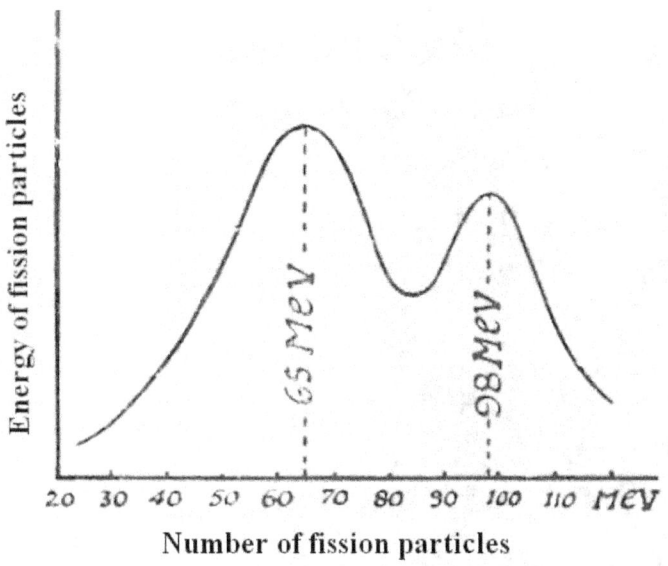

Fig. 301. Distribution-in-energy curve fission particles.

Two peaks are obtained, one at 65 MeV and the other at 98 MeV, as would be expected from probability considerations of any fission process, in general, according to which division into two exactly equal parts can occur only rarely. The sum of the energies represented by the two peaks is 163 MeV.

Jentschke (1943), measuring simultaneously with sensitive electro-meters the ionization pulses produced in argon by the two fission fragments issuing on opposite sides from a thin uranium foil bombarded with neutrons, was able to calculate the kinetic energies of the fragments on the assumption that each ion pair required 27.5 eV for its formation. He found that the total kinetic energy varied from 133 to 192 MeV with a frequent value of about 163 MeV.

Cloud chamber method

Corson and Thornton in California (1939) and Boggild, Brostroem and Lauritsen in Copenhagen have obtained fine cloud chamber photos of the tracks produced by the fission fragments by placing a thin layer of uranium inside the chamber and irradiating it with neutrons during the expansion. One such photo taken by the Copenhagen workers is reproduced here.

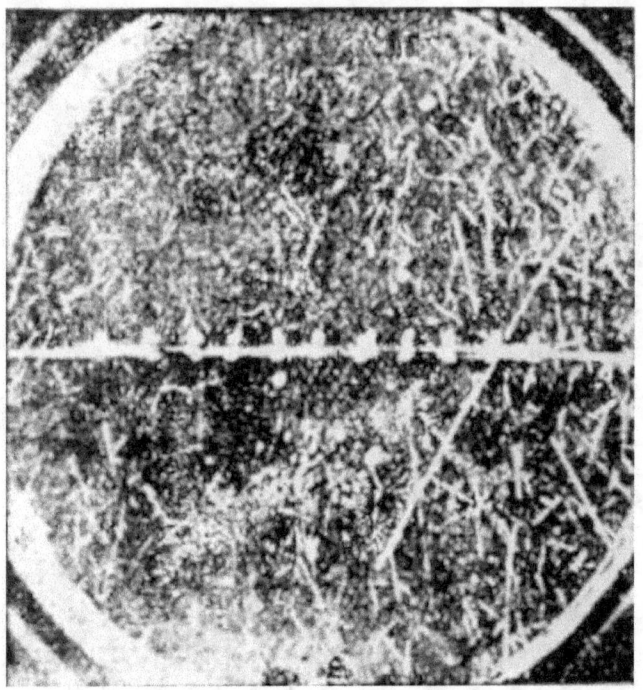

Cloud chamber photo of fission tracks. (Boggild, Brostroem and Lauritsen)

The two oppositely directed tracks, which are much thicker than numerous others forming a background throughout the chamber and produced by α-particles of non-fissioned uranium, correspond to the two fission fragments. The energies can be estimated from the ranges of the tracks. Measurements show that the energies arrange themselves into two groups, one of about 100 MeV and the other of about 72 MeV.

Calorimetric method

Henderson (1910) has measured the total average energy released per fission of uranium, by noting the total number of disintegrations and the total energy developed in a sensitive calorimeter. The value obtained is 177 MeV per fission, in fair agreement with those of other methods.

It may be noted that the experimental value is somewhat less than that derived from mess considerations. This is to be expected since the measured kinetic energy of the fragments does not represent the whole of the energy released. Part of the total energy may be carried off by high-speed neutrons which are emitted in the fission process or may be used for exciting the fragment nuclei, which would then appear in subsequent radioactivity.

(5) Production of secondary neutrons in fission
No matter how the uranium nucleus divides, the fission products have an excess of neutrons over the number contained in the corresponding normal nuclei. It has already

174

been stated that one method by which this excess can be removed is by direct neutron emission.

Many workers, having looked for these neutrons emitted at fission, have obtained positive results. Thus, in 1939, Joliot, Kowarski and Von Halban were the first to discover that the average number of neutrons emitted per fission of uranium was 3.5 + 0-7. Anderson, Fermi and Hanstein obtained the result of two neutrons per fission in uranium. Szilard and Zinn with an elaborate and carefully devised experimental arrangement determined the number of fast neutrons emitted in the fission of uranium by slow neutrons. The slow neutrons were produced from a (Ra + Be) source placed inside a paraffin block, A helium-filled ionization chamber connected to a linear amplifier was used as a detector of the secondary fast neutrons. Using uranium oxide and performing experiments with and without Cd screens, they were able to determine the number of fast neutrons emitted as 2 per fission. These results indicate that *each fission fragment from uranium emits on an average one neutron.* Theoretical considerations show that the average number of neutrons per fission fragment would increase with the mass of the fissioned nucleus.

Hence in the fission of elements heavier than uranium (say, for instance, plutonium) the average number of neutrons per fragment may be expected to be greater than one.

The *neutrons generated at fission are found to be fast neutrons, possessing energies up to 1 MeV.*

It has been observed also that *some of the emitted neutrons issue after a measurable time delay.* Roberts Meyer and Wang found that neutrons continued to come from irradiated uranium even after the cessation of bombardment,

Snell, Nedzel and Ibser (Smyth's Report —1945) have been able to measure the time delay of the emitted fast neutrons in the fission of uranium with slow neutrons. They found that about 1% were delayed by at least 0.01 sec and about a.07% were delayed by as much as a minute, which is enormously greater than the time taken for the actual fission process, estimated at 10^{-15} sec. This is to be expected, as the emission of the neutrons is really an after-effect of fission and is due to the two fragments that begin their existence in a state of rather violent vibrations, which though unable to cause a secondary nuclear fission, yet are strong enough to cause the ejection of some nuclear structural units.

(6) Chain reaction

The discovery of the emission of secondary neutrons in the fission process is a crucial fact, for it renders immediately obvious that a *self-propagating chain reaction is possible,* since the neutrons produced in the fission of a nucleus can cause fission, in their turn, in other neighboring nuclei, producing more secondary neutrons and more fissions resulting thereby in a rapid geometric increase of fissions until the whole of the fissionable material is disintegrated. This would liberate a tremendous amount of energy in a very short time since each fission releases a very high energy of about 200 MeV. It has been calculated that one cubic meter of uranium oxide might develop 10^{12} kilowatt hours in less than 0.01 sec Thus it is realized that the phenomenon of fission can be put to practical use as a source of power, far superior to the actually existing ones, nay, even as a super-explosive (the atom bomb).

The common sources of power are obtained by combustion processes. Now combustion is always self-propagating: *e.g.,* the lighting of a fire with a match liberates sufficient heat to ignite the nearest fuel, which liberates more heat which ignites more fuel and so on. Although, as a rule, nuclear reactions, unlike chemical reactions, are not self-propagating, yet in the particular case of nuclear fission, on account of the omission of particles of the same kind as those which started the fission, a self-propagating reaction is obtained involving so many nuclei that the whole mass would explode, thus producing enormous energy, many million times that of ordinary fuels.

But to *secure an efficient chain reaction is not easy* in practice, because there exist several causes that prevent the progressive neutron breeding, essential for the self-sustaining fission reaction. The general condition for the minimizing of these causes which militate against the occurrence of the chain reaction is usually expressed by a quantity known as the *multiplication factor* (k) and defined as the ratio of the number of fresh neutrons produced by fission at any place in the material to the number of free neutrons originally present at that place. *It is essential that (k) must be at least equal to 1 for the maintenance of the chain reaction.*

The two chief sources of wastage of neutrons which lead to the collapse of the chain reaction are:

(*i*) *leakage of neutrons from the system,* and
(*ii*) *presence of non-fissionable material in the system, which absorb the neutrons.*

The first of these may be reduced by a suitable choice of the size and shape of the fissionable material. The *size must be large* enough to keep the neutrons within the system so that the probability of their escape before they could hit nuclei and produce fission is rendered small. A *spherical shape* is also very conducive to obtain the desired result, since the *escape of neutrons is a surface effect* dependent on the surface area and hence on the square of the radius, while *fission is a volume effect* dependent on the volume and hence on the cube of the radius.

Thus the ratio of the rate of escape of neutrons from the system to the rate of their production varies inversely as the radius, decreasing with increasing size. As the radius is increased from a small value, a *critical size is* reached, beyond which more neutrons are produced than are lost, so that the chain reaction can progress rapidly. This critical size is of great importance, below which the fissionable material is completely stable and perfectly safe, but above which the system becomes unstable and explodes spontaneously. The critical size depends on the nature and shape of the surroundings as well as on the speed of the reacting neutrons.

The second source of loss of neutrons by absorption, non-productive of fission, may be reduced either

(*a*) *by carefully purifying the fissionable material* from other non-fissionable impurities that absorb neutrons
or
(*b*) *by neutralizing the disturbing action of the non-fissionable materials without actually removing them.*

Uranium presents a very instructive and practical example that illustrates the various conditions stated above that must be realized for a successful chain reaction. The two isotopes of uranium important in this connection are U^{238} and U^{235}, the former being 140 times more abundant than the latter.

The heavy isotope (U^{238}) has a fission cross-section of about 0.5×10^{-24} cm^2 for fast neutrons of energies greater than 1 MeV; fission ceases below 0.35 MeV; the probability of nonfission capture of slow neutrons is vanishingly small except for a pronounced resonance at neutron energy of about 25 eV with an effective cross-section of 1200×10^{-24} cm^2 as estimated by Meitner, Hahn and Strassmann.

The light isotope (U^{235}) manifests a very high probability of fission capture for neutrons of any energy and the fission cross-section increases with decrease of neutron speed reaching a high value of 400×10^{-24} cm^2 for thermal neutrons.

It is to be noted that the resonance non-fission capture by U^{238} produces a new transuranic element called plutonium, ($_{94}Pu^{239}$) which can be chemically separated from uranium; the fission properties of plutonium should closely resemble U^{235} according to the Bohr-Wheeler theory.

With these data the possibilities of the chain reaction in uranium can be studied:

In natural uranium a chain reaction is not possible

On account of the very high probability of fission capture of neutrons of any energy in the case of U^{235}, this light isotope by itself should be most effective in producing a chain reaction even in a lump of natural uranium. But this U^{235} chain reaction is normally hindered almost completely by the presence of the 140 times more abundant U^{238}. For, most of the neutrons produced in the fission of U^{235} will be slowed down by elastic collisions to energies less than 1 MeV and therefore rendered incapable of producing fission in U^{238} . These slowed down neutrons can, of course, produce fission in U^{235} But on account of the small amount of this isotope, the probability of collision of U^{235} nuclei with these neutrons is much smaller than , in the case of the heavy isotope. Hence the neutrons will escape to the exterior before they can produce further fission with U^{235} nuclei, unless the amount of material is large.

If the mass is large and extremely pure, collision with U^{238} nuclei will reduce the energy of the neutrons slowly. The energy lost in a single collision is very small because of the large mass difference and the neutrons will therefore linger for a long time in any given energy interval. The probability of capture into the resonance level of U^{238} at about 25 eV thus becomes large. There is likely, therefore, to be too great a loss of neutrons by absorption which does not lead to multiple fission required for the development of a progressive chain reaction, however large the mass of material may be. Thus a chain reaction is rendered impossible in natural uranium.

It is interesting to note that only because of the preventive action of the heavy isotope U^{238} that the highly fissionable atoms of U^{235} still exist in nature, since otherwise they would have all been destroyed long ago by a fast chain reaction among them.

A *chain reaction can be secured with uranium in the following two ways:*

— *If* U^{235} *can be separated from* U^{238} in sufficiently large quantities, greater than the critical size, which has been estimated to be 100 kilograms, a spontaneous chain reaction will result, initiated by any stray neutrons, such as those produced either by cosmic ray action or by fission itself.

The separation of U^{235} from U^{238}, however, presents the very complicated technical problem of the separation of isotopes, already described. In the present case two additional difficulties exist, *viz.,*

First the very small relative abundance of U^{235}; to separate the 100 kgms about (demanded by the critical size), 10,000 tons of natural uranium are required;

Second owing to the small difference of mass between the two uranium isotopes the separation cannot be achieved in one single process, but requires a large number of repetitions which lead to a gradual concentration of the light isotope, until finally reasonably pure samples of U^{235} can be obtained.

A much more ingenious method first suggested by Fermi *consists in forcing a chain reaction in natural uranium itself by the use of certain artificial devices known as* **moderators** which reduce effectively the adverse absorption effect of the heavier isotope and thereby permit the lighter isotope to pursue its natural chain reaction. The essential function of the moderator is to slow down the fast neutrons of fission very rapidly by elastic collisions to thermal energies, so that the chance of non-fission resonance capture of U^{238} becomes small, and consequently the probability of fission of U^{235} increases, as illustrated in Fig. 302.

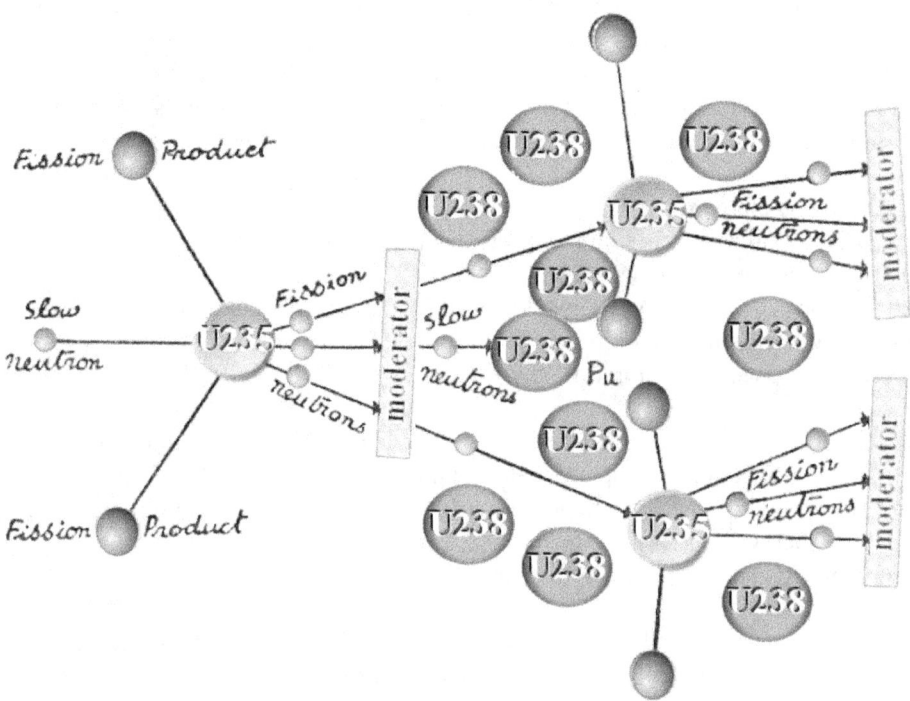

Fig. 302. Principle of action of the moderator.

The moderator must necessarily possess the property of itself not capturing neutrons readily, since otherwise it will augment, instead of diminishing, the collapse of the chain reaction.

Elements of low atomic weight are particularly suited to serve as moderators, since they reduce considerably the energy of the neutron at each collision. It was first thought that hydrogen would be the best moderator, since, having nearly the same mass as the neutron, it could reduce the energy of the neutron by about half at each collision. Calculation showed that in about 25 collisions, hydrogen could reduce the energy of a neutron from 1 MeV to thermal values 0.038 eV). But there are several disadvantages in the use of hydrogen as moderator. Being a gas, it will take up too much space.

Water (H_2O) might be used, but it contains oxygen which complicates matters. Moreover, hydrogen captures neutrons in inelastic collisions to form deuterium, while a good moderator should slow down neutrons rapidly without capturing them. Deuterium, helium, beryllium and carbon are considered as possible good moderators. Carbon in the form of graphite and deuterium in heavy water (D_2O) have been actually used.

This method has the additional advantage of producing at the same time plutonium, which when separated and used pure, is as good, perhaps oven better, than U^{235}, as a fissionable material. This new artificial element results from the resonance capture of some of the neutrons in the process of slowing down by U^{238}, as already stated.

Both the methods of developing an efficient chain reaction have been tried with successful results, in connection with the use of atomic energy for practical purposes. It may be noted that the *time factor* is very important in the production of atomic power for industrial applications. Fission after fission must occur rapidly before the neutrons could escape into the air or be absorbed by the materials without leading to fission.

(7) Transuranic elements

Although the initial researches on the bombardment of uranium by neutrons conducted by Fermi and others gave more or less clear indications of the formation of transuranic elements in the process, recognition of the more startling fission reaction diverted the attention of scientists from looking for further evidence in confirmation of the existence of such nuclei heavier than uranium.

It was only in 1940, that the existence of transuranic elements was definitely established. At present, ten transuranic elements of atomic numbers 93, 94, 95, 96, 97, 98, 99, 100, 101 and 102, called respectively neptunium (Np), plutonium (Pu), americium (Am), curium (Cm), berkelium (Bk), californium (Cf), einsteinium (E), fermium (Fm), mendelevium (Mv) and nobelium (No), are known.

McMillan and Abelson discovered, in May 1940, the first transuranic element *neptunium* ($Z = 93$), produced in the non-fission resonance capture of neutrons by U^{238}. The radioactive isotope of uranium ($_{92}U^{239}$) formed by the capture of a neutron disintegrates by β-emission with a period of 23 mins and produces neptunium according to the reaction

$$*_{92}U^{238} + _{0}n^{1} \rightarrow *_{92}U^{239} + h\nu$$

$$*_{92}U^{239} \rightarrow *_{93}Np^{239} + e^{-} \qquad\qquad (T = 23 \text{ mins})$$

McMillan and Abelson were able to identify this new element by chemical processes carried out on a *"tracer"* scale, i.e., using minute quantities. When latter produced in greater amount, it was found that

neptunium was also radioactive and transformed by β-decay with a period of 2.3 days into a second transuranic element, now called *plutonium*, which is an α-emitter with a long period of about 25,000 years.

The nuclear reactions involved are

$$*_{93}Np^{239} \rightarrow *_{94}Pu^{239} + e^{-} \qquad\qquad (T = 2.3 \text{ days})$$

$$*_{94}Pu^{239} \rightarrow *_{92}U^{235} + _{2}He^{4} \qquad\qquad (T = 2.5 \times 10^{4} \text{ yrs})$$

The two important features of this Pu^{239} isotope are:

(i) it is a different chemical element from uranium and so can be separated from uranium by chemical means;

(*ii*) it is readily fissionable with both fast and slow neutrons.

Another isotope of plutonium (Pu^{238}) was discovered shortly after, in the same year 1940, by G.T. Seaborg, McMillan and co-workers in the bombardment of uranium with 16 MeV deuterons, as a (d, $2n$) reaction

$$*_{92}U^{238} + {}_1D^2 \rightarrow *_{93}Np^{238} + 2\ {}_0n^1$$

$$*_{93}Np^{238} \rightarrow {}_{94}Pu^{238} + e^- \qquad (T = 2\ \text{days})$$

${}_{94}Pu^{238}$ was found to be an α-emitter with a period of about 50 years:

$$_{94}Pu^{238} \rightarrow *\ _{92}U^{234} + {}_2He^4 \qquad (T = 50\ \text{years})$$

In 1942, Wahl and Seaborg discovered a second isotope of neptunium (Np^{237}) produced in the following (n, $2n$) reaction

$$*_{92}U^{238} + {}_0n^1 \rightarrow *_{92}U^{237} + 2\ {}_0n^1$$

$$*_{92}U^{237} \rightarrow *_{93}Np^{237} + e^- \qquad (T = 7\ \text{days})$$

${}_{93}Np^{237}$ is an α-emitter with a very long period:

$$*_{93}Np^{237} \rightarrow *_{91}Pa^{233} + {}_2He^4 \qquad (T = 2.25 \times 10^6\ \text{yrs})$$

At present, the known isotopes of neptunium are those with mass numbers 231, and 233 to 239. They are all radioactive. The longest lived isotope Np^{237} is the representative of element No. 93 in the Periodic table.

The known isotopes of plutonium are those with mass numbers 232, 234 to 237, 238, 239, 241 and 242, all radioactive. Many of them have long periods, Pu^{238} - 89 yrs., Pu^{239} - 25,000 yrs., Pu^{240} - 6,580 yrs., Pu^{241} - 10 yes., Pu^{242} - approximately one million years.

Being the longest lived isotope, Pu^{242} is the representative of element No. 94; it decays with the emission of an α-particle and becomes U^{238}.

In 1945, J. G. Hamilton, G. T. Seaborg and collaborators were able to detect by the "tracer" method two more transuranic elements of atomic number 95 (*americium*) and 96 (*curium*) resulting from the bombardment of U^{238} and Pu^{239} respectively with 40 MeV helium produced in the 60-inch Berkeley cyclotron.

In 1950, S.G. Thompson, A. Ghiorso and G.T. Seaborg obtained *berkelium* (${}_{97}Bk^{243}$) and *californium* (${}_{98}Cf^{242}$) by bombarding ${}_{95}Am^{241}$ and ${}_{94}Cm^{242}$ respectively with 35 MeV helium produced in the same 60-inch cyclotron.

The reactions are:

$$*_{92}U^{238} + {}_2He^4 \rightarrow *_{94}Pu^{241} + {}_0n^1$$

$$*_{94}Pu^{241} \rightarrow *_{95}Am^{241} + e^- \qquad (T = 14 \text{ years})$$

$$*_{94}Pu^{239} + {}_2He^4 \rightarrow *_{96}Cm^{242} + {}_0n^1$$

$$*_{96}Cm^{242} \rightarrow *_{94}Pu^{238} + {}_2He^4$$

Americium and Curium have already been found to have several isotopes
e.g. americium eight (237, 238, 239, 240, 241, 242, 243, 244),
curium seven (238, 240, 241, 242, 243, 244, 245),

Am^{241} has a period of 475 yrs., while Am^{243} 7,600 yrs.; this is the longest-lived isotope of Am, hence representative of the element 95.

Cm^{243} has a period of 35 yrs., Cm^{244} 19 yrs., while Cm^{245} 20,000 yrs., and hence represents the element 96 in the Periodic table.

For berkelium and californium the reactions are:

$$*_{95}Am^{241} + {}_2He^4 \rightarrow *_{97}Bk^{243} + 2\,{}_0n^1$$

$$*_{97}Bk^{243} \rightarrow {}_{96}Cm^{243} \quad (K\text{-capture})$$

$$*_{96}Cm^{242} + {}_2He^4 \rightarrow *_{98}Cf^{244} + 2\,{}_0n^1$$

$$*_{98}Cf^{244} \rightarrow {}_{96}Cm^{240} + {}_2He^4$$

Berkelium is at present known to have the isotopes of masses 243, 244, 245, 249 and 250. The period of Bk^{249} is approximately one year. Californium has many more isotopes -- 244, 246, 248, 249, 250, 251; 252 and 253. The period of Cf^{249} is nearly 500 years.

The discovery of the elements 99 and 100, in 1953-1954 is of special interest: A. Ghiorso, B. Rossi, B. G. Harvey and S. G. Thompson, under the direction of G. T. Seaborg, bombarding U^{238} with nitrogen ions ($N^{14+++++++}$) accelerated with the 60-inch cyclotron were able to produce the element 99 of atomic mass 247, half-life 7.3 mins, emitting α-particles of 7.35 MeV energy. Other products of the bombardment include Cf $^{(244, 246, 247, 248)}$ and Bk$^{(243, ...)}$. The new element, identified by the usual chemical separation methods of precipitation and ion exchange, has properties similar to holmium (67) of the rare earth series and has been named *einsteinium* (E). Other isotopes of this element already identified are of masses, 252, 253, 254 and 255.

B. G. Harvey, S. G. Thompson, A. Ghiorso and G. R. Choppin, again under the leadership of G. T. Seaborg, discovered element 100 by bombarding Pu^{239} with neutrons from an atomic reactor. This marked the first time of production of *synthetic transuranic elements by means of atomic reactors,* since previously those elements were prepared only by the use of projectiles from the cyclotron.

Transmutation of an atom of Pu to element 100 is a complicated process in which 15 neutrons are absorbed by the Pu atom. During this process, six p-particles are emitted, while the atom of Pu with 94 protons and 145 neutrons is changed into atom (100) with 100 protons and 154 neutrons.

Chemical identification was made by the usual precipitation and ion exchange methods. Many of the properties of the new element were found to be similar to those of erbium (68) of the rare earth series. The element 100 thus discovered has a period of 3 hours and decays by the emission of α-particles of about 7.2 MeV, energy. It has been called *fermium,* (Fm), which has isotopes of masses 254, 255, 256.

In April 1955, G. T. Seaborg, A. Ghiorso and collaborators discovered element 101, by bombarding element (99^{253}) with an intense beam of 48 MeV helium ions from the 60-inch cyclotron. The new element has properties similar to thulium (69) and is an α-emitter. The name *mendelevium* (Mv) has been suggested for it.

Very recently (1957), the Nobel Institute of Physics, Stockholm, obtained element 102 by bombarding curium (96) with carbon ions accelerated by the cyclotron of the Nobel Institute. The name *"nobelium"* (No) has been suggested. for it, which is a solid .α-emitter, with a period of 10 to 12 mins.

Now that the ball has been set rolling as regards elements heavier than uranium hitherto considered as the heaviest, one wonders where the new periodic table will end. Seaborg from an intensive analysis of the chemical properties of the newly formed transuranic elements has suggested, as early as 1945, that the elements of Z> 88 might form another transition group (*actinides*) similar to that of the rare earths (*lanthanides*). He expects that if the transuranic elements could be extended up to 103 the second half of the actinide group will be completed.

The identified transuranic elements decay forming a series of radioactive transformations analogous to those of the three series of natural radioactive elements. All the four series have now been classified under a general law which may be stated thus:

All members of a particular series are characterized by having (4n + q) nucleons, *where q is* one of the integers 0, 1, 2, 3, characteristic of a given series, and n may be any integer, as shown in Fig. 303.

Thus the thorium series is characterized by (4n + 0),
The transuranic neptunium series by (4n + 1),
The uranium-radium series by (4n + 2) and
The actinium series by (4n + 3).

Evidently (4n +q) gives the mass number A of the members of the series.

The *thorium series* (A = 4n) begins with $_{99}Th^{232}$ having a period of 1.4 x 10^{10} yrs., and the end product is $_{82}Pb^{208}$. The artificially produced isotopes Pa^{228} and U^{228} begin two other branches of this series.

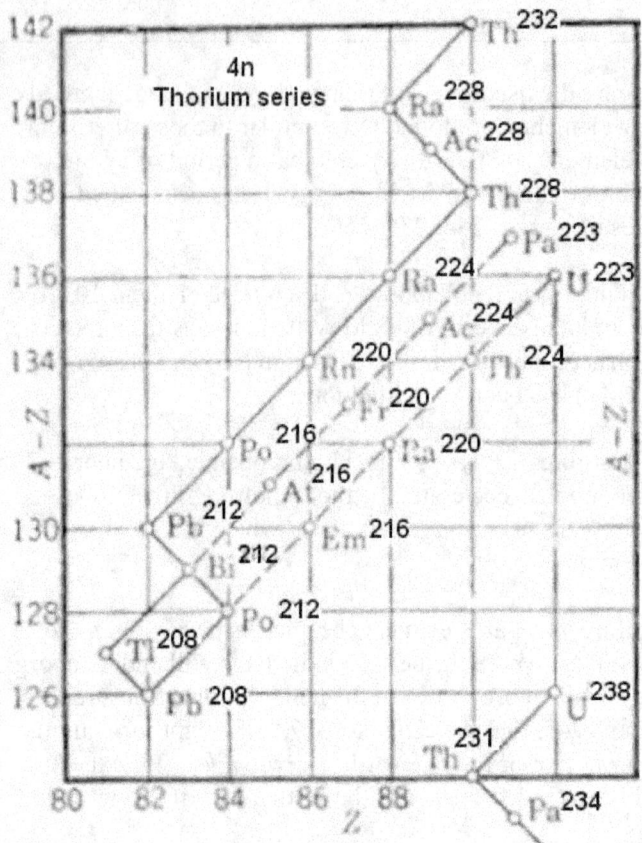

The recently discovered *neptunium. series* (A= 4n + 1) is named after Np^{237} , the longest•lived member in it, having a period of 2.2 x 10^6 yrs. All numbers of this series are either unknown or extremely rare in nature. They include isotopes of the elements *francium* Fr^{221} and *astatine* At^{217}, only recently identified. This series has a branch, beginning at artificially produced U^{229}, which transforms into Po^{213} after four alpha decays. The end product is Bi^{209} and not an isotope of Pb, unlike in the other three series.

184

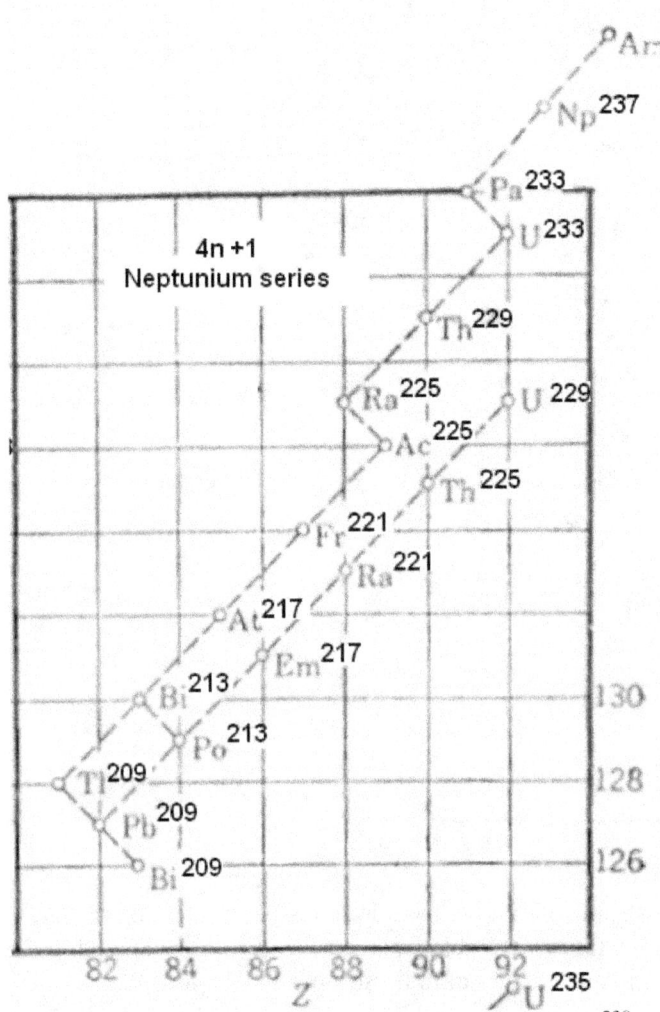

The *uranium-radium series* ($A = 4n + 2$) starts with U^{238} and ends in stable $_{82}Pb^{206}$. It has two other branches starting with artificially produced Pa^{230} and Pa^{226}. The first of these joins the main series at $_{84}Po^{214}$, while the second at $_{83}Bi^{210}$.

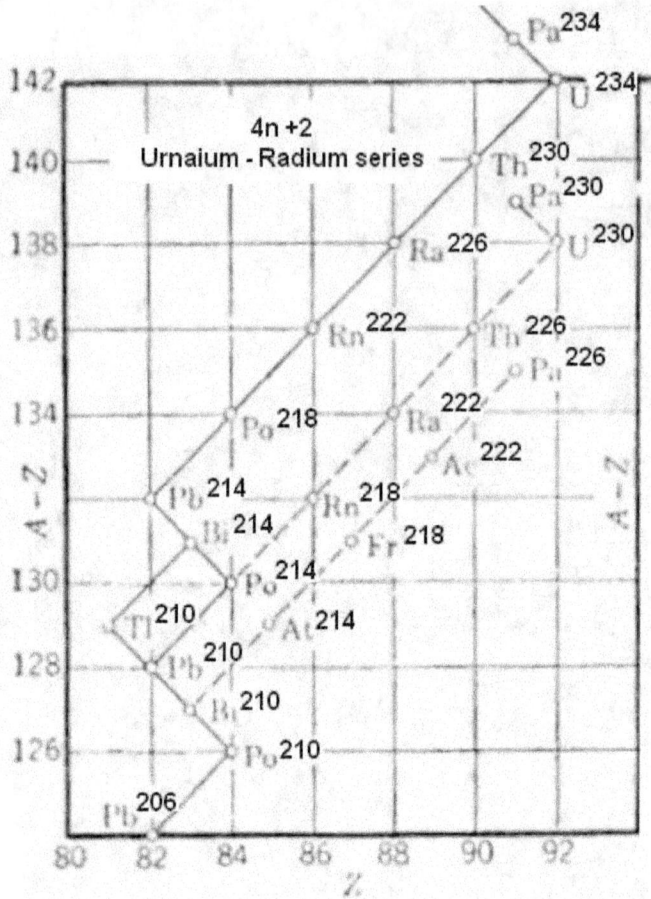

The actinium series (A = 4n + 3) starts with U^{235}, includes an isotope of actinium $_{89}AC^{227}$ which gives it its name and terminates at the stable isotope of lead $_{82}Pb^{207}$. This series has also two other branches that begin at the artificial radioisotopes Pa^{227} and U^{227}.

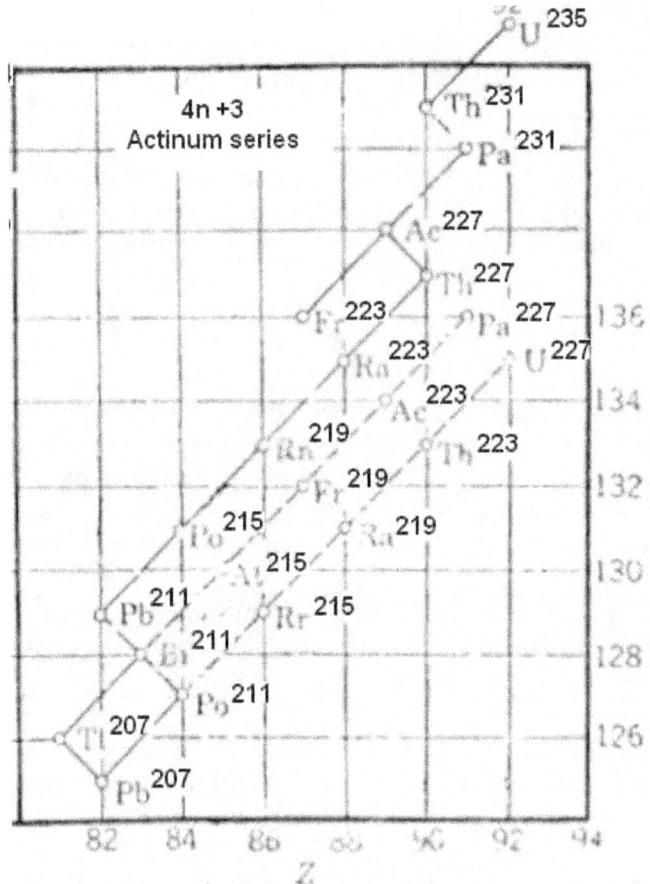

This classification brings out the great abundance of members in the different series having a large number of parents and several side branches.

Practical applications of nuclear fission

The enormous energy liberated in nuclear fission coupled with the possibility of practical realization of rapid chain reaction readily suggests the two most important practical applications of the process, viz.

(i) The *super-power plant*
and
(*ii*) The *super-explosive* or *the atom bomb.*

Both of them have been successfully attempted, in spite of complicated techniques involved. The atom bomb project has yielded better results than that of the atomic power plant. The general principle that discriminates the two applications is as follows:

For industrial purposes, a graded constant production of power which gives a temperature high enough for the efficient operation of heat engines has to be taken into consideration, while for the atom bomb a high speed cumulative effect is to be realized, where within less than a micro-second the number of fissions is multiplied to such an extent that a considerable portion of the material is affected and the temperature and pressure rise enormously causing a terrific explosion.

Accordingly, two different techniques have been developed, one a *slow neutron reacting system* called the *reactor* or *pile* suited for industrial purposes and the other *a fast neutron reacting system* for the production of atom bombs.

THE REACTOR OR PILE

Principle
The pile works on the *slow neutron chain* which is achieved with *natural uranium,* by the use of graphite as *moderator* to slow down the fission neutrons to thermal energies which thus become capable of producing further fissions, chiefly on the small percentage of U^{233} and hence of giving atomic power at a controllable rate. It consists of a great number of uranium blocks or slugs disposed in a calculated "lattice" throughout a huge mass of graphite.

Fission neutrons which escape from one of the uranium blocks, in which there is a very small chance of capture, pass into the graphite moderator, where by successive collisions they lose energy very rapidly, about one-sixth at each collision.

The chance that they encounter a U^{238} nucleus before they are slowed down to energies below the resonance level is greatly reduced and when the neutrons diffuse to a uranium block again they are readily captured by the U^{235} nuclei and give rise to further fissions.

For the chain reaction to take place the size of the system must be evidently greater than a certain critical value, when the rate of escape of the neutrons from the surface is not greater than the rate of production. This *critical size* depends upon the precise arrangement of the uranium and graphite and upon whether the surroundings reflect back any of the neutrons which escape from the surface. The size of the pile required to obtain a slow neutron chain reacting system in uranium must be obviously very large, on account of the very small amount of U^{235} in natural uranium and the great quantity of moderator to be used.

The *fact that some of the fission neutrons are delayed in time is utilized to control the operation of the pile.* If the pile is made just above the critical size, the exponential increase in the rate of fission takes place rather slowly due to the delayed neutron emission.

Now by inserting into the pile rods of a material like boron or cadmium which absorb slow neutrons strongly, the multiplication of the fissions can be prevented altogether and the pile will not operate.

This device is known as *arrestor or control rods.* Slow withdrawal of these rods allows the reaction to begin and there is plenty of time to re-insert them if the multiplication rises too rapidly. Adjustment of the rods can be carried out automatically by the use of an equipment like a thermostat, which sets the position of the arrestors so that the system runs continuously at any desired energy level.

The fission energy degenerates by collisions into heat and the uranium blocks become hot. This heat may be extracted by cooling the rods, either by blowing gas, such as air, hydrogen or helium, or by surrounding them with a tube through which water circulates.

The size of the pile may be reduced by replacing the graphite by *a better moderator* such as *heavy water,* which is more effective in slowing down neutrons. A further reduction of size and increased efficiency can be obtained by using either uranium enriched in isotope 235 or plutonium as the active agent in the pile. This is known as the *enriched reactor* or *pile.*

Piles in use

The first successful pile was constructed and put in operation, in December 1942, in the University of Chicago under the direction of Fermi. It is illustrated schematically in Fig. 304.

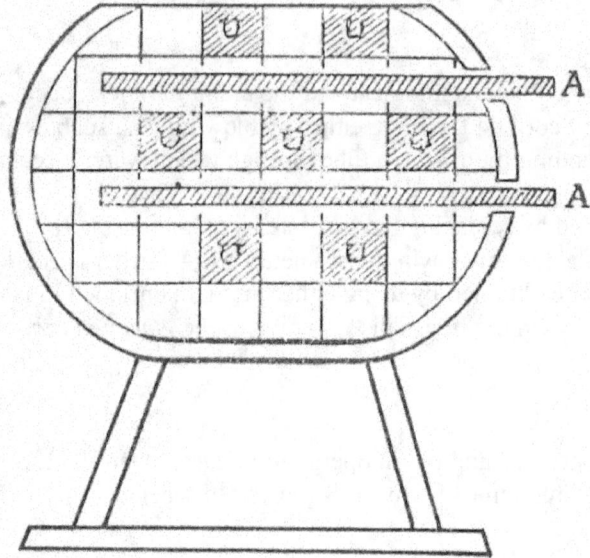

Fig. 304. The uranium pile.

The shape of the pile was an oblate spheroid flattened at the top.
It was built up in layers of pure graphite bricks.
In alternate layers were embedded lumps of uranium U, U, U, in a cubic lattice arrangement.
The arrestors, A, A, were cadmium rods that could be inserted in or with drawn from slots passing through the pile.
During the building up of the pile, the neutron density was measured as successive layers were added, in order to check the approach of the critical size. With no arrestors the required size was reached before the last layers were in place.

Great care was taken lest the chain reaction should get out of hand and cause an explosion. The pile was surrounded by a thick layer of graphite so as to reflect back into it as many neutrons as possible. But even so, in order to make up for the surface loss of neutrons, the size of the pile was large, its diameter being 6 to 7 feet.

Six tons of uranium and some hundred tons of graphite were used. When the pile was not in operation several arrestors were inserted in a number of slots which brought down the multiplication factor below 1.

To operate the pile, all but one of the arrestors were removed and the remaining one was slowly pulled out, when the intensity of the neutrons emitted began to increase rapidly.

By a careful adjustment of the single arrestor, the pile was made to work at a constant power level. This first pile produced a low power of 200 watts only, but this was enough to check up theoretical expectations. Other piles were soon constructed which gave much greater power output and could be operated for months at a time. Thus for instance, at

Clinton, Tennessee, a gas-cooled pile, having a continuous power output of 1,000 kilowatts and more, was put in action in 1943. At Handford, Washington, piles, extremely large, complex and costly, utilizing liquid cooling and capable of supplying power at thousands of kilowatts, were constructed in 1944. Since then, a few more piles have been built in Canada, England, Russia and France. India had three at the Atomic Energy Establishment, Trombay: the "Apsara" which has been operating since 1956, the "Zerlina", which started functioning in 1961 and the "Canada-India reactor".

The nuclear reactors in actual use may be classified as follows:

(*i*) *Uranium-graphite reactors*

Following the success of the pile at the University of Chicago, several others were built on the same model, but further improvements to obtain greater power output and larger neutron flux. Such are:

- The Oak Ridge reactor at Tennesse (1943) operating at 3,800 KW,
- The "Gleep" and "Bepo" at Harwell, England (1947-1948) working at 6,000 KW and
- The Brookhaven reactor (1951) delivering 30,000 KW.

Efficient air-cooling system has enabled some of these reactors to be operated at a moderately high temperature of 200 to 300°C.

(*ii*) *Heavy water reactors*

The use of heavy water as moderator in a nuclear reactor has the advantage that neutrons are reduced to thermal energies after about 25 collisions with deuterons as compared with some 115 collisions required with carbon nuclei in graphite. Heavy water has the additional advantage that it absorbs fewer neutrons than graphite.

Thus for a heavy water reactor a small volume of moderator is required, and the neutron flux is larger than for a graphite reactor at the same power level. Heavy water can be used at the same time as a cooling liquid for the reactor.

To his class of reactors belongs the first one built in 1944 at the Argonne laboratory, operating at rather low power level of about 300 KW in the original design, but completely remodeled in 1954 with the use of enriched uranium to deliver 1,000 KW and a large neutron flux.

Others built on the same model are the Canadian NRX reactor constructed in 1947 working at a maximum power output of 30,000 KW, the "Zoe" in France (1950), the "Jeep" in Norway, the "Zerlina" in India, etc. These last mentioned reactors operate at low powers of 10 and 100 KW respectively and are chiefly meant for laboratory researches.

(ii) *Homogeneous nuclear reactors*

In contrast to the reactors in which the fissionable material is concentrated at regularly spaced intervals throughout the moderator, it is possible for reactors to have fuel and moderator intimately mixed, The first reactor of this type was constructed in 1945 at Los Alamos, using enriched uranium salts (one part in six of U^{235} as sulphate or nitrate) dissolved in ordinary water as moderator.

Although ordinary water cannot be used as moderator with natural uranium because too many neutrons are captured by hydrogen nuclei, it, is possible to use it when uranium sufficiently enriched in U^{235} is employed. In the Los Alamos reactor the enriched U 235 solution is enclosed in a stainless-steel sphere of 1 ft. diameter. Helical coils through which cooling water is circulated are arranged inside the sphere, so that the reactor can be operated at about 45 KW. Traversing the sphere is a hollow cylinder into which samples may be inserted for irradiation in the neutron flux. The sphere is surrounded by a cube of beryllium oxide that acts as a neutron reflector (*tamper*) into which the cadmium control and safety rods are inserted.

An additional graphite reflector surrounds the BeO. The radiation shield around the entire assembly consists of 4 inches of lead 1/32 inch of cadmium and 5 ft. of concrete. The shield is pierced for provision of a thermal column as a source of slow neutrons. Between the graphite and the thermal column is a slab of bismuth 8 inches thick, which is a very efficient γ-ray absorber but passes neutrons without great loss.

Other reactors built on this model are the Oak Ridge Swimming pool type (1950), working at 10 KW, the M.T.R. Idaho (1952), with a power rating of 30,000 KW and the Apsara at Trombay, India (1956) with a maximum power output of 1,000 KW.

The operation of these reactors is beset with many practical difficulties: *e.g.,*

(*i*) Uranium is very active chemically and is easily oxidized in contact with air or water. To prevent this, the uranium blocks are coated with pure aluminum. But then it is necessary to keep the temperature of the air or water in contact with the pure aluminum below 100°C in order to prevent corrosion of the aluminum itself.

(*ii*) The pile warms up very much and the cooling problem is a great one, while the heat gathered up by the cooling system is actually allowed to go to waste on account of the low efficiency demanded by the low temperature operation.

(*iii*) The amount of heat which may be extracted from the U slugs directly is limited also by the low heat conductivity of the metal.

(iv) Materials inside the pile become so strongly radioactive that methods of handling them from a considerable distance become essential.

(v) The neutron radiation from the pile is so great and dangerous that no operator can approach the pile which has to be enclosed in thick concrete walls.

Applications of the reactors

The possibility of using nuclear energy as a *super-power plant* clearly depends upon the solution of engineering and metallurgical problems of extraction of the fission heat at a temperature high enough for the efficient operation of heat engines. There exists also the question of economy of the process as compared with other fuels in actual use. These points are being tackled, but have not yet been fully solved.

The great advantage of nuclear power plants is that the production of energy involves only a comparatively very small consumption of fuel and the combustion of the fuel does not require any additional material, such as oxygen, for instance.

Hence the harnessing of atomic energy enable powerful engines to be built, which will be capable of working for long periods of time without refueling or taking on supplies of other materials.

Two nuclear power submarines, the *Nautilus* and the *Sea Wolf* had been built in 1958 and put in operation in America. These are fuelled with enriched U^{235}; the heat generated by fission is removed by ordinary water which at the same time serves as moderator; the water circulates between the reactor and the boiler heat exchanger unit, where it transfers its heat to the water of the secondary circuit maintained at about 500°C required for the steam turbine. The turbine develops 8,000 H.P. and drives an electric generator which feeds the engines that turn the propellers. These submarines to do the round-the-world voyage without refueling.

As of end 2009 the total electricity production since 1951 amounts to 64,600 billion kWh. The cumulative operating experience amounted to 14,174 years by September 2010.

Country	In operation		Under construction	
	Number	Electr. net output MW	Number	Electr. net output MW
Argentina	2	935	1	692
Armenia	1	375	-	-
Belgium	7	5,926	-	-
Brazil	2	1,884	1	1,245
Bulgaria	2	1,906	2	1,906
Canada	18	12,569	-	-
China				
Mainland	13	10,048	27	27,230
Taiwan	6	4,980	2	2,600
Czech Republic	6	3,722	-	-
Finland	4	2,716	1	1,600
France	58	63,130	1	1,600
Germany	17	20,490	-	-
Hungary	4	1,889	-	-
India	20	4,391	5	3,564
Iran	-	-	1	915
Japan	54	46,823	2	2,650
Korea, Republic	21	18,665	5	5,560
Mexico	2	1,300	-	-
Netherlands	1	487	-	-
Pakistan	2	425	1	300
Romania	2	1,300	-	-
Russian Federation	32	22,693	11	9,153
Slovakian Republic	4	1,792	2	782
Slovenia	1	666	-	-
South Africa	2	1,800	-	-
Spain	8	7,514	-	-
Sweden	10	9,303	-	-
Switzerland	5	3,238	-	-
Taiwan	6	4,980	2	2,600
Ukraine	15	13,107	2	1,900
United Kingdom	19	10,137	-	-
USA	104	100,747	1	1,165
Total	**442**	**374,958**	**65**	**62,862**

Nuclear power plants world-wide, in operation and under construction, as of Jan 19, 2011 (http://www.euronuclear.org/info/encyclopedia/n/nuclear-power-plant-world-wide.htm)

The construction of "portable" type of nuclear engines which can be used in automobiles and planes is not yet feasible. It may be noted that the British reactor Bepo is used to heat laboratory buildings by circulation of the air in its cooling system and the American E.B.R. operates a steam turbine and generates electric power sufficient to supply the laboratory in which it is housed.

Most of the *reactors now in operation are used chiefly to produce plutonium and other radioactive materials* required far research purposes. As the reactor operates, plutonium is being formed constantly in the uranium blocks. When the action has proceeded sufficiently, this fissionable plutonium may be separated chemically from the parent uranium.

To produce one gram of plutonium per day the reactor must operate continuously at about 1,000 kilowatts. From this it is seen that the production of plutonium means **a** reactor of gigantic size developing thousands of kilowatts power and a chemical separation plant to extract the plutonium. Further, great precautions must be taken by the operators on account of the strong radioactivity of the fission products.

Reactors designed to produce plutonium are known as *breeder reactors.* In America, the A.E.C. put into operation in 1952 an experimental breeder reactor (E.B.R.) for the conversion of U^{238} and Pu^{239} based on the fast neutron reaction. Fast Breeder Reactors (F.B.R.) have been built and operated in the USA, the UK, France, the former USSR, India and Japan. An experimental FBR in Germany was built but never operated. There are very few breeder reactors actually used for power generation, there are a few planned, and quite a few are being used for research related to the Generation IV Reactor Initiative. In many countries, nuclear power has been opposed politically and thus many breeder reactors have been shut down, or are planned to be shut down, with various justifications.

The radioactive fission products can also be recovered in large quantities, which find wide application in several branches of applied science. If an aperture is made at the top of the reactor, an intense uniform beam of neutrons can be obtained, which can be used for the production of radioactive isotopes and further experiments in physics, medicine, biology, etc.

It has been reported that a small enriched heavy water reactor with a few pounds of active agent in the form of a solution surrounded by a neutron reflector constitutes the most compact but most intense source of neutrons well suited for laboratory researches such as neutron diffraction, neutron-electron interaction, etc. A spectacular example of Cerenkov radiation is observed when the "hot" fuel rods of a nuclear reactor are immersed in water. The γ-rays from the fission products produce many high-speed electrons in the water and these travel with sufficient velocity to excite Cerenkov radiation, whose color is of an unsaturated blue.

THE ATOM BOMB

Principle

The atom bomb, which must necessarily be an explosive type of reaction, works on the *fast neutron chain* that is accomplished by bringing together very rapidly into intimate contact two pieces of a fissionable material, such as U^{235} or Pu^{239} each of mass somewhat greater than half the critical value. The two pieces, when they are kept separated, are perfectly stable and safe, since they are smaller than the critical size. But when they are put together by a mere mechanical operation, such as, for instance, a sudden jerking out of a thick cadmium screen inserted between them, the total mass becomes greater than the critical value and in consequence will explode violently.

No other mechanism for artificially inducing the explosion is required, since there are always some stray neutrons in any mass of metal produced by cosmic rays, and these suffice to start the chain reaction.

It is essential that the chain reaction develops so rapidly that a considerable portion of the active material is affected before the explosion.

Technical details

(i) The actual production of the atom bomb involves essentially the *preparation of* U^{235} Pu^{239} in *sufficiently large quantities*. This is evidently an extremely complicated and costly project, but it has been achieved under the impetus of World War II as a major national effort in the U.S.A.

Plutonium is obtained by the use of the uranium pile and of chemical separation plants, as already stated.

The technique used for the concentration and separation of U^{235} forming a very small part of natural uranium is still more involved and difficult. But it has been achieved, of course at prohibitive costs, using chiefly the gaseous diffusion method and employing very probably the only known gaseous compound of uranium, UF_6. The gigantic and complicated installations involved in this work may be considered a unique achievement in the history of science. The whole project was highly speculative, until the first test bomb exploded in New Mexico desert according to theoretical predictions.

(ii) As pure fissionable materials are used in the atom bomb without any moderator, the *size of the bomb* is bound to be *far smaller than that of the* pile; which eliminates transport difficulties to a great extent. The critical size depends upon the shape of the bomb and the nature of the material in which the bomb is enclosed.

(iii) The two pieces to be detonated are enveloped in a cover of a very dense substance, the *tamper,* which serves the double purpose of reflecting back into the bomb neutrons which might otherwise escape into air and of delaying the expansion of the bomb until the temperature and pressure of a high explosive is built up.

Atom bomb in action

The atom bomb is the most terrific and devastating explosive so far made by man, although a more powerful competitor, the so-called "hydrogen-bomb", has been realized so after (cr. below). The ruins of Hiroshima and Nagasaki bear witness to its destructive power, estimated to be about that of 10,000 tons of trinitrotoluene (T.N.T.) In the actual explosion which occurs in a couple of microseconds after the detonation and for which only a portion of the active material is effective, the energy liberated is sufficient to raise the temperature to the order of 10,000,000°C and more and to produce a pressure of several millions of atmospheres.

Since radiation is proportional to the fourth power of the temperature, the explosion will be accompanied by a violent and intense blast of visible, ultra-violet, X-ray and γ-ray radiations causing a blinding flash. A large quantity of radioactive matter is also released both by the fission products and by the action of the emitted and neutrons. Under the conditions of such high temperature and pressure, the active material itself will get quickly vaporized and expand, causing a fall in the density, diminution of neutron absorption and the consequent collapse of the chain reaction. The time for which the explosion lasts is very small and depends on the rate of expansion of the active material. The radioactive substances formed are carried away by air currents and will die away if they have short period, or get deposited on surrounding objects.

Note:

(1) Source of stellar energy

With the knowledge gained in studying various types of nuclear reactions, it has now become possible to provide a satisfactory explanation of the origin of the energy radiated by the stars. Considering, for instance, the sun which is one of the innumerable stars forming the Milky Way, we know that it is constantly radiating energy and yet its temperature is sensibly constant, 6000°C at the surface, while in, the interior much higher, having the tremendous value of 20,000,000 °C at the center. (The surface temperature can be readily estimated by measuring the *solar constant* and applying Stefan's law of radiation, while the temperature at the center can be calculated from the surface temperature and from the heat conductivities of the gases constituting the material of the sun). The constancy of the solar temperature implies that some processes are taking place in the sun which make up for the loss of energy by radiation.

Hans Albrecht Bethe (July 2, 1906 – March 6, 2005) Germany

Carl Friedrich Freiherr von Weizsäcker (June 28, 1912 – 28 April 2007) Germany

Bethe and Weizsäcker were the first, in 1939, to propose a *thermonuclear origin for stellar energy, i.e.,* the production of energy in the sun and the stars is due to nuclear reactions caused by the very high temperatures in their interior. A nuclear reaction being really a fast collision, at the actual internal temperatures of the stars, collisions, chiefly of light atoms which have enormous velocities, become nuclear disintegrations. These thermonuclear processes in which light atoms are involved correspond to the phenomen of effusion of two small liquid drops of the Bohr-Wheeler theory. In such cases the speeds of the interacting particles must be great enough to overcome the electrostatic repulsion, as we have already noted. Such speeds are reached in the extremely hot interior of the stars. In these processes, very large amounts of energy are released which are transformed into radiation. The energy radiated from the surface being thus constantly replenished by the nuclear energy produced in the interior, the state of the star will remain constant, as long as the nuclear reactions can go on.

Combining the information concerning the central temperature (about 20×10^6 °C) and the hydrogen content (about 35%) of the sun with known facts about the rates of various nuclear reactions, Bethe showed that a *"carbon-nitrogen cycle" was mainly responsible for the production of solar energy.*

This cycle consists of *a chain of nuclear transformations in which hydrogen is converted into helium, with the aid* of *carbon and nitrogen* as *catalysts, the highly energetic thermal protons constituting the projectiles in the successive reactions.* A proton collides with ordinary carbon (C^{12}) and forms the lighter isotope of nitrogen (N^{13}), in a (p, γ) reaction. The nucleus N^{13} being radioactive, emits a positron and is transformed into the stable heavier isotope of carbon (C^{13}). Being struck by another proton, this carbon is transmuted into ordinary nitrogen (N^{14}) in a (p. γ) reaction again. The nucleus N^{14} collides with still another proton and produces the unstable oxygen isotope O^{15} which decays with positron emission to form the stable nitrogen isotope N^{15}. Finally, this N^{15} reacting with a fourth proton breaks up giving a carbon nucleus (C^{12}) and a helium nucleus (He^4):

$$_6C^{12} + {}_1H^1 \rightarrow * {}_7N^{13} + hv$$
$$* {}_7N^{13} \rightarrow {}_6C^{13} + e^+ \quad (T = 9.9 \text{ mins})$$

$$_6C^{13} + {}_1H^1 \rightarrow {}_7N^{14} + hv$$
$$_7N^{14} + {}_1H^1 \rightarrow * {}_8O^{15} + hv$$

$$*{}_8O^{15} \rightarrow {}_7N^{15} + e^+ \quad (T = 2.1 \text{ mins})$$
$$_7N^{15} + {}_1H^1 \rightarrow {}_6C^{12} + {}_2He^4$$

It will be noted that these reactions form a *closed chain* or cycle, returning to the starting point after every six steps. The chain of reactions can start with either C^{12} or N^{14} and end in the reproduction of the same nucleus with which the chain is initiated. Carbon and nitrogen therefore remain unaffected, enabling the cycle to continue and hence they may be considered to act as mere catalysts. The net result of the chain is the consumption of four protons to form a helium nucleus. The period of a complete cycle is estimated to be about 5 million years.

The *amount of energy liberated in this chain of reactions* can be readily evaluated from the mass deficiency involved.

The mass of 4 protons = 4 x 1.00813 = 4.03252 m.u.
The mass of a helium nucleus = 4.00386 m.u.
Therefore,
Mass that has disappeared = 0.02866 m.u.

Hence the energy liberated = 0.02866 x 931 = 26.7 MeV, according to Einstein's mass-energy relation. Considerations of the actual hydrogen and helium content of the sun show that this amount of energy is enough to keep the sun going at its present internal temperature for about 30 billion years.

Assuming that the sun must have been formed not much before the formation of our earth, the ago of the son must be more or less the same as that of the earth, *viz.,* about 3 to 4 billion years. Hence the sun must be considered still young and will continue to shine with approximately the present intensity for billions of years to come.

According to modern views of stellar evolution, the above-mentioned carbon-nitrogen cycle comes at a rather late stage in the life of the stars. A star is formed by the condensation of a large amount of matter at a point in space. Under the influence of gravitational attraction the amassed matter gradually contracts, resulting in a rise of temperature until the central part is about 200,000°C. At this stage, there is sufficient thermal energy for a *proton-proton interaction* to form a deuteron as given by

$$_1H^1 + {}_1H^1 \rightarrow {}_1D^2 + e^+ + v$$

which is followed by two additional proton reactions resulting in an α-particle, as

$$_1D^2 + {}_1H^1 \rightarrow {}_2He^3 + \gamma$$
$$_2He^3 + {}_1H^1 \rightarrow {}_2He^4 + e^+$$

These reactions are, in the main, responsible for the luminosity of the *stars fainter than the sun, e.g.,* the so-called *red-dwarfs.*

After about a million years, when all the deuterons are exhausted, the star shrinks again and grows hotter and the core temperature rises to about 5,000,000°C. The thermal velocities of the protons, are now great enough to allow them to interact with heavier elements, such as Li, Be, B again forming He as the final nucleus. The star is now said to pass through the *"giant stage",* accounting for the luminosity of the so-called *"red giants".*

When all the above three types of nuclei have been destroyed, further contraction sets in, with the increase of temperature to 20,000,000°C. The carbon-nitrogen cycle begins at this stage, supplying energy for the most important radiating life of the star. When the proton content is finally exhausted the star undergoes a rapid contraction and is almost at the end of its career. The so-called *"white dwarfs"* belong to these last phases of stellar life. The energy liberated during the final contraction soon exceeds the thermonuclear energy resulting sometimes in catastrophic explosions as evidenced in what are known as "novae" and *"supernovae"*. As the sun maybe considered to belong to the category of stars which are still young, although its internal temperature corresponds to stars in mature life, the production of solar energy might involve both the *proton-proton* and *carbon-nitrogen cycle interactions.*

(2) The hydrogen bomb. The *mechanism* of this bomb which is talked about so much at the actual hour, as an explosive 1,000 times more powerful than even the atom bomb, must be similar to that of the production of stellar energy, viz., the *fusion of hydrogen or other light atoms into heavier* ones, *with release of enormous energy.* Hence it works just the opposite way to that of the atom bomb, where very heavy atoms break down by fission. The essential conditions required for the operation of the hydrogen bomb are the extremely high temperatures and pressures that are present in the interior of the stars promoting thermonuclear reactions. To reproduce these in terrestrial matter is not easy. But such conditions have been realized by using the atom bomb as a *"primer"* which by first exploding could provide the very high temperatures and pressures necessary for the successful working of the hydrogen bomb.

The superiority of the hydrogen bomb over the atom bomb is probably due to the fact that it has, not, in principle, the limitation of a *critical size,* unlike the atom bomb. If the active material, in the atom bomb, exceeds the critical mass spontaneous explosion results. On the other hand, any amount of active material can be used in the hydrogen bomb, which would not start exploding until part of it had been heated to the ignition temperature. The first hydrogen bomb, prepared in America and nicknamed "Mike", was exploded in November 1952 at Eniwetok in the Marshall Islands.

Ivy Mike (Eniwetok-Atoll - 31. October 1952)

The explosive violence of Mike was equivalent to 3 million tons of T.N.T. The fuel used for the thermonuclear reaction was probably liquid deuterium and was fired with all atomic bomb "trigger". Mike weighed about 65 tons and was not a transportable bomb.

First Soviet test of a thermonuclear device.
Originally from USSR Nuclear Weapons Tests and Peaceful Nuclear Explosions: 1949 through 1990; The Ministry of the Russian Federation for Atomic Energy, and Ministry of Defense of the Russian Federation; ed. V. N. Mikhailov; 1996

In August 1953, the Russians exploded a thermonuclear bomb, which had a less explosive power, about 1 million tons of T.N.T. only, but was technically more advanced than Mike, since it was a transportable one and very probably the fuel used was other than liquid deuterium. Of the following possible thermonuclear reactions:

$$_1D^2 + {_1H^1} \rightarrow {_2He^3} + hv$$
$$_1D^2 + {_1D^3} \rightarrow {_1H^3} + {_1H^1}$$
$$_1H^3 + {_1H^1} \rightarrow {_2He^4} + hv$$
$$_1H^3 + {_1D^2} \rightarrow {_2He^4} + {_0n^1}$$

only the last one develops with sufficient speed (of the order of 10^{-6} sec.). Hence deuterium and tritium should be ordinarily employed in the construction of the hydrogen bomb.

A pure fusion bomb will not produce any bad after-effects, since the end products are not radioactive; but the radioactive fission products of the 'primer' atom bomb will always be present. In order to produce a *clean bomb, i.e.,* a fusion bomb without radioactive end products, attempts have been made to replace the atom bomb primer by a chemical trigger, *i.e.,* to produce the required temperature of a million degrees by firing a chemical explosive contained in some specially designed container (principle of shaped charges). However the possibility of the large number of high energy neutrons liberated during the explosion inducing radioactivity in the surrounding air and dust cannot be neglected, but compared to the fission the effect will be very small.

The so-called *N-bomb* (neutron bomb) is a clean bomb, produced by tailoring the energy of a fusion explosion (H.-bomb), so that, instead of heat and blast, its primary product is a burst of neutrons. Such a burst would operate as a king of death rays. It would do next to no physical damage; it would result in no contamination; but it would immediately destroy all life in the target area. This, of course, would make it an ideal battlefield weapon. A fission atom bomb is inherently *dirty*.

(3) The rigged H-bomb
It has been found that when the hydrogen bomb is encased in a special metallic sheath, still worse effects are produced by the giving off of lethal radiations and this, in some cases, for months and years. Such a contrivance is popularly called the *hell bomb*. The so-called *cobalt bomb is one* of this kind, which consists in encasing the hydrogen bomb in a sheath of metallic cobalt. When the hydrogen bomb is exploded, the neutrons that are emitted act on the cobalt cover and render it intensively radioactive, due to the formation of Co^{60} which is some 320 times more powerful than radium.

If, for example, the bomb contains one ton of deuterium, 250 lbs of neutrons are produced. This acting on cobalt would produce 15,000 lbs (7.5 tons) of radioactive cobalt (Co^{60}), equivalent to 4,800,000 lbs (2,400 tons) of radium. The half period of Co^{60} is about 5 years, *i.e.*, after 5 years it will be reduced to 1,200 tons, after 10 years to 600 tons and so on. During the explosion, the cobalt will be naturally pulverized and converted into a gigantic radioactive cloud that will kill everything in the area it blankets. Nor will it be confined to a particular area, since the winds will take it to distant parts, thousands of miles away, carrying death and devastation wherever it reaches, and this for many years.

The transportable 'thermonuclear bomb', built by the Americans and exploded in March, 1954, is perhaps another of the rigged H-bomb type. This bomb had an explosive power of 15 to 20 million tons of T.N.T. According to available reports, the fuel used was lithium deuteride (a compound of lithium and deuterium). When the bomb was exploded by the use of an improved form • of atomic bomb as trigger, tritium (H^3), an unstable heavy isotope of hydrogen, was produced on the spot by the capture of neutrons, released in the fission of the core, by Li^6. The fusion of tritium and deuterium by thermonuclear fusion process produces fast neutrons in great quantities. These are absorbed in an outer sheath of U^{238} which thereby undergoes 'fast-fission' and releases enormous quantities of energy as well as intensely radioactive fission products. Hence it is a case of *"fission-fusion-fission,"* type of bomb, where thermonuclear reaction is of secondary importance.

The use of fast fission of U^{238} as the primary process makes this bomb much more dangerous than a purely thermonuclear bomb from the point of view of radiation damage. These are generally referred to as *'high yield'* bombs in comparison with ordinary atom bombs, which are of *'low yield'* type.

The high yield bomb is essentially a *'radiological'* weapon, in the sense that the damage caused to human life by the radioactive end products outweighs the damage caused by its blast and heat radiations. Of the long lived radioactive end-product the isotope Sr^{90} (strontium) presents the greatest danger. For its period is 28 yrs and it is chemically similar to calcium, the most important constituent of our bones. When Sr^{90} falls on the ground it is assimilated by the plants along with Ca in the soil and gets into our system directly through the grains and vegetables we eat and indirectly through cattle (milk, meat, etc.) which live on grass. Once deposited in bone Sr^{90} stays there for years and can cause bone cancer, leukemia and other diseases of the blood. The danger of Sr^{90} is greater to children and undernourished people because of their large need of calcium.

(4) Controlled thermonuclear reaction

A means of producing useful power from the boundless supply of hydrogen in ocean water was discovered in 1960 by Dr. F. H. Coengsen and collaborators at Berkeley, California.

With a system of powerful magnets, a cubic inch of hydrogen was squeezed until it reached a temperature of 35 million degrees. In this state, some hydrogen atoms were added which formed the light variety of helium (He^3) with the production of a cloud of neutrons and a release of more than 3 MeV energy per atom as by-product.

The reaction lasted 1/1000th of a second, which though far too short to be practical was much longer than the result of previous experiments ($1/10^6$ sec.). The reaction looks like **a** real thermonuclear reaction, because of

(*i*) the power interval, which means that hydrogen was contained in the magnetic grip for a significant space of time
and
(*ii*) the reaction produces 10,000 helium atoms from 20,000 heavy hydrogen atoms, which means that neutrons should be ejected in the formation of He^3 .

Neutron counting instruments around the magnet recorded 10,000 neutrons coming through the machine's walls during the thousandth of the second, which is considered as a further evidence that fusion has taken place.

CHAPTER 15

COSMIC RAYS

Introduction

Cosmic rays are extremely penetrating radiations, coming from where we do not know for certain as yet, but doubtless from regions very far away from the earth, and continually bombarding it on all sides with more or less uniform intensity. In the domain of modern physical researches, these rays occupy a unique place on account of the extreme complexity of the phenomenon, the delicateness of observations, the adventurous excursions of the experimenters, the subtlety of analysis and the ingenuity of deductions. In fact, the study of cosmic rays has proved itself to be a subject vast enough to captivate the closest attention of hundreds of workers all over the world and difficult enough to tax the keenest wits of many great theoretical and experimental physicists. During the past century, since the first experiments were made, hundreds of articles have been published and many books written on cosmic rays; yet one is forced to confess that our knowledge of these mysterious rays remains still far from being complete.

All the same, information already gathered from the incessant researches of eminent scientists such as Millikan, Neher, Compton, Anderson and Neddermeyer in America, Hess, Bothe, Kolhörster, Bethe and Heitler in Germany, Auger, Ehrenfest and Leprince Ringuet in France, .Blackett, Occhialini, Powell and their co-workers in England, Clay and his associates in Holland. Piccard, Cosyns and Lemaitre in Belgium, Rossi in Italy, Regener in Switzerland, Bhabha and Gill in India and a host of others, have definitely established the importance and interest attached to the study of cosmic rays.

The chief reason for such an enthusiastic and almost universal scientific study of these radiations of extra-terrestrial origin may be briefly stated as follows:

Louis Leprince-Ringuet (27 March 1901-. 23 December 2000)

Although the investigations have been directed, in the main, towards the understanding of the precise nature and origin of these rays, yet it has become clear that their careful exploration can add considerably to the knowledge of nuclear structure. Thus, two new particles, the *positron* and the *mesotron,* which are as fundamental to the nucleus as electrons, protons and neutrons have been discovered in cosmic ray researches. Moreover, our present understanding of the mechanism of the internal structure of nuclei arises chiefly from the departure of energetic particles from nuclei as in radioactivity and the penetration of fast projectiles into nuclei as in artificial transmutation.

These phenomena, important as they are, do not take us far towards unraveling the mysterious forces that bind the nuclear particles into stable units, because the projectiles involved in these processes, are not energetic enough to probe fully into the interior of the nuclei. In cosmic rays, on the other hand, particles have been found with fantastically large amounts of energy, of the order of 10^{12} eV and more, which are capable of exploring the interior of nuclei with much greater effectiveness and thereby lead to very important results concerning their internal state. The study of cosmic rays, therefore, naturally constitutes an important section of nuclear physics.

Discovery

The discovery of the cosmic rays may be considered to have originated in a simple observation made by C.T.R. Wilson, Elster and Geitel as early as 1900. They found that electroscopes exhibited a small residual leak, in spite of the best insulation. This could have been caused only by ions about the instrument. But the surrounding air could not have been the source of these ions as its conductivity was found to be constant. Hence, it was concluded that there must be some *external agent* other than air, which continually made good the loss of ions due to the residual discharge. This surmise was further confirmed by Rutherford and Cooke, in 1903, who showed by the use of absorbing screens of iron and lead that a considerable part of the permanent ionization was due to a *penetrating radiation coming from outside the instrument.*

It was thought for a time that radioactive matter occurring in small quantities everywhere, in the earth, surrounding atmosphere, etc., might be responsible for this penetrating radiation, if this were so, the rate of discharge of an electroscope should diminish considerably at very great altitudes, far removed from such influences. This inference was tested by balloon observations by three German scientists. In 1909, Gockel measuring the rate of discharge with a Wulf electroscope at different altitudes up to a maximum height of 4,500 meters, found that the rate of discharge diminished much more slowly with height than could be explained by the hypothesis that the penetrating radiation was of terrestrial or atmospheric origin.

In 1911-1914, Hess and Kolhörster, using sealed ionization chambers, extended the balloon observations to much higher altitudes than Gockel had done and found that above a few hundred meters the intensity of radiation increased with height continuously up to 9,000 meters, as much as live to ten times the value at ground level, which, of course, could not happen if the radiation was due to radioactive contamination of the earth, since it would then be almost completely absorbed in going up through a few hundred meters of atmosphere. Hence it appeared certain that the radiation had its origin outside the earth and its atmosphere. Furthermore, the fact that

the radiation produced effects at the earth's surface and was able therefore to penetrate the entire atmosphere showed that it was far more penetrating than even the γ-rays from radioactive substances.

Kolhörster's experimental data proved that the new radiation was absorbed only to about 1/10 the amount of the hardest γ-rays. Hence these authors concluded that an *extremely penetrating type* of *radiation whose origin, was entirely beyond our atmosphere,* fell upon the earth almost uniformly in all directions. They called it by the special name *"hohenstrahlung* or *"ultrastrahlung",* (*i.e.,* radiation from high above or beyond our atmosphere). All later investigations have supported this interpretation of the phenomenon, which may therefore be considered as the discovery of cosmic rays.

The first World War (1914-1918) interfered with the continuation of the researches on the newly found radiation and it was only in 1922 that further investigations were restarted by Kolhörster and his co-workers in Germany and Millikan and his associates in America. In 1923-1924 Kolhörster worked at the mountain station, Jungfraujoch, in Switzerland, in order to obtain a reasonable intensity of ionization, and found that the intensity of the incoming radiation was the same during day and night, even during a solar eclipse. This was disproved later. But Kolhörster concluded from his observations that the *origin of the radiation could not be the sun, nor even the stars, but should be somewhere for beyond our galaxy.*

Millikan and his associates, in 1925-1926, studying the absorption of these rays in the snow-fed (and hence non-radioactive) water of high mountain lakes and finding that the absorption in water agreed with that in air confirmed the opinion that the origin of the rays was outside the atmosphere. Further, from the data obtained from these absorption experiments, Millikan made bold to affirm that

the *main part of these penetrating rays might consist of energetic photons of* a *complex nature,* (*i.e. electromagnetic radiation),* emitted in the process of the building of helium nuclei from hydrogen in interstellar space.

However, as the evidence has become overwhelming that most of these rays are electrically charged and of a higher order of energy than was supposed, it has become necessary to abandon Millikan's first suggestion. As regards the real existence of highly penetrating radiation coming from outer space, however, there is no doubt, and the name cosmic rays, first employed by Millikan to designate them, has come to be universally used. Thus the discovery of cosmic rays was finally established.

RESEARCHES ON COSMIC RAYS

From the time of their definite discovery, extensive investigations have been carried out on these radiations, chiefly with a view to know more about their nature, composition and properties. The existing atomic tools have been modified and adapted to suit the needs of the study of cosmic rays which are characterized by the smallness of the effects to be observed. The experimental data gathered are vast and often overwhelming. We shall limit ourselves here to a brief survey of the main aspects of these studies in order to bring out the salient features of the cosmic rays.

APPARATUS USED IN COSMIC RAY RESEARCHES

A. The cosmic ray ionization chamber
For the study of cosmic rays, the ionization chamber has been improved chiefly in two directions in order to measure as accurately as possible the small ionization current produced by them. The Neher's electroscope and the Compton's Carnegie model "C" cosmic ray meter may be cited as typical examples of these modifications.

The **Neher's electroscope** is a very *rugged, light-weight, tilt* and *vibration-free instrument,* devised by Neher (1936) and well suited to the study of the altitude effect. It is represented diagrammatically in Fig. 305.

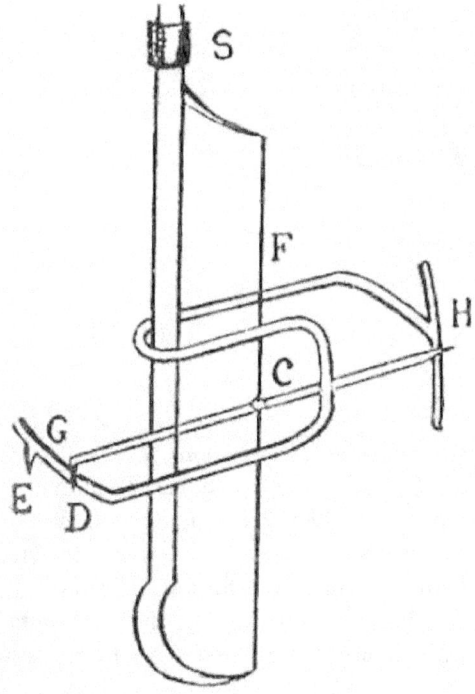

Fig. 305. Neher's electroscope.

It is made entirely of fused quartz, consisting of a movable cross arm, C, 30μ thick, fixed to a 5μ stretched fiber F. The cross-arm is bent at right angles at one end D where it is drawn out to 10μ thickness, and is well balanced. The image of D, magnified ten times by means of a lens, is cast on a recording motion picture film. The fiber E serves as a reference mark and the parts G and H combine to give a linear scale over practically the whole range of discharge. A permanent twist of about 30° is given to the torsion fiber F so that no motion takes place except at a starting voltage of 250 volts. The electroscope is charged periodically to a constant potential by a charged condenser, contact being made through a platinum sleeve S, below which everything is sputtered with gold so as to be conducting. The slope of the discharge curve recorded on the moving film is a measure of the intensity of ionization produced in the chamber. The whole apparatus, *viz.,* the ionization chamber, the electroscope, the barometric and thermal recorders, weighs only four pounds about. The electroscope part is not only free from tilt but also insensitive to vibration on account of the large ratio strength to weight.

Compton's model "C" cosmic ray meter which has been employed by Compton and Turner in their extensive researches on the latitude effect (1937) is schematically represented in Fig. 306.

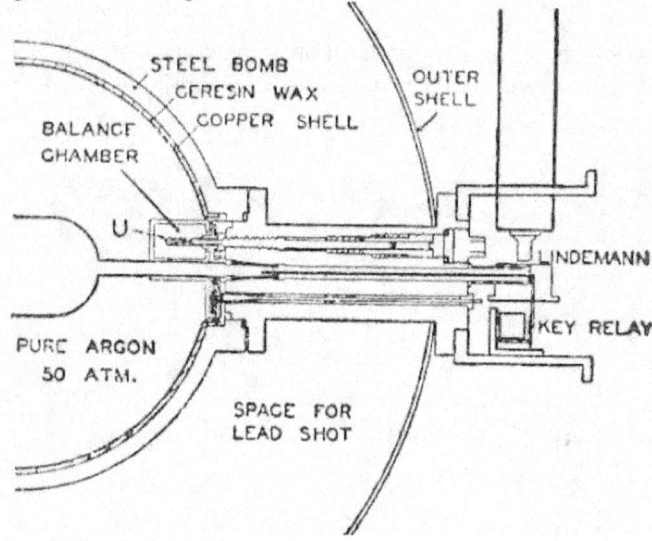

Fig, 306. Model "C" cosmic ray meter.

The *ionization, chamber is very large, made up of a spherical steel bomb of capacity 20 litres filled with pure argon at a high pressure of 50 atmospheres.* The use of argon at high pressure provides an ionization nearly 70 times that which is produced in air at atmospheric pressure. The cathode of the chamber is a copper sheet lining inside the steel bomb, well insulated from the latter with wax, while the anode is a fork at the center of the sphere. The chamber is shielded from local radiations by surrounding the steel bomb with an outer shell filled with lead shots to an equivalent thickness of about 11 cms. solid lead.

A *method of balancing* is used in the measurement of cosmic ray intensity by the following device. The current to compensate the ionization current due to cosmic rays is supplied by the ionization produced within a small auxiliary chamber (*balance chamber*) by the β-rays from a uranium source U. The compensating current is adjusted to approximate equality with the mean cosmic ray ionization current, so that small changes in cosmic ray intensity rather than the total intensity is recorded. A great advantage of this method is that it automatically compensates for pressure or temperature changes of the gas.

A *Lindemann electrometer* connected to the central electrode records the changes in the net compensated intensity, the displacement of the shadow of the electrometer needle being projected through a compound microscope on to a continuously moving strip of photographic film. The readings of a barometer and a thermometer are also recorded on the same film.

In operation, the central electrode is grounded for a period of 3 mins at the beginning of every hour by a clockwork mechanism; it is then left insulated for 57 mins to gather the net charge in the system and communicate it to the electrometer needle. At the end of every four-hour period an automatic sensitivity calibration is performed by the application of a known voltage to the central electrode.

B. The coincidence counters

The Geiger-Muller counter with its associated electronic circuit is a very efficient instrument for detecting the passage of single ionizing particles, as we have already seen, and should therefore be a very useful tool in the investigation of cosmic rays, where one has to deal with relatively small numbers of ionizing particles and in many cases even with single particles. But one great drawback of this apparatus is that it is equally sensitive, to all types of radiations and in consequence gives rise to an ever present *spurious background effect* due to radioactive contamination of the surroundings, which masks to a large measure the small effects due to cosmic rays. This limitation has been removed to a great extent by a simple and ingenious device known as the *coincidence counters,* first conceived and used by Bothe and Kolhörster, in 1928, and then improved by Rossi and others so that the G-M counter has now become a very powerful weapon in researches on cosmic rays.

The device consists in arranging two or more counters in such a way that a record is made only when all the counters discharge almost simultaneously. This type of discharge is known as *coincidence,* which may be due either to a single particle passing through all the counters at the same time practically, or to two or more separate particles coming by chance one to each counter at the same instant. The former is considered *true coincidence* while the latter *chance coincidence.* In cosmic ray work, the two kinds of coincidences can be differentiated, since the number of chance coincidences can be readily estimated from the separate counting rate of each counter.

Bothe and Kolhörster, using two counters arranged one above the other and recording on a moving photographic film running at 1 cm./sec. the counter discharges which were considered as coincidences if within 0.01 see of each other, obtained many more coincidences than could be explained by chance occurrences and showed that the coincidences left over after those due to chance had been deducted, were due to particles which passed through both the counters.

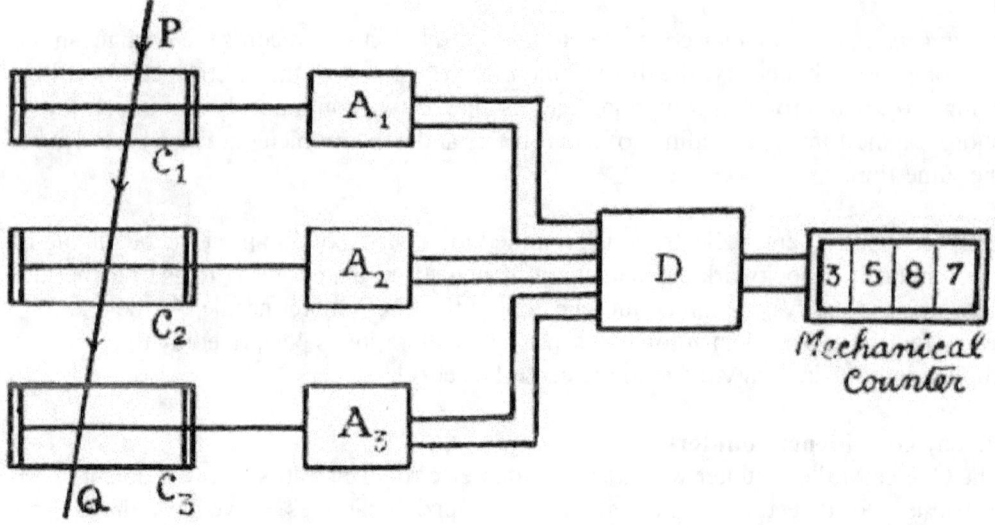

Fig.307. Coincidence counter set.(Rossi)

Rossi, in 1930, improved the coincidence counter technique in the following two ways:

(i) *Use of three counters in a row* and *recording only when all three counters respond simultaneously.* In this way the possibility of chance coincidences, which can hardly traverse the three counters, is much more effectively eliminated than in the Bothe-Kolhörster two-counter arrangement, while the genuine coincidences due to penetrating particles passing through all the three counters remain unchanged.

(*ii*) *Use of an electromechanical device, instead of photographic film, for recording the discharges of the counters.* This is accomplished by means of modern *electronic selecting circuit,* schematically shown in Figs. 307 and 308.

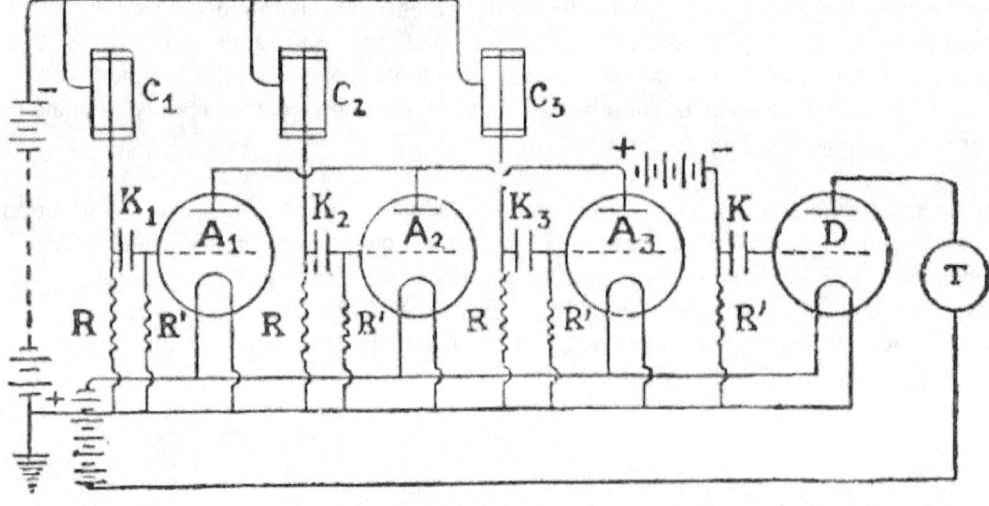

Fig. 308. Coincidence circuit for G-M counters.

Three counters C_1, C_2 and C_3, arranged one below the other in a line, are connected to the grids of three triodes A_1, A_2 and A_3 in such a way that the discharge of a counter momentarily lowers the potential of the grid of the corresponding triode and stops the flow of thermionic current in it. The plates of these three valves are connected among themselves and communicate with the grid of a detector D through the intermediary of a grid bias and a resistance. These elements of the circuit of the plates are so regulated that the grid potential of the detector is sufficiently negative so that no current passes in its anode circuit.

Now, if an energetic cosmic ray, such as PQ, traverses all the three counters almost at the same time, a discharge is produced simultaneously in all of them and the flow of current in all the three corresponding valves is completely cut off. This causes a. momentary decent rise in the grid potential of the detector and a sufficiently strong pulse of current flows in its plate circuit actuating a telephone which produces an audible click or a mechanical counter which records the coincidence. If, on the other hand, only one or two of the counters are affected, the currents in the corresponding valves are cut, but as nothing happens in the third valve, the grid potential of the detector valve is not sufficiently altered to permit a pulse of current in its output circuit.

Such an arrangement of valves acts as a *selector,* which chooses, among the various discharges in the counters, only those that occur simultaneously in all the counters, thus eliminating the non-coincident effects in the counters. The circuit is quite *symmetric* with respect to the counters as seen in Fig. 308 and allows the use of any number of them. It is to be remarked that the selector circuit has the following tasks to accomplish, *viz.,*

(a) to make the amplitude of the final pulses to be as constant as possible and their period as short as possible in order to reduce the number of chance coincidences, without however missing true coincidence and

(b) to modify the period of the final coincidence pulses to a value required to function mechanical counter.

This kind of grouping of counters is sometimes called a **cosmic ray counter telescope,** for it can be pointed in any given direction and made to record the passage of the cosmic rays in that direction. Bernardini and Johnson, using such an arrangement of counters, were able to establish an outstanding characteristic of cosmic rays, *viz.,* their incidence on the earth with uniform intensity from all directions.

Other combinations of counter are also used, depending on the nature of the researches to be made. Thus, for instance, there is the arrangement known as the *shower-counting* array, Fig. 309 (d), first used by Rossi in the study of cosmic ray showers. With three counters C_1, C_2 and C_3 arranged not in a row but as shown in the figure, chance coincidences will still be rare but a single energetic particle cannot pass through all the three. At least two separate particles are required to function this arrangement and the more particles there are in the simultaneous group, the more surely at least one will pass through each counter. Particles incident in this way in groups are assumed to be of common origin and form a *shower*.

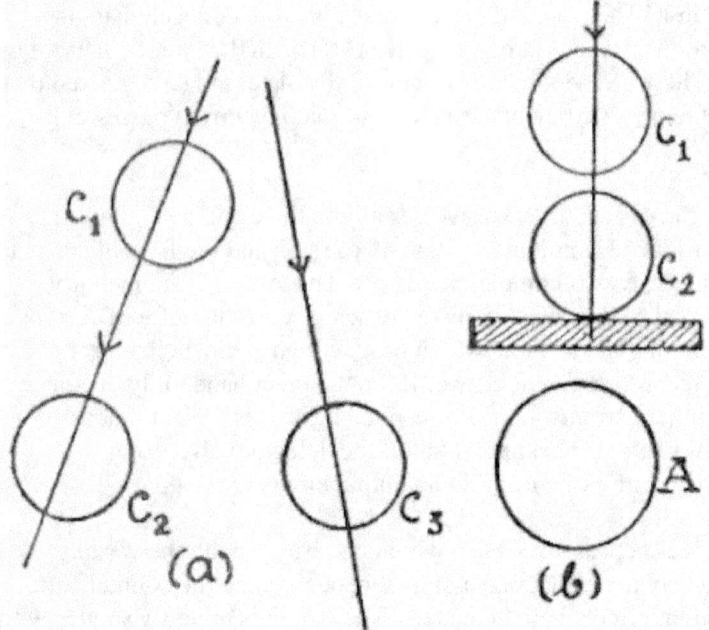

Fig. 309. Some important arrangements of coincidence counters.

Fig. 309 {*b*) represents another set-up, called *counters in anticoincidence*, which has proved a very powerful means of investigating the nature of cosmic rays. A count is recorded only when C_1 and C_2 operate, while A the anti-coincidence counter does not. A is ordinarily a group of counters so large that a particle passing through C_1 and C_2 must necessarily pass through A, unless something happens to it in between, for example, if it stops in an absorber before reaching A, as shown in the figure.

C. The scintillation counters

The scintillation counter technique, already described, has lately been employed in cosmic ray researches with greater efficiency and better results than the older G-M counter equipment. The new type of counter has decidedly two advantages over the older one:

(*i*) *Reduction of background effect:* the gas-filled G-M counter often shows larger pulses (caused by the background particles) than are due to the particles under investigation and the unwanted particles are more numerous in the ratio of, perhaps, 100: 1; this cannot happen with a scintillation counter.

(*ii*) *Comparatively short resolution time:* the G-M counter has a comparatively long resolution time so that the maximum permissible counting rate is relatively low, the 'useful counts' to be recorded thus being but a few per minute; but with the scintillation counter, data which took several months to obtain can now be recorded in a matter of days. It is to be noted that the 'telescope' arrangement, 'coincidence' and 'anti-coincidence' techniques can be readily employed with the scintillation counters also, as in the case of the G-M counters.

D. The counter controlled cloud chamber

Skobelzyn was the first, in 1927, to observe accidentally tracks of cosmic ray particles in the cloud chamber, but the magnetic field was not strong enough to deflect them.

Anderson, in 1932, placed a cloud chamber in a very strong magnetic field, investigated the tracks of cosmic ray particles and discovered the positron.

In the same year, Blackett and Occhialini made a very substantial contribution to this experimental technique in the study of cosmic rays by introducing a most ingenious and economic device of operating the cloud chamber with the coincident discharges of Geiger-Muller counters. When a cloud chamber is worked at *random*, due to the inefficiency of the apparatus during the relatively long "resetting" time on the one hand and the comparatively rare cosmic ray phenomena on the other, only a low percentage (about 10 to 20%) of the photographs taken contain cosmic ray tracks, which means uneconomical use of time and film. To overcome this difficulty, the above mentioned authors conceived and successfully put into operation for the first time the counter-controlled cloud chamber which is now generally employed.

The essential details of the apparatus are shown in Fig. 310. A *vertical* expansion cloud chamber E, well stated for cosmic ray work, of large dimensions, with a glass plate window G on the front side through which the tracks are photographed, is used. At the back, a piston P made of a light metal, such as duraluminum, is supported to the sides of the chamber by a rubber diaphragm RR which enables the piston to move easily and rapidly during expansion. It also makes the observation chamber gas-tight so that the latter may be filled with any desired gas at a pressure required by the expansion ratio. The two adjustable screws SS serve to control the total movement of the piston and hence the expansion ratio. 'The space behind P, closed by means of light, smooth and quickly operating valve, V, is filled with compressed air so that the pressure on both sides of the piston is the same. When this valve is released, the compressed air rushes out into the chamber A which is in communication with the outside or may be even evacuated if required, and the piston falls quickly completing its stroke within 0.005 sec. and causing a sudden expansion in the cloud chamber.

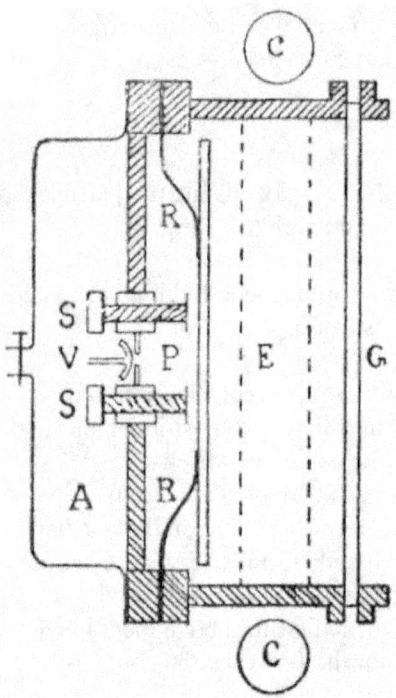

Fig..310. Counter-controlled cloud chamber.

The *modification introduced by Blackett consists chiefly in the automatic release of the value V by the simultaneous discharge of G-M counters CC placed above and below the expansion chamber and connected so as to record coincidences.* With this arrangement an ionizing particle passing through the counters must necessarily traverse also the expansion chamber between them. The coincidence discharge due to the passage of a particle through the counters, considerably amplified by means of thermionic tubes, is made to operate a system of relays controlled by a small electromagnet which triggers the valve V, sets the piston P into motion and thereby causes the adiabatic expansion in the chamber. The same amplified coincidence pulse is made to operate another mechanism which illuminates the chamber, exposes the film after a proper delay, resets the piston, winds the film and thus prepares the chamber for the next operation automatically. The tracks of the passage of particles are photographed by this device immediately after their arrival in the chamber, within about 0.01 sec. The counter-controlled cloud chamber thus makes *the particles take their own photographs.*

The great advantages of this method are evidently *economy* and *quick results.* Over 75% of photos taken contain cosmic ray tracks. Further, by a suitable combination of the control counters, it is possible to photograph tracks of cosmic rays arriving in particular directions. This method has been fruitful of very important results concerning the energy of the cosmic ray particles, the different components of the cosmic rays, the effects produced by them, such as the production of showers, the ionization of the atoms traversed, nuclear collisions with emission of radiation, nuclear disintegrations caused, by the very high energy cosmic ray particles, etc.

216

Two designs of *vertical* expansion chambers are used in this method, one *cylindrical* about 10 to 20 cms in diameter and 5 to 10 cms in width and the other *rectangular* and much bigger, about 25 to 60 cms in height and 10 to 20 cms in breadth, which is best suited for the study of very high energy particles. The photo of one such rectangular chamber, designed and built in the laboratory of Maurice de Broglie under the direction of Leprince Ringuet, is reproduced here.

Cloud chamber for cosmic ray work. (Leprince Ringuet)

It is 55 cms high and 15 cms broad to suit the pole-pieces of the huge electromagnet, 75 cms in diameter, capable of producing 50,000 gauss, installed at Bellevue, Paris, by the French Academy of Sciences. With this chamber cosmic ray tracks 40 cms long can be photographed. The two circular pole-pieces of the electromagnet are seen in the background of the picture.

E. The photographic emulsion method
This has proved itself to be a very important tool in cosmic ray researches, chiefly in the analysis of *mesons* and of *explosive disintegrations* of nuclei under bombardment of extremely high energy particles, as already stated. In cosmic ray work, special thick emulsions, stacked up in a number of layers, are used in order to obtain greater lengths of the tracks of particles passing through. Recently (1957) Dr. E.P. Ney, of the University of Minnesota, collaborating with the Bristol University, has been able to record on a "stack" of photographic emulsion an extremely high energy particle, a helium nucleus moving at

217

nearly the speed of light and possessing an energy of 10^{15} eV, whereas the man-made powerful accelerators (bevatron) can produce only 10 BeV (10^{10} eV) about.

F. The bubble chamber method

D. A. Glaser, at the Brookhaven National Laboratory, conceived and constructed, in 1952, the bubble chamber, in an attempt to eliminate the limitations and combine the respective merits of the cloud chamber and the emulsion techniques, by the use of a suitable liquid which would be dense enough to provide the frequent collisions and precise tracks of the emulsion and flexible enough to be readily influenced by magnetic fields and to give the decent-length-single-track picture of the cloud chamber, with a quick recovery after each exposure.

Principle

It is well known that in a clean, smooth-walled vessel a very pure liquid can be heated above its ordinary boiling point without boiling. When the superheated liquid does begin to boil, it erupts with great violence. Often, bits of broken glass are introduced in the liquid, which provide triggering points for boiling and thus prevent superheating. Glaser conceived the idea that if the supersaturated vapor of a cloud chamber was replaced by a superheated liquid, fast particles passing through the liquid might play the role of bits of broken glass and trigger the formation of the microscopic bubbles that start the boiling process. Further, if the particle was charged, as it passed through the liquid, it would produce small clusters of ions of the same sign (negative or positive). Pushed apart by their mutual repulsion, the ions might form small bubbles all along the path of the particle in the liquid and thus produce a visible track. For such a thing to take place actually, the liquid must possess the following properties:

- It should be non-conducting, so that the ions would retain their charges;
- It should have low surface tension, so that the force tending to collapse a bubble would be weak and
- It should have high vapor pressure, which would tend to enlarge the bubble formed in the liquid.

Putting to practical test these ideas, Glaser took ether, superheated to 285°F, in a small glass bulb; he lowered the pressure in the bulb quickly by a mechanical piston device, irradiated the bulb with radioactive cobalt and took high speed movie photographs of the happenings in the liquid. To his great satisfaction, the pictures disclosed tracks of tiny bubbles. The experiment also demonstrated that the larger the bulb containing the liquid, the shorter the time it can maintain its superheated condition, because of the greater likelihood of intercepting some stray radiation. The bulb remained quiet only for a few seconds after the pressure was reduced. Thus the bubble chamber, as a particle-trapping device, came into existence.

It was next shown that the *bubble chamber was a very sensitive recorder*. Even fast .μ-mesons, which ionize very lightly, made visible tracks in the superheated liquid. To obtain this result, the chamber was placed under 4 inches of lead and between two G-M counters arranged in 'coincidence circuit' so that, whenever a particle passed through both the counters and hence through the chamber also, the firing of the counters set off a flash illuminating the chamber for a camera whose shutter was left open in a dark room. By delaying the light flash for a given time after the counters fired, the bubbles along the track of the particle could be photographed at any stage of their growth.

Some bubble chambers in actual use

The chamber built by Glaser and his associates and used with the cosmotron at the Brookhaven National Laboratory is rectangular in shape, three inches by two in dimensions, and made of duralumin with glass windows. A diaphragm actuated by compressed air fully expands the chamber in five thousandths of a second, and the desired drop in pressure and superheated condition is maintained for about seven thousandths of a second. The liquid used is pentane. The workers haven been able to obtain hundreds of excellent pictures of tracks of particles from the accelerator. The bubbles grow so fast that the tracks can be photographed before swirling motions in the liquid distort them. The track photographs are as good as the best cloud chamber records and are about 10 times as accurate.

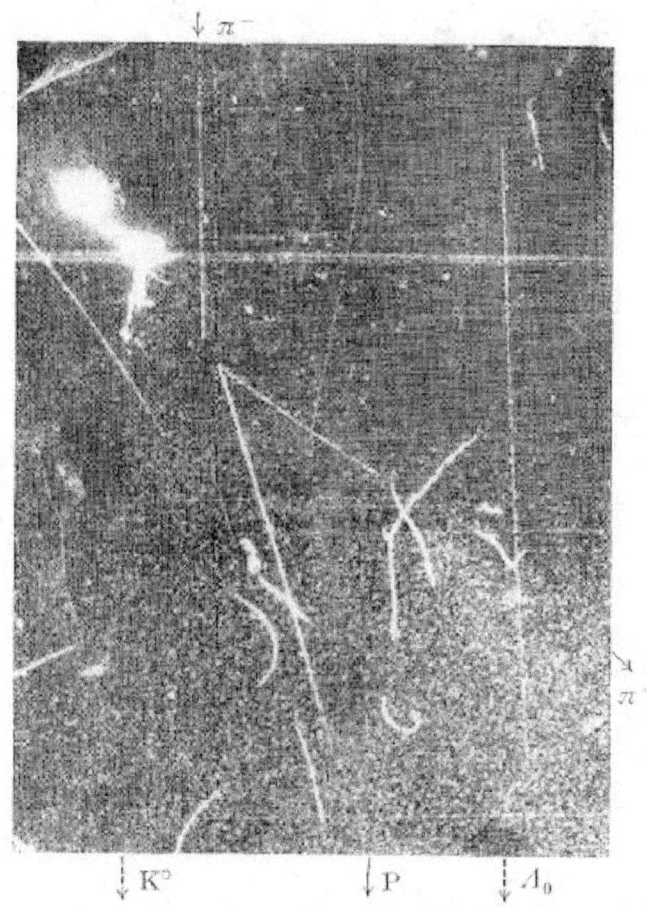

A neutral hyperon A_0 is produced in a collision between a 1.5 GeV from the Brookhaven Cosmotron and a

A proton in 20 atm of hydrogen in a diffusion cloud chamber The A_0 particle is produced at the point where the π^- ceases to ionize, its electric charge being neutralized by that of a proton. This Ao decays into a proton (p) and a π^- after living for 4×10^{-11} seconds. If only one single neutral particle is assumed to conserve momentum and energy of incident π^-, its mass would have to be 1300 electron masses and its direction of flight as indicated by (K^0). The picture was obtained by the courtesy of Dr. R. P. Shutt and earlier published. Fowler, Shutt, Thorndike, and Phys. Rev. 91 (1953) 1287).

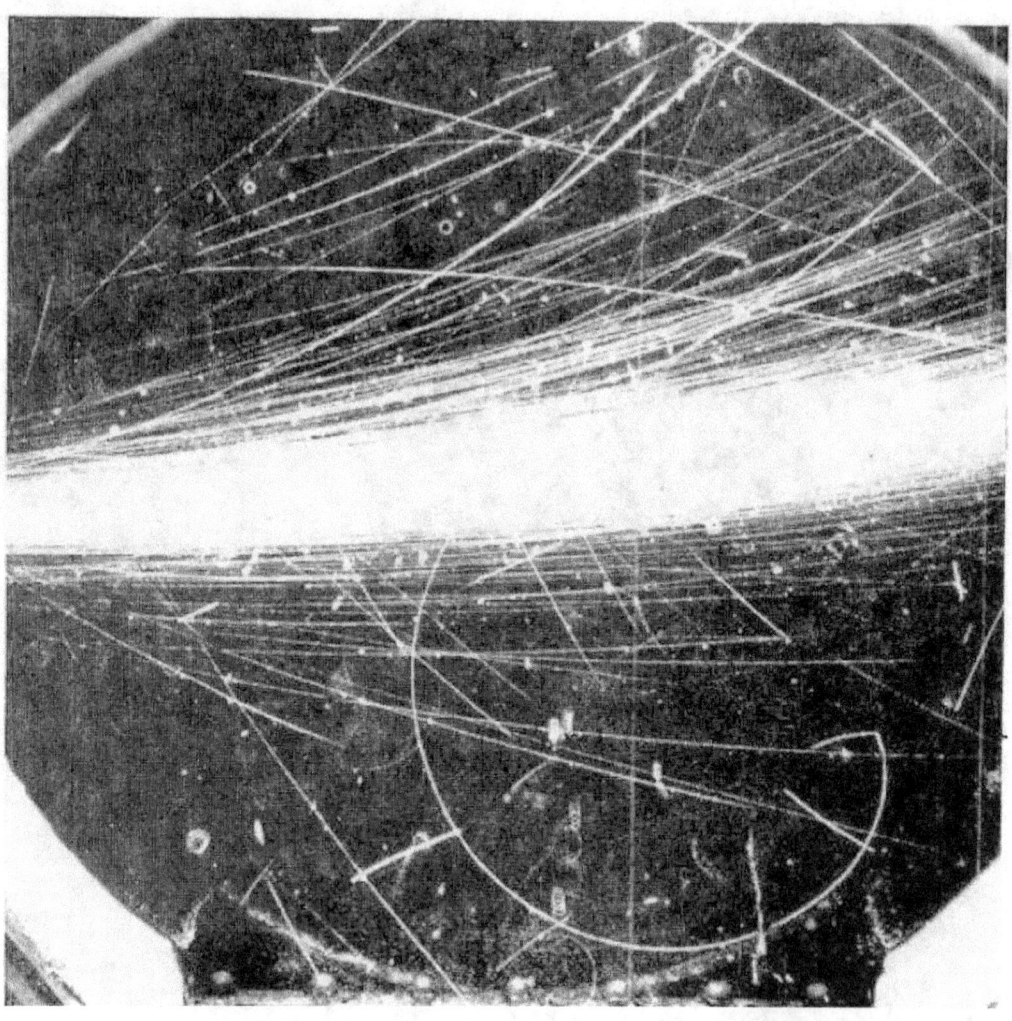

About 20 000 protons of 380 MeV energy enter the Liverpool diffusion cloud chamber. The G-shaped track is produced by a π^+-meson, which decays at the point into a μ^+-

220

meson. Field strength 11 200 oersted. The picture by the courtesy of Dr. Margaret Alston, and earlier published. Alston, Evans, von Gierke (Proc. Phys. Soc. A 69 (1956) 691).

R. H. Hildebrand and his colleagues, at the University of Chicago, have successfully operated a pentane filled, bubble-shaped chamber, using a beam of π-mesons from a synchro-cyclotron. Their chamber can be compressed and expanded once a second; it is capable of collecting data rapidly on the interaction of π-mesons with the nuclei of hydrogen and carbon atoms in the pentane. One of the photos taken by them is reproduced here, where a π-meson enters the chamber at the left, undergoes an elastic scattering and than ejects an electron (wavy track) from a pentane molecule. It clearly demonstrates that different types of particles produce recognizably different tracks.

At the University of California, Dr. L. Alvarez and his group have built a bubble chamber, circular in shape and 2.5 inches in diameter. Using liquid hydrogen at about 400° below zero Fahrenheit as the superheated liquid and exposing the chamber to neutrons, they have been able to take photos of protons in the liquid hydrogen recoiling from collisions with the incoming neutrons.

The bubble chamber technique has the following special advantages over the cloud chamber and the emulsion:

(i) *The greater frequency with which interesting events are recorded.* In the Brookhaven Laboratory, it has been shown that the *six-inch* bubble chamber catches as many events as would an ordinary cloud chamber *140 feet* long. For instance, a complete record of π – μ – e decay is obtained in a single picture, which is rarely seen in nuclear emulsions, while it is virtually unknown in cloud chambers, since the path of a μ-meson in the cloud chamber would be more than 3 meters long, unlike a 3 mm. path in the bubble chamber.

What is still more interesting is that the first 22 photos taken in 11 minutes with the bubble chamber contained eight events similar to the one stated above.

(*ii*) *Its adaptability to the different conditions of research.* The chamber may be filled with a light liquid which does not deflect particles much and therefore will permit magnetic field experiments as the cloud chamber, or with a highly dense liquid which will produce a great amount of scattering as the emulsion.

(*iii*) *The tracks are not much distorted,* unlike in the cloud chamber due to convection currents.

(iv) *The density of the bubbles might prove a better index of particle energy than the droplet density in a cloud chamber.* Above all these, the bubble chamber will certainly speed up the rate at which one can gain information about strange particles and still stranger nuclear forces.

It may be noted that, in June 1959, Alvarez has been able to build a very big bubble chamber, 6 ft. long containing 150 gallons of liquid hydrogen. He intends to shoot into that chamber antiprotons from the Berkeley 6 BeV bevatron and photograph their tracks that will reveal the innermost secrets of matter. In particular, he has been able to trace the birth, death and after-effects of an *anti-lambda* particle, the counterpart of lambda particle which is produced when high energy protons hit protons at rest.

In another series of experiments, bombarding the chamber with extremely powerful particles from the bevatron, he has been able to obtain photos of tracks made by μ-mesons knocked out of smashed atoms. A few of these tracks are found to have gaps in them that are quite puzzling at first sight. One of the photos which such gaps is reproduced on the opposite page. Alvarez has offered the following ingenious explanation, which seems to point to a new and better method of obtaining nuclear energy.

Luis W. Alvarez (June 13, 1911 – September 1, 1988)

The *μ-mesons,* liberated from the bombarded atoms and having negative charges, get attached to positive hydrogen nuclei and revolve around them as electrons normally do. Since, however, the μ-mesons are about 200 times heavier than electrons, the radii of their orbits will be about 200 times smaller than those of the electronic orbits; hence they will revolve close to the nucleus. The atoms formed in this *way* are known as *"mesic" atoms,* which are somewhat heavier than the ordinary hydrogen atoms, but still extremely small. They can therefore pass through the electron defenses of ordinary atoms and fuse with their nuclei. This is what has happened in the photo under study. Mesons form mesic atoms with the nuclei of heavy hydrogen (deuterium) present in the bubble chamber. Mesons so occupied make no bubbles, which accounts for the gaps in the meson tracks. But when a mesic deuterium atom hits an atom of ordinary hydrogen atom, the nuclei fuse together forming an atom of helium-3 and releasing 5.4 MeV of energy. The meson, ejected from the helium atom thus formed, carries away the energy as velocity. It may eventually stop and decay into an electron. In the photo, the μ-meson enters the chamber at upper right and after traveling a good distance (the long thick track slightly curved downwards), forms a mesic atom with deuterium, which drifts slightly to left forming a gap in the track.

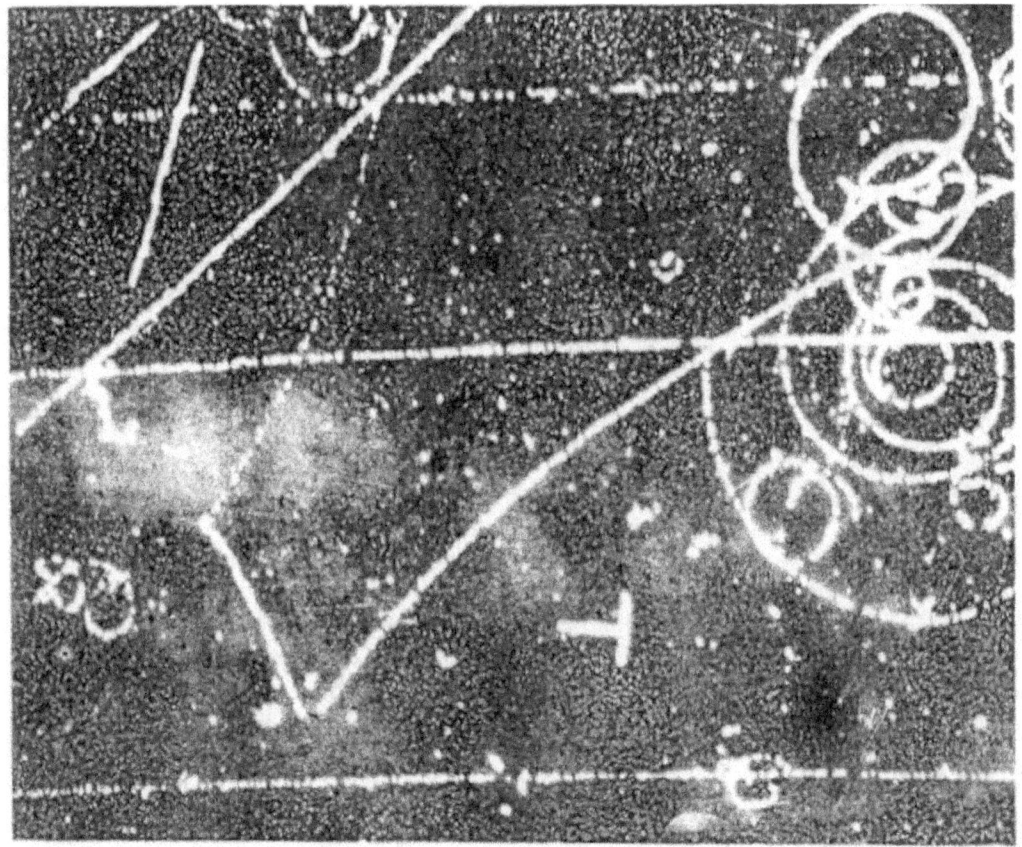

μ-meson acting as catalyst—bubble chamber photo

At the left side of the gap the mesic deuterium atom has fused with an atom of ordinary hydrogen. The meson ejected from the resulting helium (3) atom moves towards the left, stops after a while and decays into an electron, which produces thinner track curving upwards.

The meson that shoots off from the helium (3) atom may also form another mesic deuterium atom and cause it to fuse with another ordinary hydrogen atom. When mesons act in this way, they behave like catalysts, which induce chemical reactions without themselves being changed. If such catalyzed fusion could be made practical, it would have advantages over known methods of releasing nuclear energy:

It would not, require expensive fuel, as uranium fission does, and it would not create harmful radioactive fission products.
It would not require exceedingly high temperature as thermonuclear fusion reactions do.
It might burn peacefully, releasing a considerable amount of nuclear energy.

But the method does not offer as yet a direct means of tapping fusion energy in commercial amounts on account of the following difficulties:

(*i*) there is no dependable source of mesons at present except the giant machines like the Berkeley bevatron;

(*ii*) since the μ- mesons are short-lived, decaying into other particles in two millionths of a second, they have very little lime to act as catalysts.

If a longer-lived particle could be found that does the work of the catalyst, the reaction would look promising indeed. The Russian physicist, A. Alikhanian claims to have evidence that such a particle exists, but his claim has not yet been confirmed by others.

INVESTIGATIONS ON THE NATURE OF COSMIC RAYS

A new era of investigation of the cosmic rays, chiefly of their nature, *was* initiated in 1927-1928 with a remarkable observation known as the *latitude effect,* made by the Dutch physicist, Clay. Measuring the cosmic ray intensity at different latitudes in sea voyages from Holland to Java, he found a decided variation in the intensity, being distinctly less in the equatorial regions than in higher latitudes. This phenomenon was explained by him as due to the presence of charged particles in cosmic rays. For, if these rays consisted only of uncharged photons, as was till then supposed, they would not be affected by the earth's magnetic field and should therefore come as freely to the equator as to the polar regions of the earth. If, on the other hand, cosmic rays were made up of charged particles also, they would be obviously influenced by the earth's magnetic field. The particles which approached the earth at the equator, with their directions of motion, for the most part, *at right angles* to the earth's magnetic lines of force, would be bent round and deflected away from the earth (Fig; 311).

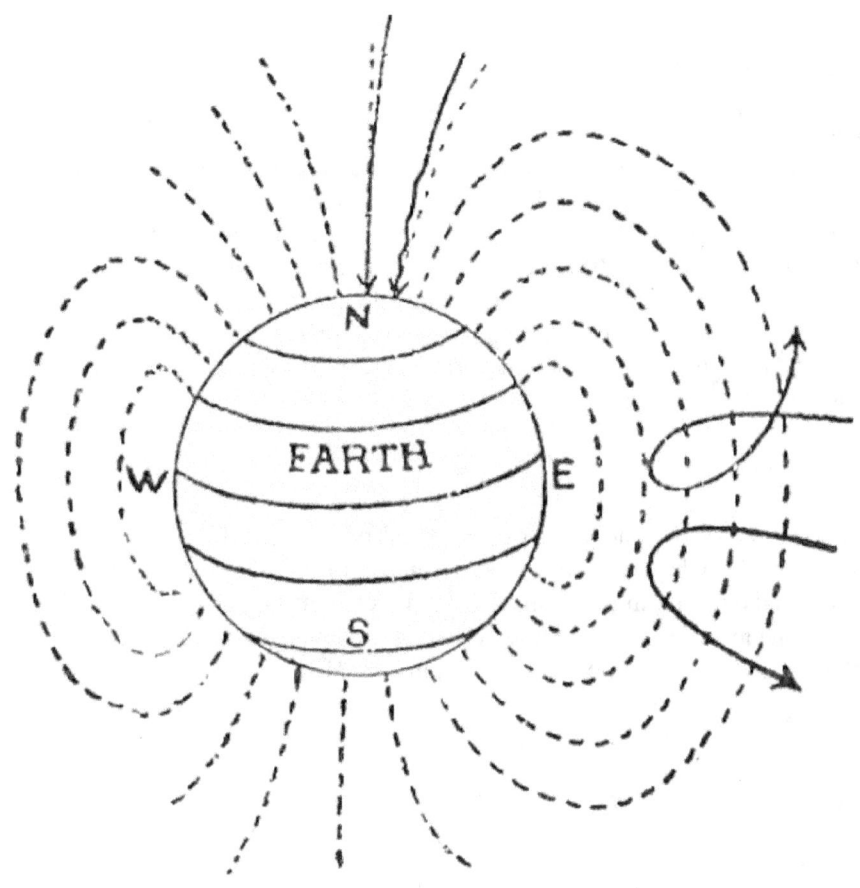

Fig, 311. The latitude effect.

But the particles approaching the earth near the poles and moving mainly *along* the earth's magnetic lines of force would have no difficulty in reaching the surface of the earth. Hence, a decrease in intensity from the pole to the equator would be expected, if the cosmic rays consisted of high-energy charged particles.

Clay's discovery of a latitude effect was, therefore, really momentous and it received a further confirmation, in 1929, by the experiments performed and Bothe-and .Kolhörster with coincidence counters and an interposed absorber, which gave a very strong. indication of the presence of charged particles among the cosmic rays observed within the earth's atmosphere. The*se* corroborative evidences for a charged-particle composition of cosmic rays were the starting point of a whole series of extensive researches, not only on the geographical distribution of cosmic rays, such as the latitude, longitude, azimuth and north-south effects, but also on other supplementary phenomena such as the variation in intensity with season and time, the altitude effect, especially at very great heights, absorption in matter, energy measurements, etc. These important investigations which have enabled, to a great extent, to unravel the complex nature of the cosmic rays, will now be briefly described.

1. THE LATITUDE EFFECT

(a) Experimental study

The variation of intensity of cosmic rays with latitude has been investigated by a great number of workers, such as Clay (1927-1932), Millikan, Neher and Bowen (1933-1936), Compton and Turner (1932-1937), Piccard and Cosyns (1936), Carmichael and Dymond (1938), Gill (1939) and Gross (1939). The effect was studied both at *sea level*, involving many sea voyages to various parts of the world from about 78° N to 45° S geographic latitudes and *at high altitudes* by observations made on airplanes, free sounding balloons and stratosphere balloon flights. Data obtained at sea level are more numerous than those at high altitudes.

Sea level observations

The investigations of Compton and Turner during their twelve expeditions across the Pacific Ocean and of Gill who made fifteen additional voyages later, with the model C cosmic ray meter as the measuring instrument, may by cited as typical examples of the amount of work done in studying the latitude effect at sea level. Some of the combined results from these various observations are shown in Fig. 312.

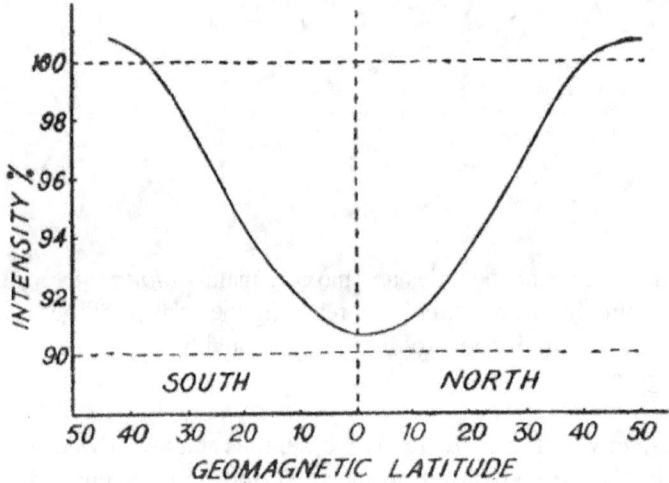

Fig. 312. Variation of cosmic ray intensity with magnetic latitude.

The curve represents the variation of cosmic ray intensity expressed as a percentage of the maximum value observed (ordinates) with the geomagnetic latitudes (abscissae). Quite from the beginning of the study of the effect by Clay, it has been clear that the observed results correlate much better with the geomagnetic latitude than with the geographic latitude. From the figure it is seen that the minimum intensity occurs at the equator, which is about 10% less than that at the terminal latitudes. For latitudes greater than 40° the variation is very small. There is, in effect, a *polar cap* extending from the poles to about 50°, where the intensity is practically constant. This critical latitude, above

226

which variation with latitudes ceases, is known as the "knee" of the intensity-latitude curve, due to a sudden bend in the curve. Different values ranging from 45° to 55° have been obtained by different workers for this "knee".

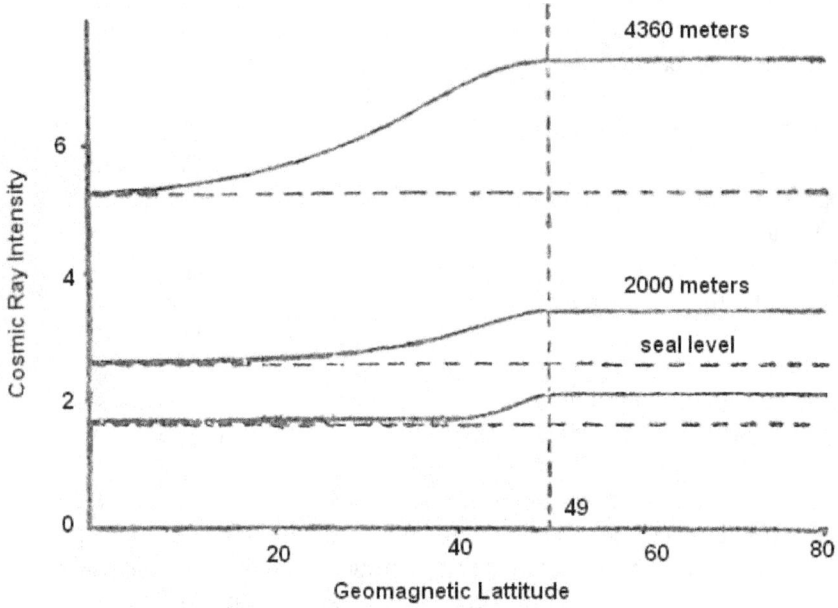

Fig. 313. Latitude effect at different altitudes.

High altitude measurements at different latitudes have been made by Compton and his co-workers up to a height of about 5,000 meters employing unmanned sounding balloons. The results thus obtained are shown in Fig. 313. Each of the three curves represents the variation of cosmic ray intensity with geomagnetic latitude at different altitudes, the lowest one referring to measurements at sea level, while the other two to those at heights 2,000 meters and 4,360 meters respectively. It is readily seen from these curves that at *all* altitudes the intensity remains constant till about 48°, at which point there is a sudden drop and then the intensity decreases up to the equator, but by different amounts according to the altitude, being about 12% at sea level, about 20% at 2,000 meters and about 35% at 4,360 meters.

Carmichael and Dymond, using unmanned sounding balloons from a point very close to the north pole (west coast of Greenland), have been able to show that the cosmic ray intensity there, is, height for height, practically the same as that in the whereabouts of latitude 50°.

Piccard and Cosyns, on the other hand, making investigations in the stratosphere over Europe up to a height of about 30 kilometers with manned balloons in the form of sealed gondolas, found that the knee of the curve was still at 40°, thereby confirming the fact that there exists *a critical latitude above which, the intensity is constant, independent of altitude and hence the same throughout the whole atmosphere* (the polar cap).

(b) Theoretical study

The existence of a latitude effect in cosmic rays receives a straightforward interpretation on the assumption that a considerable part of cosmic rays consists of charged particles which are affected by the earth's magnetic field, as already stated. Several physicists, like. Störmer, Epstein, Fermi and Rossi, Lemaitre and Vallarta, have investigated in detail the effect of the earth's magnetic field upon cosmic rays and have been able to obtain a *quantitative confirmation* of the observed variation of intensity with latitude.

The foundation for these theoretical considerations was provided by Störmer's study of the trajectories of the electrons emitted by the, sun, which produce the phenomena of the aurora in the polar regions. But the following special features involved in cosmic rays had to be taken into account:

(*i*) There are two wholly different types of influences tending to prevent the cosmic rays from arriving at the earth's surface, viz., (a) the *earth's magnetic field,* always exerting a force upon them at right angles to the direction of their motion and (b) the *absorption in the earth's atmosphere.* The former alone will be effective at the top of the atmosphere, while both have to be considered at lower altitudes and the surface of the earth.

(ii) Since the extreme penetrating power of the cosmic rays shows that the particles in them should possess tremendous energies, it may be thought that the relatively weak magnetic field of the earth can have only a negligible influence on them; but as the earth's field extends to thousands of miles, being appreciable even at a distance of 10,000 miles, the total deflection suffered by even very high speed particles may be quite large.

(iii) As the earth's field is not uniform and its direction changes from point to point, the trajectories followed by the particles coming to the earth in all directions are bound to be very complicated helical paths.

Without entering into the difficult mathematical treatment of the problem, we shall see how the important theoretical predictions confirm, even quantitatively, experimental observations.

Considering *particles having a continuous distribution of energies and coming* uniformly *from all directions,* which is presumably the case proper to cosmic rays, *very energetic* ones among them, that exceed a certain limit, will not be affected by the earth's field to any appreciable extent and so will reach the earth at any *latitude* and in *every direction.*

Less energetic particles, i.e., those of energy below the above-mentioned limit, will be so affected by the earth's field that their arrival at the earth or otherwise will depend upon both their directions of approach and their energies.

At the poles, since the particles are moving *along* the lines of force, *all of them, whatever their energy, can reach the earth;* but *at the equator,* as the particles travel *across* the lines of force, *only those with* more *than a certain definite energy can reach the earth,* all the others with lower energy being turned away from it without ever striking it; *at intermediate latitudes,* an intermediate effect will be produced, which is best studied by considering *particles of a given energy coming from all directions.*

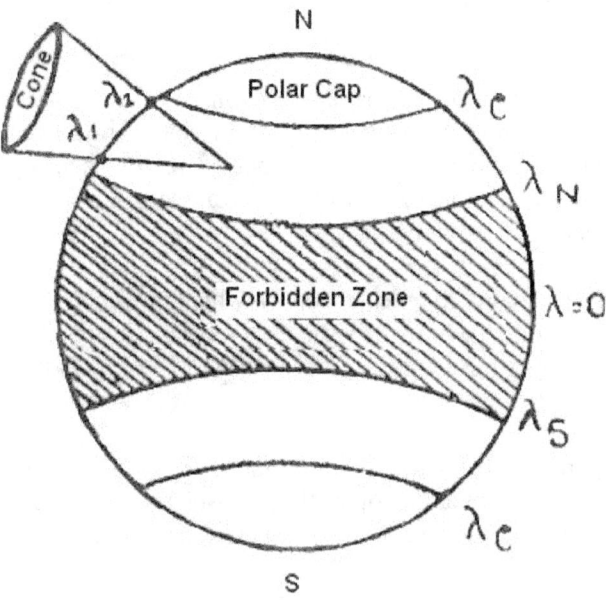

Fig. 314.

According to theory, three regions on the earth can be distinguished as shown in Fig. 314:

(*i*) *a region extending from the pole up to a minimum latitude* λ_C where all of them can came to the surface of the earth from all, directions; this is the *polar cap,* experimentally observed, where the intensity of the cosmic rays remains constant;

(*ii*) an *equatorial forbidden zone,* between λ_N and λ_S, where none of the particles can reach the earth. The extent of this zone will be smaller the greater the energy of the particles, and beyond a certain upper limit the zone will disappear;

(*iii*) a *shadow cone* with complicated boundaries of λ_1 and λ_2 latitudes, in each magnetic hemisphere, where only those particles approaching within a certain range of directions limited by λ_1 and λ_2 can strike the earth. The angular opening of this cone becomes larger the greater the energy of the particles, while it vanishes for a critical energy below which no particles can reach the earth.

The theory further shows that the *minimum energy that a particle should have so as to reach the earth at a latitude* λ is given approximately by

$$E_{min} = 1.92 \times 10^{10} \cos^4 \lambda$$

in electron volts, the exact value being somewhat larger at the equator than at higher latitudes. According to this expression, the minimum energies of the particles that reach

229

the earth at the equator ($\lambda = 0$) is 1.92×10^{10} eV. At the critical latitude $\lambda_C = 49°$, corresponding to the knee of the latitude curve, the minimum energy of the particles that can reach the earth is about 3×10^9 eV. Particles of lower energy cannot reach the earth at any latitude, not even in the polar cap, since otherwise the intensity in the polar cap will not be constant.

From these theoretical results it becomes clear that if cosmic rays consist of charged particles of different energies, a steady decrease in the intensity of the cosmic rays from the pole to the equator might be expected, as particles of lower energy are gradually prevented by the magnetic field from reaching the earth. But the actual manner in which the intensity varies with latitude depends on two other factors:

(i) The *absorption in the atmosphere* (of about 20 miles in depth), which prevents particles of less than about 2×10^9 eV from reaching the surface of the earth, as this amount is roughly the energy required by a particle to cover ionization losses in passing through the atmosphere.

Hence the minimum energy of the particles that begin to be cut out at sea level is about 5×10^9 eV, while that of the particles that are cut at the top of the atmosphere is about 3×10^9 eV. This explains why the latitude effect is more pronounced at high altitudes than at sea level.

(*ii*) *The energy distribution among the cosmic ray particles.* The existence of a well-defined "knee" in the latitude curve at 49° indicates that no charged particles with energy less than about 3×10^9 eV can ever reach the earth. As this value agreed approximately with the energy loss due to absorption in the atmosphere, it was first suggested that the "knee" in the curve merely represented the *absorption effect of the atmosphere.* If this explanation were correct, the knee would be displaced to higher latitudes at high altitudes, where the energy absorbed by the atmosphere is less.

Since, however, it is experimentally found that the value of the knee is independent of altitude, another explanation, first given by Janossy, in 1936, viz., the *sun's magnetic field* prevents particles with energy below 3×10^9 eV from reaching the earth, is now generally accepted.

Thus the study of cosmic ray intensity at different latitudes clearly proves that whatever may be the constituents of primary cosmic rays, high-energy charged particles must form an appreciable part of them. This charged particle composition applies to the primary cosmic rays, since the magnetic deflection of particles within the relatively small extension of the earth's atmosphere will be slight. The fact that at the equator a large number of particles of energy about 2×10^{10} eV reach the earth, combined with the existence of the "knee", implies that the primary radiation is mixed with an energy spectrum in the range of 3×10^9 to 3×10^{10} eV approximately.

As regards the nature of the charged particles, whether positive or negative, protons or electrons, the latitude effect, by itself, gives no indication whatever; nor does it exclude the possibility of the presence of photons or neutrons. These points have to be decided from other experiments. It may be noted that at the energies involved in the charged particles that are able to break through the blocking effect of the earth's field even at the equator, it is not possible to distinguish between an electron and a proton, since both of these at such high energies behave similarly as regards magnetic deflection and specific ionization.

The variation in intensity of cosmic rays with longitude at a fixed geomagnetic latitude and altitude was discovered, in 1934, independently by Clay and by Millikan and Neher. In the different sea voyages undertaken for the study of the latitude effect, crossing the equator at different longitudes they found that the apparent equatorial intensities varied appreciably indicating that at the equatorial belt the drop in intensity differed at different longitudes, being greater in the eastern than in the western hemisphere. The effect amounts roughly to 5% at the equator and diminishes to zero as higher latitudes are reached. These observations have been interpreted as due to *the earth's magnetic field not being symmetrical with respect to on axis passing through the center of the earth, even at great distance.* As regards the nature of the cosmic ray particles, no additional information is obtained from this effect, other than that deduced from the latitude effect.

3. THE AZIMUTH EFFECT—EAST-WEST ASYMMETRY

By the study of this effect it has been possible to decide the sign of the charge carried by the high energy particles in the primary cosmic rays, since if particles of one sign of charge should be more numerous than those of the opposite sign, the earth's field should produce an east-west asymmetry in the cosmic ray intensity. For, the magnetic lines of force *outside* the earth being from south to north, particles approaching the earth, say in a vertical direction, if positively charged, will be pushed, in accordance with Fleming's motor rule, towards the east, so that they will appear to come in a direction inclined to the west of the vertical, *as* shown in Fig. 315, where the reader is supposed to be looking north along the magnetic axis of the earth, the lines of force outside the earth being directed into the page.

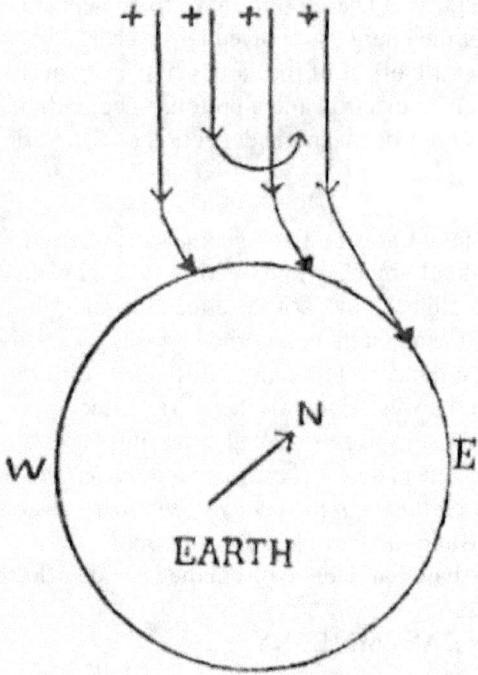

Fig. 315. Deflection of positively charged particles by the earth's field.

According to theory, to reach the equator, a positively charged particle coming from the west, needs at least an energy of 10^{10} eV, while 6×10^{10} eV would be the minimum energy if it were to come from the east. This means that the west is favored at the expense of the east, if the original particles are positively charged. Less energy is needed for these particles to reach higher latitudes. Negatively charged particles coming in vertically, will, for the same reason, strike the earth in a direction inclined to the east of the vertical. Hence, if there are more positive than negative particles, there should be a greater intensity of cosmic rays reaching the earth from the west than from the east; if more negative particles than positive, the reverse.

This is what is known as the east-west asymmetry. If there are, however, equal numbers of similar positive and negative particles, there will evidently be no east-west asymmetry and no azimuth effect. The shadow cone will appear in the west or the east according to whether the particles are positively or negatively charged.

Theoretical considerations further show that:

(*i*) the asymmetry should be a maximum at the equator falling off rapidly at higher latitudes,

(*ii*) the asymmetry should increase continuously with zenith angle and

(*iii*) there is an increase in the asymmetry with increasing altitude.

232

It is to be noted that the greater part of the azimuth effect must be produced before the rays enter the earth's atmosphere, since the atmosphere is so thin relative to the earth's dimensions that any effect produced within it can have very little influence on the large scale distribution of the intensity over the earth. Hence the effect will give important information about the nature of the *primary* radiation.

The azimuth effect, first predicted theoretically by Lemaitre and Vallarta and by Rossi, in 1930, was looked for and finally discovered, in 1933, independently by Johnson of the Bartol Foundation and Alvarez and Compton of the University of Chicago. To the former goes the credit of having investigated the effect in very great detail for a number of years (1933-1940).

The experimental technique consists of two or more coincidence counters arranged on a vertical frame one above the other and disposed in the north-south direction. This effective "cosmic ray telescope" can be tilted to different angles with the vertical, either towards the east or towards the west, and the corresponding coincidence impulses recorded. Thus thin intensities of cosmic rays coming from the east and from the west at different zenith angles can be measured. Observations have been made at various latitudes and altitudes.

The results obtained by Johnson in one of his experiments (1935) at Cerro de Pasco, Peru, situated on the magnetic equator at an altitude of 4,300 meters, where the effect is bound to be appreciable on account of both the equatorial region and high altitude chosen, are given below:

Zenith angle θ	Counts-West (W)	Counts-East (E)	Asymmetry %
15°	52,295	55,049	8·4 ± 0·4
30°	39,601	35,418	12·5 ± 0·5
45°	11,024	9,764	13·9 ± 0·8
60°	6,760	6,028	17·4 ± 1·5

The percentage of asymmetry was taken to be given by the expression:

$2 \times 100 \times (W - E) / (W + E)$

It is seen from the above table that the intensity of cosmic rays coming from the west exceeds that from the cast for all zenith angles. This fact indicates definitely that there is *an excess of positively charged particles in the primary cosmic radiation.*

Although the asymmetry increases with increase of the zenith angle, it has been experimentally found that it is greatest between 45° and 60°, decreasing for greater angles. This decrease, not to be expected from theory, is due to the absorption effect in the atmosphere. Rays coming from near the horizon have to penetrate a greater amount of atmosphere than those coming from above. Hence, for large zenith angles, the low-energy particles are absorbed to a greater extent than the high energy particles, which cause the actual decrease of asymmetry beyond 60°.

Johnson was able to confirm also the other predictions of theory. Thus he proved that the east-west effect which was maximum at the equator decreased to about 2 or 3% at latitudes of about 40°, becoming scarcely perceptible at latitudes greater than 50°, *i.e.,* in the polar cap. The earth's magnetic field acts of course on charged particles inside the polar cap also, but the uniform intensity inside the cap means that the field there causes as many particles to be bent into a given angle of incidence on the earth's surface as out of it.

It is possible in principle to estimate the relative abundance of the positive and negative charged particles in the primary radiation by using the results of the azimuth effect in combination with those of the latitude effect. Johnson has investigated this possibility in great detail, proving that the *positive excess is very considerable.* Thus a large part of the primary radiations consists of positively charged particles. These *cannot be the mesotrons* which form the penetrating component of cosmic rays observed at sea level, for various reasons that will be given later; nor *can they be positive electrons,* since then they would certainly have with them an equivalent number of negative electrons required for the maintenance of electrical equilibrium of the earth. Most *probably they are protons* as suggested by several investigators, a conclusion agreeing closely with the recent finding about the production of mesotrons in the upper atmosphere.

Johnson also discovered first that near sea level at intermediate latitudes the east-west asymmetry increased with altitude, in consonance with theory, but later found that it almost vanished at high altitudes. This absence of the effect at high altitudes has been interpreted to indicate the existence of a *softer* and more easily absorbable component in the radiation at high altitudes consisting of nearly equal number of positive and negative electrons, over and above the hard component of positives which produce the penetrating mesotron radiation. More recently (1945) Arley has used the absence of the effect at high altitudes to give support to the hypothesis of *negative protons.*

4. THE NORTH-SOUTH EFFECT

If the cosmic rays come from beyond our own galaxy, then according to the Lemaitre-Vallarta theory, an asymmetry between the north and south unidirectional intensities is to be expected. The intensity observed in the northern hemisphere should be greater than that observed in the southern hemisphere. The effect is small amounting to about 0.5% only, at high latitudes. Early experiments by Compton, Johnson and others appeared to bear out this theoretical expectation. But a more recent and complete treatment of experimental data, where allowance is made for seasonal and diurnal changes, indicates that there is no appreciable difference between the intensities at equal north and south latitudes. Hence, no argument can be made from this effect which must be negligibly small, if it exists at all, for the origin of cosmic rays outside our own galaxy. Nay more, there are now good reasons to believe that they may originate within our galaxy itself.

5. SEASONAL AND DIURNAL CHANGES IN COSMIC RAY INTENSITY

The existence of a *small seasonal change in cosmic ray intensity* was definitely established, in 1937, by Compton and Turner from their experimental data on the latitude effect. They found that the shape of the intensity-latitude curve depended slightly upon the season in which the data were taken. They were able to resolve the observed intensity-latitude variation into two components, *via.,* the *atmospheric effect* and the *pure magnetic effect* and show that while the latter was free from seasonal changes, the former, particularly apparent at high latitudes, being 1 to 2% of the total intensity, showed seasonal variations.

More recent researches by Gill (1938), by Forbush (1938), by Millikan, Neher and Smith (1939) and by Duperier (1942-1946) have not only confirmed the existence of the seasonal effect, but also established its important special features.

Forbush in America has made a continuous record of cosmic ray intensity over a period of more than 18 months at four different places covering the latitudes from 48° S to 50° N. Duperier in London, working with a triple coincidence counter telescope made up of several large counters at each row, thus presenting a big receiving area, has been able to record continuously the cosmic ray intensity over a much longer period of four years. From these data, the following results concerning seasonal variations have been obtained:

(*i*) The effect has an amplitude ranging from zero at the equator to 1 or 2% of the total intensity at a latitude of 50°;

(*ii*) it is opposite in the two hemispheres;

(*iii*) it is definitely correlated with atmospheric temperature, being a maximum always during the colder part of the years—in January in the northern hemisphere and in July in the Southern.

The cause of the seasonal variation, according to the explanation now generally accepted and first proposed by Blackett, in 1938, is the temperature variation in the mesotron-producing layer of the upper atmosphere. The mesotrons, which constitute the highly penetrating component of cosmic rays and are produced in the upper atmosphere, are radioactive with a very short period of a few microseconds so that some of them decay before reaching the surface of the earth.

Now, during summer, the air-layers in which mesotrons are produced, are shifted to higher altitudes and are less compact than in winter, so that the mesotrons have a larger chance of decaying before they reach the recording apparatus in summer than in winter.

There exists also *a diurnal change, i.e.,* a variation of cosmic ray intensity with *solar time* of the day, as clearly established from the data gathered over long periods of many years by different workers such as Messerschmidt (1932) in Germany, Compton and Turner and Forbush in America and Duperier in England. Experiments were conducted without and with lead screens surrounding the measuring instrument in order to study separately the soft and hard components of cosmic rays. The results obtained are:

(i) There is a small diurnal variation in the *softer* component, the intensity reaching a maximum at about midday and a minimum at about midnight. On the other hand, the intensity of the penetrating component is practically constant.

(ii) This variation of intensity with solar time appears to be independent of both latitude and altitude.

The actual cause of the variation of cosmic ray intensity with solar time is not known. But there are several evidences to show that the sun might be responsible for these changes. Over and above the periodic variation per 24 hours with maximum intensity at noon and minimum at midnight, a 27-day cycle, corresponding to the moon's revolution, and chiefly a marked change in intensity, as much as 10%, during the solar bursts associated with sun-spot activity, which cause major magnetic storms in the earth, clearly indicate the influence of the sun on cosmic ray intensity. But there appears to be no mechanism correlating these phenomena, though it has been suggested that the *magnetic, field of* the *sun* might be responsible for the observed small diurnal changes of intensity in cosmic rays.

6. THE ALTITUDE EFFECT

The study of the variation of cosmic ray intensity with altitude, by means of which the discovery of cosmic rays was first made was undertaken anew, in greater detail, in order to arrive at a proper knowledge of the nature and origin of these rays. The chief aim was to measure very carefully the ionization produced by the cosmic rays at higher and higher altitudes, reaching up practically to the top of the atmosphere. For this purpose, *mountain stations* at altitudes of *2 to 3 miles,* such as Jungfrau in Switzerland, Mount Blane in the Italian Alps, Pie du Midi in the Pyrenees, Bolivian Andes in South America, Mount Evans in the U.S.A., have been used. These permit the use of massive equipments and long periods of continuous observation.

Airplanes up to altitudes of 6 or 7 miles, and *stratosphere manned balloons* (first devised by Piccard) *reaching up to 10 to 11 miles* have also been used. This technique has the advantage of allowing the observer to keep watch on what is happening. Still higher altitudes have been attained by *unmanned sounding balloons* (initiated by Regener) that can *ascend to about 20 miles* which is almost the top of the atmosphere, and record for hours; with very delicate apparatus the cosmic ray ionization in that region. Recently, *rockets* carrying ionization chambers, counters, photographic emulsions and even cloud chambers, have pushed the inquiry to altitudes of *more than 100 miles,* although their flight-lime is very short, three to four minutes. With the introduction of space *rockets* it is now possible to reach out to several thousands of and more.

As regards the investigations directly concerned with the altitude, effect, *the experiments performed during the years 1933-1938 by Neher and Haynes* with sounding balloons at different magnetic latitudes, such as Saskatoon, Canada (60°N), Omaha, Nebraska (51°N), San Antonio, Texas (38°N) and Madras, India (3°N) Are illuminating and hence will be described as typical exam.

Self-recording electroscopes of the Neher type were sent up by means of a number of sounding balloons of small size connected together, similar to those used for weather observations. After reaching a certain high altitude, some of the balloons burst, while the others brought back the recording instruments to the earth without serious damage. With groups of ten balloons, heights of 60,000 ft. corresponding to within less than 2% of the top of the atmosphere were reached. It may be noted that Korff and his collaborators (1938), replacing the Neher electroscope by small counter tubes for measuring cosmic ray intensity and using suitable transmitting valves, which periodically sent records of the actual pressure and counting rates to earth by radio waves, thereby eliminating the uncertainty of recovering the recording apparatus, were able to obtain data to an altitude of 116,000 ft., *i.e.,* within ½ % of the top of the atmosphere.

When the data gathered from these experiments are plotted, *i.e.,* cosmic ray intensity as a function of the altitude, curves of the form shown in Fig. 316 are obtained.

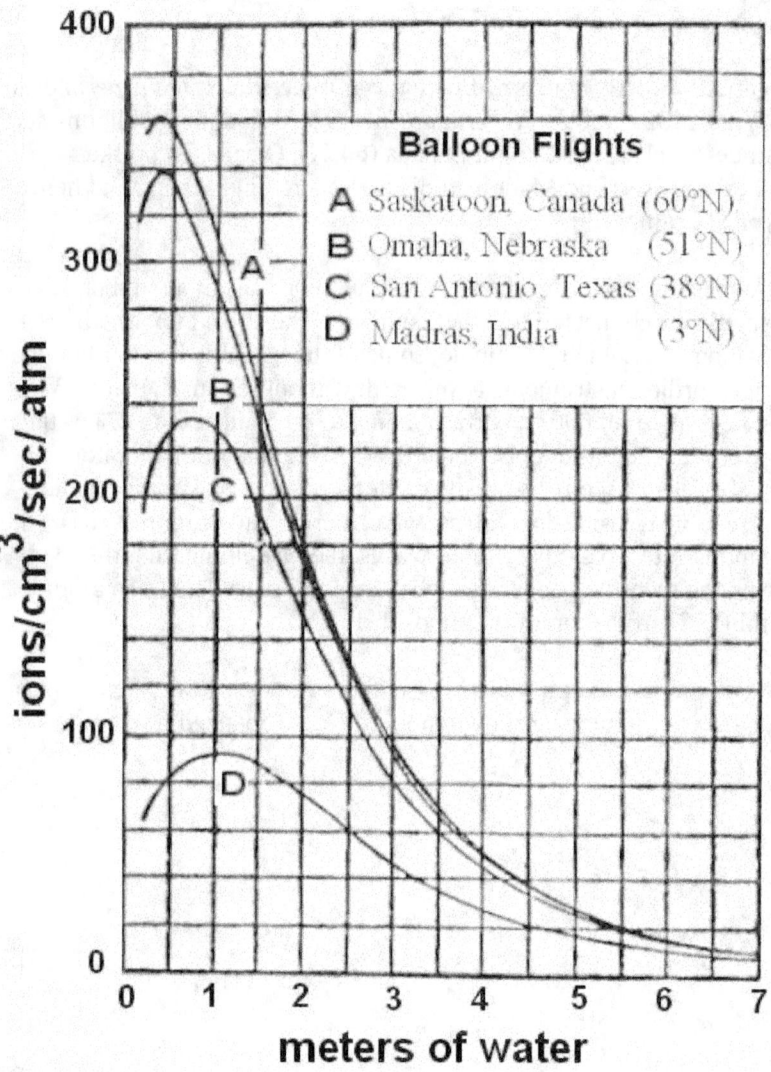

Fig. 316. Altitude curves obtained at four different latitudes.

The *abscissa* represents the altitude, conveniently expressed in terms of equivalent meters of water below the top of the atmosphere. (On the assumption that a layer of air absorbs the same amount of the radiation as a layer of water having the same mass per unit area of surface, the absorption in a given layer of the atmosphere can be expressed in terms of an equivalent layer of water and it can be shown that the entire atmosphere is equivalent to 10 meters of water approximately).

The *ordinate* represents the number of ion-pairs per per second that would be produced by the cosmic rays in. air at standard density; which serves as a measure of their intensity. Each curve represents the results obtained at a different latitude as designated. All of them have the same general form, reaching a maximum, not at the top of the atmosphere,

238

but well below it, at a level of about 0.7 meter of water from the top. From the maximum point the intensity decreases almost exponentially. But the range of intensity variation is quite different for the curves referring to different latitudes, the observed intensity at high altitudes; say at the maximum point, being much greater at higher latitude. Likewise, the altitude at which the maximum ionizations occurs varies slightly for the different curves, being shifted closer to the top of the atmosphere at higher latitudes.

The following important conclusions can be drawn from these special features of the experimental curves:

(i) The *position of the maximum* indicates that the cosmic ray intensity at the top of the atmosphere is less than that occurring within about 5% below it. This might, at first sight, mean that the recording apparatus had actually ascended to the *source* of the rays, which might thus, after all, be *within the atmosphere.* But there is no need to draw such a conclusion, since admitting that the primary cosmic rays originate entirely beyond the atmosphere, as everyone believes to-day, it is still possible to account satisfactorily for the occurrence of the maximum well below the top as follows. The energetic primary rays on entering the atmosphere, react with the attenuated top layers and produce secondaries whose ionizing power is greater than that of the primaries. This process progressively increases the total number of ionizing particles, traveling downward, in spite of absorption, until an equilibrium state is reached represented by the maximum point. Thereafter the ionization decreases, as both primary and secondary rays lose energy in collisions with air molecules in the more dense layers of the atmosphere.

There are also indications to the effect that as the top of the atmosphere is reached within 1/2 % the intensity drops very rapidly to about 1% of the maximum value. This means two things:

(a) *most of the cosmic ray phenomena observed are secondary effects produced near the top of the atmosphere;*

(b) *the primary rays are fewer in number as compared to the secondaries,* being only about 1/5 of the secondaries at the maximum point.

(ii) The greater intensity variation and the slight shift of maximum towards the top of the atmosphere at higher latitudes are evidently due to the action of the earth's magnetic field on the cosmic rays approaching the earth. Millikan and his associates, combining the experimental data of the differences between the curves at the different latitudes with the maximum possible energies of the particles able to reach the earth at those latitudes (as given by the geomagnetic theory of Lemaitre and Vallarta) and the energy spent in ionization, were able to construct a curve representing the energy distribution among the primary cosmic rays entering the atmosphere and show that

(a) *the whole cosmic ray energy comes in as a relatively sharply limited band and*

(b) particles having energies 6 x 10^9 eV are most numerous, although there exist some particles with lower energies and many with considerably higher energies of 17 x 10^9 eV and more.

It has been possible also to estimate the *total number of primary cosmic ray particles* that fall in unit time upon the top of the whole atmosphere from the experimental curves, taking into account the geomagnetic latitude effect. It is found to be 2 x 10^{18} particles per second.

Likewise, *the number of particles striking each sq. cm. of the top of the atmosphere* per minute at the various magnetic latitudes has been estimated to be 1.9 at 3°, 6.5 at 39° and 21.8 at 52°.

The *number* of particles, including both primary and secondary, *that arrive per minute per sq. cm. of the earth's surface,* is still smaller, about 1.5 particles per sq. cm.per minute at sea level, as indicated by cloud chamber experiments.

Hence many *of the primaries and a large number of secondaries formed high in the atmosphere are unable to penetrate the entire atmosphere.*

(iii) Analyzing more closely the *almost exponential decrease of intensity* from the maximum downward,

Millikan was forced to recognize the complicated nature of absorption of cosmic rays and thereby reject his original interpretation of attributing an electromagnetic nature and admit instead a *charged particle nature* for these rays.

He was further able to distinguish *two components* in them, one very rapidly absorbed, mostly effective in the upper part of the atmosphere down to a depth of a third of the atmosphere (3.3 meters of water) and the other a more penetrating component effective even in the *lower* part of the atmosphere.

The total radiation at sea level is made up of about ¼ soft and 3/4 hard components and has an intensity of a little more than one *particle per sq. cm. per minute.*

7. THE ABSORPTION OF COSMIC RAYS IN MATTER

The study of the absorption of cosmic rays in air, water, rocks, of the earth's crest, in lead and other metals has led to very important results concerning the complex and powerful nature of these mysterious radiations. Although early altitude-depth experiments of Kolhörster and Milliken established that the cosmic rays were extremely penetrating radiation, no accurate data of their absorption in matter was available, since their extra-terrestrial origin was still not quite evident. When once this point was definitely established by the latitude effect, more reliable conclusions regarding the rate of absorption in the atmosphere could be drawn. But the fact that the intensity- altitude curve depended very much upon the magnetic latitude showed that the distribution of cosmic ray intensity in the atmosphere was not one determined by absorption alone.

Hence it was found necessary to measure directly the rates of absorption in media other than the atmosphere. This was undertaken by several investigators using different methods.

(i) Absorption in water

Millikan and Cameron were the first, in 1926, to make a systematic study of the absorption of cosmic rays in water. They mounted electroscopes in water-tight containers and immersed them meter by meter in mountain lakes, such as Muir Lake (altitude 3,595 meters) down to 20 meters, which was later (1931) extended to 72 meters. From the electroscopic records of cosmic ray intensity at various depths they were able to arrive at the following conclusions:

(a) The effect was still appreciable at the greatest depths they were able to reach,
(b) there was a gradual hardening of the radiation as it passed through mere and more matter and
(c) a considerable portion of the rays reaching the earth could be classified into two groups, with different absorption coefficients, one of them much more penetrating than the other.

Very accurate and extensive observation of cosmic ray intensity under water was made by Regener in 1928. Using a large, strongly built ionization chamber of 30 litres capacity, filled with CO_2 at a pressure of 30 atmospheres, in order to increase the intensity of ionization, lie lowered it in Lake Constance (altitude 395 meters) down to a depth of 280 meters. From the data obtained he was able to plot a curve showing the manner in which the intensity continuously falls off with increase in depth. The observed intensity at the bottom a of the lake was still one per cent that at the surface.

From the observed rate of absorption, he found three different absorption coefficients indicating three different components of the cosmic rays. The less penetrating components were filtered first, leaving mainly the hardest component after great depths of water had been traversed.

More recently (1938) Clay and his associates have extended these underwater studies to a depth of 440 meters. The residual ionization at such great depths of water indicates the presence of very highly penetrating particles among the cosmic rays.

(ii) Absorption in rocks and coal mines.

V. C. Wilson, in 1938, employing a four-tube coincidence counter cosmic ray telescope, made measurements down to a depth of about 384 meters in a mine. Measurable radiation penetrated this nearly one quarter of a mile of rock, equivalent to about 1,100 meters of water or 100 meters of lead. The intensity was of the order 1/10,000 that incident upon the rock. By tilting the apparatus so that it recorded particles that came obliquely passing through a greater thickness of the rock, he could still detect the radiation at a depth equivalent to about 1,400 meters of water.

Gernert (1938) was able to detect radiation at a depth of 610 meters in a coal mine, equivalent, to 1,600 meters of water. Some of the cosmic rays are thus extremely penetrating although only 1/20,000 of the radiation incident on the earth actually penetrates to these depths.

(iii) Absorption in lead and other heavy metals

Bothe and Kolhörster were the first, in 1929, to attempt a direct measurement of the absorption of cosmic ray particles in metallic screens using two coincidence counters with an interposed block of gold about 4 cms thick. They were able to prove the existence of ionizing particles of mass absorption coefficient of 3×10^{-3} cms^3 gm^{-1}, which was of the same order of magnitude as the absorption coefficient of cosmic rays in air at sea level, deduced, from Millikan and Cameron's experimental data.

Rossi, in 1932, using three counters in coincidence arranged in a vertical line studied the absorption in lead of cosmic ray particles which had already been filtered through 7 cms of lead. He was able to establish the existence of particles capable of traversing one meter of lead placed between the counters. Assuming the absorption to be exponential through each successive layer of lead, the following values were found for the mass absorption coefficient:

in the first 10 cms, $\mu/\rho = 1 \ 8 \times 10^{-3}$;
in the next 15 cms, $\mu/\rho = 0.5 \times 10^{-3}$; and
in the last 76 cms, $\mu/\rho = 0.55 \times 10^{-3}$.

This indicates the existence of a *soft* and a *hard* component in the cosmic rays. Since the absorption coefficient is practically constant for rays that have traversed 15 cms of lead, whatever passes through this thickness of lead may be considered as the *hard* component.

Auger, Leprince Ringuet and Ehrenfest, in 1936, using a cosmic ray counter telescope of the Rossi type studied the absorption of the *soft* component in different metallic screens such as lead, tin, copper and aluminum and found a considerable variation of the mass absorption coefficient from one material to another. The absorption of this soft component appeared to obey a Z^2 law, where Z is the atomic number of the absorber.

Alocco, in 1935, had already shown that the *hard* component obeyed an entirely different absorption law, where the absorption coefficient varies roughly as Z.

Sittkus, in 1938, measuring the rate of absorption of the hard component, obtained by filtering the radiation through 25 cms of lead, in different materials such as lead, iron, aluminum, copper and paraffin, showed that the mass absorption coefficient of these hard cosmic rays remained approximately constant, irrespective of the material used as absorber.

8. MEASUREMENT OF THE ENERGY OF COSMIC RAYS

Study of the latitude, azimuth and altitude effects as well as the absorption in matter gives *indirect* and *average* estimates of the energy of the cosmic ray particles, as we have already seen. Thus, the latitude effect leads to the existence of an energy spectrum in the primary radiation ranging from 3 x 10^9 eV to 3 x 10^{10} eV approximately, while the azimuth effect indicates even higher energies up to 6 x 10^{10} eV for some of the primary particles. The altitude effect referring chiefly to the secondly effects produced near the top of the atmosphere, shows that the majority of the particles have an energy of about 6 x 10^9 eV, while a good number much higher energies of about 2 x 10^{10} eV. Observations made on the absorption in matter, though less reliable, point to the existence of cosmic ray particles of perhaps still greater energy. There arises, therefore, the need of checking up these values by a *direct measurement of the energies of the individual cosmic ray particles*. This has been successfully achieved during the years 1933-1938 by several physicists in different countries, such as Leprince Ringuet and Crussard in France, Blackett and Brode in England, Anderson and Neddermeyer in America and Kunze in Germany, employing the cloud chamber technique of photographing the tracks of cosmic, ray particles, bent by a very strong applied magnetic field.

The **principle** of the method is simple. If the radius of curvature ρ of the track and the intensity H of the magnetic field are measured, it is possible to evaluate the energy of the particle responsible for the track. Ordinarily the following method of calculation is used in the case of very high energy cosmic ray particles, provided it is made certain that one is dealing with electrons and not more massive particles.

The action of time magnetic field causing the fast moving particle to describe a circular are is represented by the usual relation

$mv^3 / \rho = H\ ev$ or $H \rho = mv / e$

Applying relativistic correction,

$H \rho = (m_0 / e) / (1 - \beta^2)^{1/2}$,

where $\beta = v/c$. H is ordinarily measured in gauss and ρ in cms. Since 1 gauss = 3 x 10^{10} e.s.u.,

$H \rho = 3 \times 10^{10} (m_0 / e)\ v / (1 - \beta^2)^{1/2}$ e.s.u.,

Since c = 3 x 10^{10} ,

$H \rho = (m_0 / e)\ v c / (1 - \beta^2)^{1/2}$ e.s.u.,

As *v* is very high, approaching *c*, we may replace *v c* by c^2 as a first approximation. Therefore,

$H \rho = (m_0 / e)\ c^2 / (1 - \beta^2)^{1/2}$ .... (1)

Considering the total energy W of the particle, according to the theory of relativity,

$W = m\ c^2 = m_o\ c^2 / (1 - \beta^2)^{1/2}$ ergs

Since 1 erg = 300 /e   electron volts,

$W = 300 \times m_o\ c^2 / (1 - \beta^2)^{1/2}$ electron volts .... (2)

Comparing equations (1) and (2) we have

$W = 300\ H\rho$

where W is the energy of the particle in electron volts, provided H is expressed in gauss and ρ in cms

This relation simplifies the method of evaluation of the energy of the particle from experimental data. The approximation made (viz., $v\ c = c^2$) is justified for most cosmic rays which are electronic in nature and of very high speed. Taking a concrete example, if H = 14,000. gauss, ρ= 24 meters,

$W = 300.x\ 14 \times 10^3 \times 24 \times 10^2$ eV
$= 3 \times 14 \times 24 \times 10^7 = 1.008 \times 10^{10}$ eV

The momentum ρ of the particle is also used in the analysis of cosmic ray tracks and it can be shown that
$\rho = 300\ H\rho$ (eV) / c
with no approximations being made, unlike in the case of the energy W.

Experimental details
In order to measure the energies of the cosmic ray particles passing through the cloud chamber, the following points must be carefully attended to in practice:

(*i*) A *vertical cloud chamber of large dimensions, counter-controlled* for its operation as the one described earlier is used. Such a chamber is well suited for the study of high energy cosmic ray particles coming in *vertically*. It has the further advantage of making the particles themselves take good photos of their own tracks with great economy and quick results, since the chamber functions only when a cosmic ray passes through the chamber and the counters in coincidence simultaneously and the inertia of the mechanical parts is reduced to a minimum, so that the track is photographed within an extremely short time after expansion, before the ions produced diffuse to any appreciable extent.

(ii) The chamber is placed between the pole-pieces of a *large and powerful electromagnet, giving a horizontal strong field* of the order of 5,000 to 25,000 gauss and uniform over the whole area of the chamber, The adjacent photo was obtained with a vertical cloud chamber, designed by Leprince Ringuet placed between the huge pole-pieces of the powerful electromagnet at Bellevue, Paris. Some of the incident cosmic ray particles are so energetic that even high fields fail to deflect them into an observable curvature.

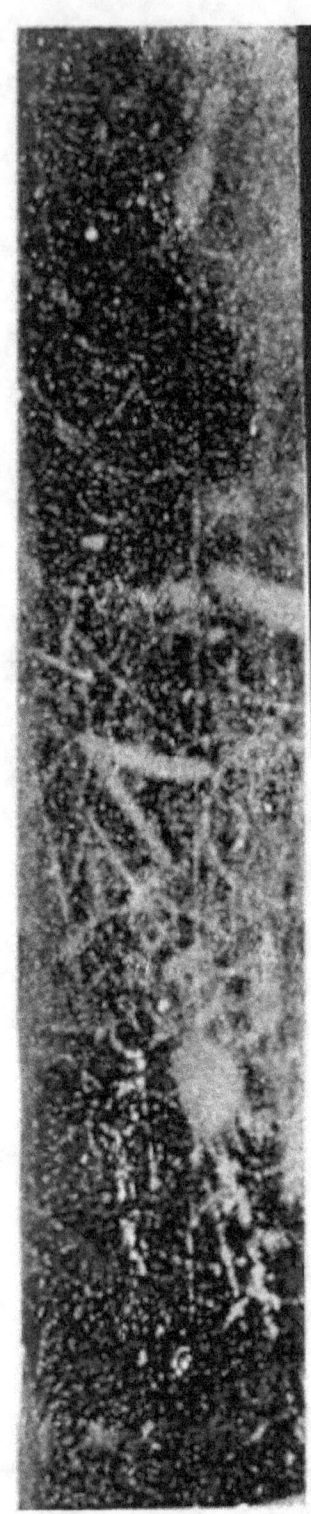

A cosmic ray event with a vertical cloud chamber (Leprincd Ringuet)

Anderson overcame this difficulty to a certain extent by inserting a metal plate absorber across the center of the chamber, which reduced the energy of the incoming particles so that the tracks that entered the plate with a slight curvature emerged with a sharper curvature.

(iii) *The tracks are photographed stereoscopically,* which give precious information about the direction and orientation in space of the passage of the particles through the chamber. For this purpose, either a stereo-camera is used and the photo of the image of the chamber reflected by a mirror set at 45° is taken or an ordinary camera is fitted into a hole in the pole-piece along with a mirror suitably placed at the side of the hole and the two views of the chamber, one direct and the other reflected by the mirror, are recorded together on the same film.

(iv) *Special precautions have to be taken to avoid distortion of the tracks,* arising chiefly from two causes, known as *chamber* and *optical distortions.* Chamber distortions are due to:

(a) the *cosmic* ray *going through the chamber before the expansion* in a counter-controlled arrangement, which makes the original track to be subjected to a strain by the sudden expansion,
(b) the *motion of the condensation droplets through the gas of the chamber* which renders impossible photographing without distortion later than about 0.25 sec after expansion,
(c) *convection currents* caused by the rapid fall of the central mass of cold gas in the chamber which makes it necessary that the chamber be illumined powerfully already before the start of the growth of the drops and the photographs taken very soon after the end of the expansion and
(d) the *presence of a metal plate* across the center of the chamber, which causes considerable distortions especially in the neighborhood of the plate, that could he reduced only by a careful temperature control.

Optical distortions are due to:

(*a*) *photographic lenses which* must be of large aperture in order to photograph from a direction nearly parallel to the magnetic field and hence at right angles to the illuminating beam, where the light scattered is very small and

(*b*) the *glass plate of the chamber* through which the photographs are taken.

In practice, to secure best results, in spite of the several distortion factors which are unavoidable, *curvature corrections* are applied with the help of a distortion curve drawn from measurements made on tracks in zero magnetic field.

(v) The *radius of curvature* of the very slightly bent cosmic ray tracks have been measured by the following two methods:

(*a*) *The method of coordinates,* employed by Leprince Ringuet, consists essentially in estimating the displacement of the center of the circular are, in which the bent track lies, by means of a graph plotted on a magnified scale, the coordinates of different points on the track measured with a traveling microscope. From the curvature thus graphically obtained the radius of curvature ρ is readily measured by using the relation $\rho = l^2/8d$, where l is the length of any chord on the curve and d the distance between the center of the arc cut off by the chord and the mid-point of the chord.

(*b*) *The optical null method,* more accurate than the previous one, chiefly when the length of the track is not great, has been used by Blackett. It is based on the fact that it is possible even with the naked eye to judge the straightness of a line with remarkable accuracy, if it is viewed very obliquely along its length. This allows the development of a null method, in which the curvature to be measured is compared by some optical device so as to reduce the actual measurement to a judgment of straightness. A convenient practical way of doing this is to project the image of the curved track on a white screen and then to place in front of the lens of the projector a suitable curvature-producing device. To compensate very small curvatures, an inclined glass plate with parallel sides can be used, while for large curvatures achromatic prism is convenient. The apparatus is calibrated by the use of lines of known curvature, which have been ruled with a beam compass on the emulsion side of a photographic plate, so that the radius corresponding to the measured curvature can be readily obtained.

(vi) *To measure the energies of the highly penetrating components* of the cosmic ray particles, a sufficient thickness of lead block is interposed between the two coincidence counters that control the functioning of the chamber, the counter system with the lead block in between being placed either above or below the chamber. Under these conditions extremely high energy particles alone can enter the chamber and hence their tracks can be photographed almost individually.

Results
Anderson and Neddermeyer using a cylindrical cloud chamber 17 cms in diameter and 4 cms in width between the pole- pieces of an electromagnet capable of giving a magnetic field of 15,000 gauss and photographing cosmic ray tracks 12 cm long, studied the energy distribution of cosmic ray particles at sea level by drawing the *energy-spectrum curve* from experimental data, *i.e.,* by plotting the number of tracks in a given energy interval against the energy of the particles. Their energy measurements went up to values of 5 x 10^9 eV.

Blackett, using a larger cylindrical cloud chamber 26 cms in diameter and 3 cms in width in a magnetic field of 14,000 gauss, with which cosmic ray tracks 17 cms long could be photographed, was able to extend the energy spectrum of cosmic rays up to 10^{10} eV.

Leprince Ringuet and Crussard with the very large rectangular chamber, 55 cms high and 16 cms wide and a magnetic field of 13,000 gauss obtained with the big Paris magnet were able to measure energies up to 2 x 10^{10} eV of the extremely penetrating cosmic rays which could pass through 7 to 14 cms of lead absorber and yet produce a track of about

40 cms in length. Each team of these workers gathered their experimental data from several hundreds of tracks. The results obtained may be summarized as follows:

(i) Energy distribution and sign of charge of-the very high energy cosmic ray particles near sea level

Anderson, discarding tracks under 3 MeV as likely to be produced by low energy phenomena, found that 75% of the tracks had energies under 4×10^9 eV.

The photographs further showed tracks of opposite curvature, proving that both kinds of charged particles, positive and, negative, were present in the cosmic rays up to energies of 6×10^9 eV. *The distribution of energy in the two kinds of particles was about the same and their number were in equal proportions, although the positives seemed to be slightly in excess.*

Blackett's results substantially confirmed, those of Anderson, except for the additional finding that the *particles with energy greater than 10^{10} eV seemed to be mainly positive.*

Leprince Ringuet working with very high energy particles obtained by a 7 cms lead filter found that, while in the energy range of 1 to 2×10^9 eV particles of both signs occurred in nearly equal numbers, *particles of higher energies up to 2×10^{10} eV were predominantly positive.* Although 2×10^{10} eV appears to be the limit up to which cosmic ray energies can be measured with the cloud chamber, there are strong indirect evidences for some of the particles having energies even as high as 10^{18} eV.

(ii) The "soft" and "hard" components

The penetrating power of the cosmic radiation into an absorber was also studied in these experiments by measuring the loss of energy, given by the difference of curvature of the cosmic ray track before and after it had passed the absorber placed inside the chamber. Different heavy metal sheets of Cu, Pb, Pt and Au of varying thicknesses were used. Analysis of the data thus gathered showed the existence of two components in the cosmic rays at sea level, one highly penetrating, i.e., the *hard* component, represented by certain isolated tracks which passed almost straight through large thicknesses of the absorber, thereby indicating very little loss of energy in the absorber, and the other less penetrating, *i.e.,* the soft component, represented by the more frequent positive and negative tracks, which were fairly easily absorbed, losing a good amount of energy in the absorber. These findings are in accord with the results obtained by the study of absorption of cosmic rays in matter, described in an earlier section.

(iii) Nature of the particles in the cosmic rays at the sea *level*

Although there are positively charged particles in time cosmic rays observed in the cloud chambers, there are good reasons to believe that *all the rays studied at sea level are secondaries produced in the atmosphere itself and do not contain the primary positives* (detected by the east-west effect) that enter the atmosphere from above, as already indicated in connection with the altitude effect,. *The excess of positives among the very high energy particles at sea level,* whose number appears to augment with increase of energy, according to the results obtained by Leprince Ringuet, may *be due to the fact that high energy negatives are more easily absorbed* in *lead than are high energy positives on account of the strong positive charge of the nucleus in the absorber.* It may also be due to the emission of high energy positive particles by nuclei disintegrated by cosmic rays, chiefly the single isolated positive tracks observed in the chamber. It is to be noted that while the primary positives are probably protons, almost all the particles met with in the cloud chamber have been shown not to be protons, although at the high energies involved proton tracks would be indistinguishable from electron tracks.

The *soft component* part of the observed cosmic rays *apparently consists of free electrons, negative and positive, in equal numbers,* as established by the discovery of the positron and the study of the phenomenon known as cosmic ray showers. The *particles constituting the hard component,* first suggested as possibly protons by the French scientists on account of their finding a positive excess in this range of energy, *have been conclusively shown* by more recent experiments *to be a new type of particles known as mesotrons,* whose mass is intermediate between that of electrons and protons and whose greater penetrative power is due fundamentally not to their greater energy but rather to their different nature, distinct from electrons, and protons.

The cloud chamber studies of cosmic rays have thus been extremely fruitful, leading up to discoveries of far-reaching significance, such as two new particles, the *positron* and the *mesotron,* and the so-called *shower* phenomenon. Those important aspects of cosmic rays will be considered in the following pages.

9. THE POSITRON

The positron is a fundamental particle which is the *exact counterpart of the electron, i e.,* identical with the electron, except for the positive sign of its charge. The possible existence of such a particle was predicted by Dirac in connection with his relativistic wave mechanical theory of the electron, long before it was actually discovered.

Carl David Anderson (3 September 1905 – 11 January 1991)

Discovery

The credit of the discovery of the positron goes first to Anderson and next to Blackett who, in their researches on cosmic rays with the cloud chamber, arrived independently, in 1932, at the same conclusion about the existence of this new particle. Among the cosmic ray tracks photographed with the cloud chamber in a strong magnetic field, Anderson found a certain number which were exactly like those produced by electrons, as regards general aspect and density of track, but bent in a direction opposite to that of a negatively charged particle.

This meant that either the particles which produced such tracks came from *below,* in which case they ought to be also *electrons,* the inversion of curvature being due to the change of direction of motion, or they came from *above* and then they ought to carry a *positive* charge.

The first alternative was very improbable in the actual arrangement of the apparatus, chiefly since cosmic rays of great energy came from above. In any case the sign of the charge of the particle could be definitely fixed only when the direction of motion of the particle was known. In order to decide this point, Anderson took photographs of tracks with a lead sheet placed across the center of the chamber. One of these historic and highly important photographs, which definitely showed that the sign of the charge of the particles which made positive curvature in a magnetic field was undoubtedly positive and thereby led to the discovery of the positron, is reproduced here.

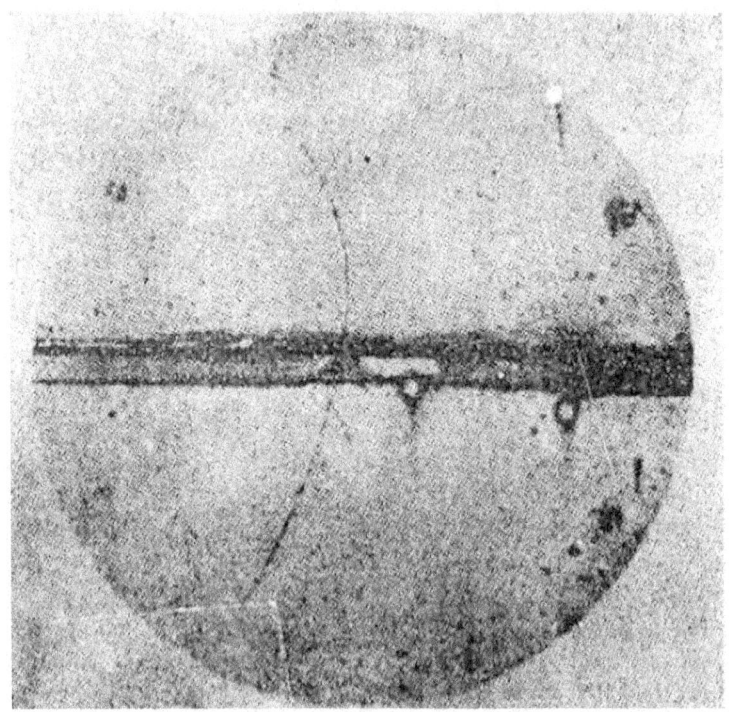

Discovery of the positron.(Anderson)

The track obviously represents the path of a single particle which traverses the lead plate of 6 mm. thickness.
Its general appearance closely resembles that ordinarily produced by an electron.
It is less curved below the lead sheet than above.
The difference in curvature must be evidently due to a loss of energy by the particle in the lead plate.
The direction of motion being assumed to be from the side where the track is less curved to that where it is more curved indicating a diminished speed.
Measurement of the radius of curvature of the track, below and above the plate, gives the energy of the particle as 63 MeV and 28 MeV respectively.
The loss of energy in the absorber is therefore 40 MeV, which is approximately the amount of energy one would expect such a particle to lose in a lead plate, 13 mm thick.

Hence, the particle must have been moving upward. If this conclusion is correct, then the sense of the curvature of the track and the known direction of the magnetic field show definitely that the particle carries a *positive charge.*

The fixing up of the sign of the charge of the particle as positive does not, however, prove at once *that the particle is a positive electron.* It might very well be a proton, even if the general appearance of the track resembles that ordinarily caused by an electron. For, at the high energies involved the relativistic mass of an electron approaches that of a proton and in consequence there exists very little difference in the specific ionization (directly proportional to the actual mass of the ionizing particle) produced by them so that their tracks are no more easily distinguishable.

But, Anderson, continuing his analysis of the track, was able to show that *it was definitely not produced by a proton.* A proton leaving a track having a curvature of the amount seen above the plate would have an energy of only 300,000 eV and a proton of such low energy is known to ionize much more heavily than what is indicated by the track.

Moreover, the range of such a proton can be only about 5 mm, whereas the actual track shows a range of 5 cms and more. Hence, *he concluded that the track was caused by a positively charged particle which must have a mass much less than that of a proton, but about the same as that of the electron.* Thus a hitherto unknown particle was discovered, which Anderson called the **positive electron** or **positron.**

The discovery of such a new particle would have remained doubtful—for, after all, a few photos where the tracks are curved inversely might be attributed to some parasitical complications—had not other observations made by other workers soon followed which placed beyond doubt the existence of positrons.

One such very important confirmatory evidence was the phenomenon met with in cosmic rays themselves and known as *cosmic ray showers.* Skobelzyn was the first to indicate the existence, in his cloud chamber photos, of several tracks of cosmic rays arising simultaneously from about the same point. Anderson himself had noted groups of such rays, bent in opposite directions and radiating from a common place.

But it was Blackett and Occhialini who, with their ingenious counter controlled cloud chamber, were the first to place in clear evidence this interesting cosmic ray phenomenon of "showers". One of the earliest photos of such showers, obtained by Blackett, is reproduced here.

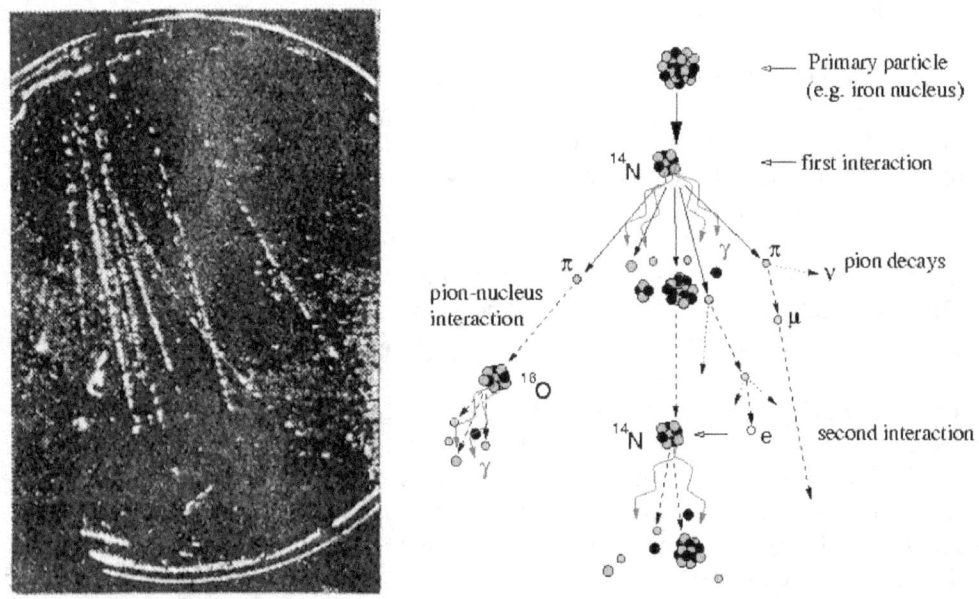

Cosmic ray shower (Blackett)

It consists of two groups of tracks bent in opposite directions, originating from a common point, usually in the massive metal parts of the apparatus, such as the mass of copper wire surrounding the chamber. The two groups consist of an approximately equal number of tracks, thereby indicating that a shower contains positively and negatively charged particles in about equal numbers. The mean aspect is the same for the negative as well as the positive tracks, suggesting a perfect symmetry between the two kinds of particles which would thus have the same intrinsic mass with equal charge but opposite in sign.

From the measurement of energies and amount of ionization, it can be shown that *all the tracks* in *a shower are due to particles which are electronic in nature.* The common spot from which a shower is found to originate decides the direction of motion of the shower particles, *i.e.,* outward and not inward. Under these conditions, the sense of curvature of the tracks unambiguously fixes the sign of the charge.

A more detailed study of the cosmic ray showers will be considered later. For the present, it is enough to remark that the "shower" phenomenon proves the existence of positrons in a very convincing manner.

Other sources of positrons

Though it was in association with cosmic rays that positrons were first discovered, other experimental sources of positrons were soon found, which not only confirmed the results obtained by Anderson and Blackett but also threw light on the different possible mechanisms involved in the production of positrons. The two important methods by which positrons can be readily produced and which have been extensively studied are:

(*A*) *materialization of radiant energy;*
(*B*) *artificial or induced radioactivity.*

A. Materialization of radiant energy

Chadwick, Blackett and Occhialini in England, Curie and Joliot in France, Meitner and Phillip in Germany and Anderson and Neddermeyer in America have studied this source of positrons in great detail.

The **principle** of the method, as already stated, is that when hard γ-rays of energy greater than $2m_0c^2 \approx 1$ MeV are absorbed by matter, the quantum of radiation is sometimes completely absorbed with a simultaneous production of an *electron-positron pair*.

The **experimental procedure** consists in irradiating different elements with γ-rays emitted either from a (Po + Be) source with an energy of about 5 MeV or from ThC'' of energy 2.62 MeV. A cloud chamber containing the element on which the γ-rays act and maintained in a magnetic field is used to photograph the tracks of the electron-positron pairs produced. To exclude from the chamber the photoelectrons and Compton electrons which might confuse the issue, the γ-rays are filtered with thick lead screens and the chamber itself is well protected with similar screens. Two photographs, one obtained by the English workers and the other by the French, are reproduced here as typical illustrations of the phenomena studied.

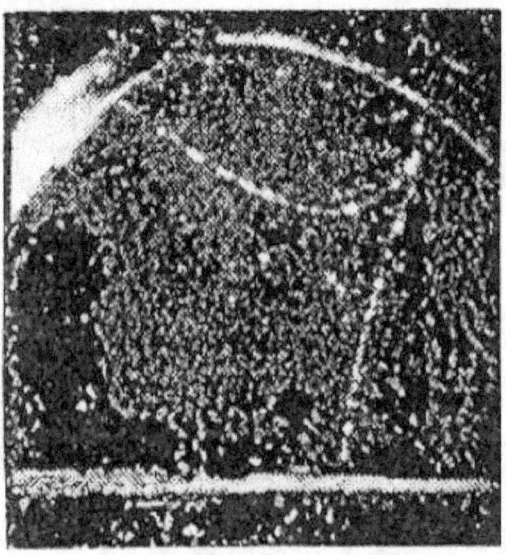

An electron-positron pair produced by γ-rays.(Blackett)

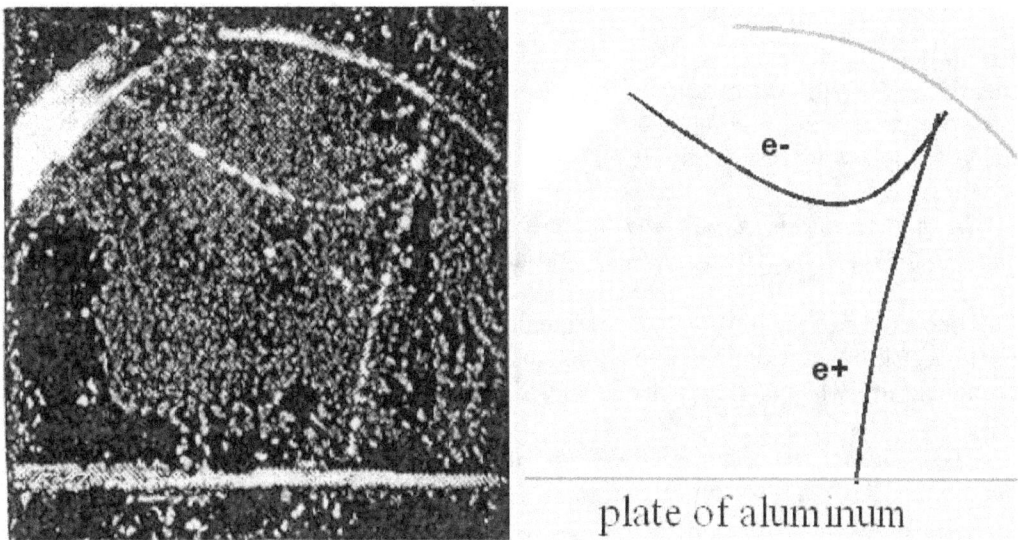

plate of aluminum

An electron-positron pair produced by γ-rays.(Blackett)

In the first one, the tracks of the positron and electron starting from a common point in the lead absorber surrounding the chamber and bent in opposite directions under the influence of the applied magnetic field are seen. The straighter of the two is the track of the positron. The horizontal bar is a plate of aluminum about 1 cm. thick, which the positron has not penetrated.

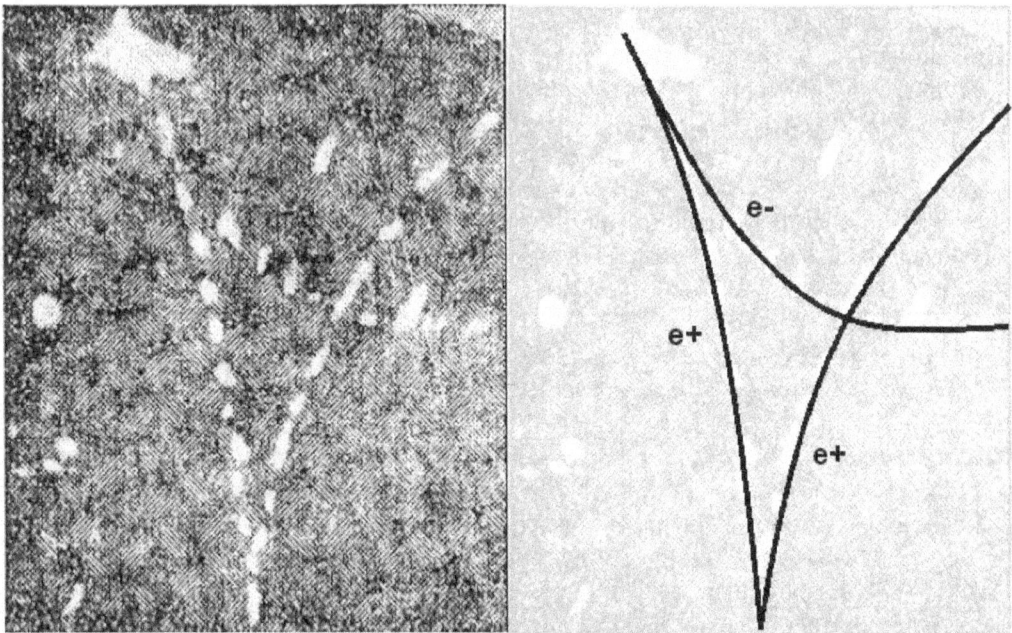

A pair production in the gas of the cloud chamber.(Joliet)

In the second (given below), we have a positron-electron pair produced by a γ-ray in the gas of the cloud chamber.

Results
Measuring the radii of curvature of the tracks, the kinetic energies of the two kinds of electrons can be readily computed. From a statistical study of a great number of such tracks the following results have been obtained:

(*a*) *The number of positrons produced increases with the energy of the incident quantum and the atomic number of the element irradiated.*

This shows that of the tree methods of interaction of radiation with matter, *viz.,* photoelectric effect, Compton effect and pair production, the last one assumes a prominent role when *high energy* quanta and *heavy elements* are involved.

(*b*) *The phenomenon occurs only when the energy of the incident radiation is greater than 1 MeV, but never for lower energies.*

Furthermore, *the kinetic energies of the pair particles are related to the energies of the incident quantum* according to the expression

$hv = E^+ + E^- + 1.02$ MeV,

where *hv* is the energy content of the radiation, E^+ and E^- the measured kinetic energies of the positron and electron respectively.

Since it can be shown that 1.02 MeV is the energy corresponding to the rest-masses of the electron and positron, viz., $2m_0c^2$. (their masses assumed equal) and since the pair originate simultaneously from a common point, one is forced to conclude in the first place *that this phenomenon of pair production is a veritable materialization of radiant energy unknown to classical physics.*

Secondly, as experimental data verify the above relation in every case studied, the assumption involved about the rest-masses of the positron and electron must be considered valid so that one is compelled to admit *that the mass of the positron should be the same as that of the electron.*

(*c*) *The observed maximum energy of the positrons produced by a given radiation is independent of the nature of the absorber.*

This suggests that *pair production does not take place inside the nucleus,* since, if the process had a nuclear origin a variation of energy of the positron with the kind of nucleus would be expected. This is further confirmed by calculation made by Blackett from experimental data of the effective cross-section for pair production, which is found to be nearly ten times greater than the diameter of the nucleus
. Hence *pair production is not essentially a nuclear phenomenon, but extranuclear,*

occurring in the Coulomb field, outside the nucleus, although the presence of the nucleus seems to be necessary as a sort- of catalyzer of the materialization of energy, required for the conservation of momentum.

(d) The energy distribution curves of the positron and electrons are similar, but *in several cases the positron energy is greater than the electron energy* ($E^+ > E^-$).

This is explained by the fact that

the pair is created in the neighborhood of the nucleus. Assuming that both the particles are produced with the same amount of kinetic energy, the positron, being repelled by the nuclear field, gains kinetic energy, while the electron, being attracted, loses it.

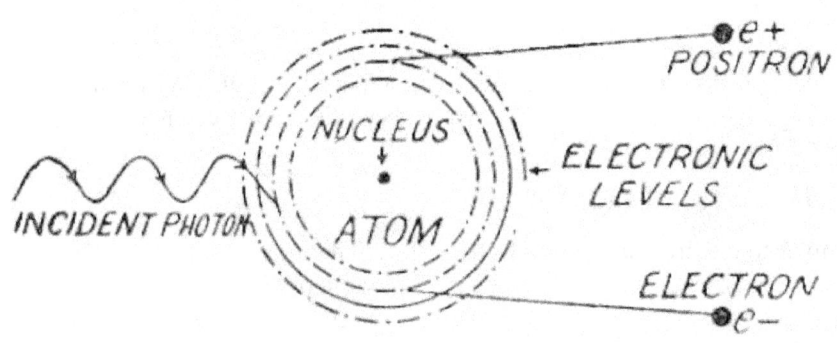

Fig. 317. Pair production.

Theoretical interpretation of results

These experimental results are *in accord with the general atomic theory,* according to which the positrons have no room inside the nucleus, and must therefore be born outside the nucleus. Such being the case, to conserve electric charge, an equal negative charge must also be born simultaneously, since the incident quantum has no charge. Conservation of angular momentum demands, in its turn, that *the positron should have a spin* 1/2 like the electron, as there is evidence that the incident photon cannot have a half integral spin associated with it. This is the phenomenon of pair production whose mechanism is illustrated in Fig. 317.

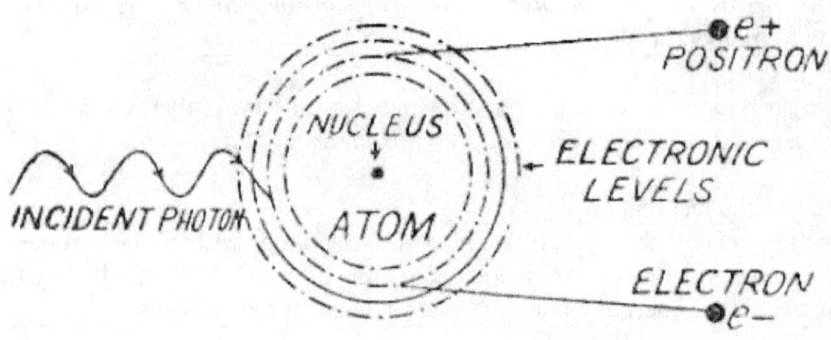

Fig. 317. Pair production.

The experimental discovery of the positron has also been a triumph for Dirac's theory, according to which, pair-production occurs when a high energy quantum hv removes an electron from a negative energy state — $[E_1]$ and places it in a positive energy state

$$+ E_2 = hv - [E_1]$$

in a manner similar to the ordinary photoelectric effect. Then it is observed that a negative electron of energy E_2 and a positive electron of energy E_1 are created, the light quantum being completely absorbed. The converse process of *annihilation of matter* predicted by the theory of Dirac in which an electron from a positive energy-state falls into a hole in the negative energy state — $[E1]$ thereby causing the union of the positron and electron with their consequent annihilation as material particles and production of radiant energy $\{E_2 + [E_1]\}$ he also been experimentally observed, as we shall presently.

Bethe and Heitler's theory of pair production

On the basis of Dirac's theory, Oppenheimer and Plesset (1933) and Bethe and Heitler (1931) have estimated the probability of pair production and obtained theoretical expressions for the effective cross-section for pair formation between a γ-quantum of energy hv and a nucleus of charge Z. It is found that the phenomenon cannot be expressed by a simple formula for the whole energy range, since different approximations are required. Furthermore, the theory is expected to break down when

$$hv \approx 137 \approx 70 \text{ MeV}.$$

With this restriction, the theoretical formulae show:

(1) *that the probability of the process of creation of pairs is directly proportional to the square of the atomic number* (Z^2) *of the absorber* and

(2) *that it increases rapidly first with increasing energy of the incident quantum, reaching an asymptotic value for very high values of hv.*

The manner in which the total available energy ($hv - 2m_oc^2$) is shared by the two particles of the pair has also been investigated and it is found that the *distribution would be symmetrical except for a distortion due to the fact that since the positron is repelled and the electron attracted by the nucleus, the former has on an average a somewhat higher energy than the latter,* the difference being proportional to the atomic number Z though negligible for small values of Z.

Theory predicts also that the two particles of the *pair tend to come off within a small angle of the direction of the incident quantum,* being ejected in a nearly straightforward direction for high energies of the photon. These consequences of the theory are qualitatively at least in agreement with the experimental results stated above. Direct study of the absorption of hard γ-rays of energy 2.62 MeV in heavy elements by Chao (1932), Meitner and Hupfield (1932), Gray and Tarrant (1934) and others show that the absorption coefficient is larger than can be accounted for by the Compton scattering and photoelectric absorption and that the additional absorption is proportional to Z^2 in consonance with theory.

The absorption due to pair production may assume a preponderant importance over that due to Compton effect, so that the total absorption of γ-rays passes through a minimum and then becomes very great for radiations of very great energy. Hence, if at all there exists electromagnetic radiations in cosmic rays, it is probable that they cannot reach the surface of the earth, being absorbed rapidly in the atmosphere. Measurement of the energies of the pair particles ejected in gases, such as argon, krypton, xenon and methyl iodide by Simons and Zuber (1937), Grosev and Frank (1938) and others confirm qualitatively the theoretical prediction about the difference between the average kinetic energies of the particles of the pair.

As regards the angular distribution, experiments performed by Adams (1937) lead to results in fair agreement with theory. In conclusion we may say that the theory is correct, even quantitatively, for energies of the order of 3 to 10 m_oc^2. Cosmic ray studies indicate that the theory is probably valid even up to energies very much higher than 137 m_oc^2.

Several other processes in which energy materializes itself with consequent pair-production have been theoretically predicted. Some of these have received sufficient experimental confirmation, some very little, and finally others none at all.

Thus, for instance, Nedelsky and Oppenheimer (1933) have analyzed *pair formation by the internal conversion of the γ-rays given off after a nuclear transmutation.* When light elements are bombarded with α-rays, pair production sometimes results. Joliet and Curio using (Po + Be) found that when this source of γ-rays was placed near the orifice of the cloud chamber without any absorber at all, the source itself directly emitted positron-electron pairs. The energy of the pair particles was of the order of some millions of electron-volts. The amount of production was too high to admit that they were produced by materialization of the γ-rays with the atoms of Be, an element of small atomic weight. Hence, the relatively intense emission was interpreted as resulting from the internal materialization of the γ-rays as they came out of the product nucleus C^{12} formed in the transmutation of Be by the α-rays of Po. The probability of the process calculated from theory appears to be in good agreement with experimental results, and it seems to *diminish rapidly as the atomic number of the element increases,* which fact differentiates this internal conversion mechanism of pair formation from the one considered above.

Francis Perrin (1933) has suggested another process where *a pair is generated when a photon interacts with an already existing electron.* In such a case, conservation of momentum demands that the energy of the incident photon should be greater than $4m_oc^2$, hence double the energy necessary in the case of a nucleus. If the process can be experimentally observed, say in a cloud chamber, there will be three tracks of electrons, two negative and one positive, directed in the forward direction, emerging from the point where the invisible track of the photon, which has disappeared, has stopped. In a magnetic field, two of the tracks will have negative curvature, while the third positive curvature. The individual energies of the particles will be different and their inclinations with the direction of the photon also different. The probability of this process will be Z times smaller than in the case of the nucleus, since the cross-section for materialization near a nucleus of charge Z, being proportional to Z^2, will be Z times less relative to any one of the Z peripheral electrons.

Certain cloud chamber photos taken by Curie Joliot and specially the one obtained by Ogle and Kruger (1945) with the 2.68 MeV γ-rays from artificial radioactive Na^{24} and reproduced here appear to confirm the existence of this type of materialization of energy.

Furry and Carlson (1933) have worked out the theory for *pair production, by the interaction of a high energy electron with an atomic nucleus,* which assumes that even forms of energy other than the radiant type, and in particular *kinetic energy, can also materialize itself.* A primary electron or positron of high energy passing through the Coulomb field of a nucleus may lose part of its energy in the production of pairs. This process is similar to the materialization of radiant energy, except for the fact that the created pairs receive only a small portion of the energy of the primary electron, unlike in the other case where all the energy of the incident quantum is absorbed. A few cloud chamber photos obtained by Skobelzyn. and Stepenowa (1934) where an electron suddenly stops at a common point of separation of two apparently paired tracks might correspond to this type of materialization of energy. But these workers affirm that the phenomenon observed by them has nothing in common with the mechanism considered by Furry and Carlson, but that it refers to the case of ejection of a positron alone when different elements are bombarded with electrons, where the probability of the process decreases with increase of atomic number of the element, contrary to theoretical expectations.

A more recent careful study by Crane and Halpern (1939) has shown that there is no evidence whatever for creation of either pairs or positrons by electron bombardment.

Other mechanisms of materialization of energy, such as collision of two electrons, one of which has an energy greater than $6m_nc^2$ and collision of two high energy photons have been considered, but have received no experimental confirmation so far.

B. Emission of positrons by artificially produced radioelements

Artificially induced radioactivity has provided another excellent source of positrons, as we have already seen. There are more than 60 of the artificial radioelements that emit positrons, some of them having quite long periods. Thus convenient and copious supplies of positrons can be had in the laboratory today as by-products of artificial radioactivity.

The idea of interpreting the ejection of positrons from artificial radioelements by the general mechanism of materialization of energy, in one form or another, say internal conversion of γ-rays, is both simple and attractive. But there are very serious difficulties against such a hypothesis.

In all materialization processes there ought to be a simultaneous formation of pairs, *i.e.,* positrons and electrons, while in the case of artificial radioactive substances, *only positrons or electrons and not pairs are ejected.*

Furthermore, the *emission takes place for a considerable time,* very much like in natural radioactivity, while one would expect an instantaneous emission in the materialization process.

No satisfactory solutions have yet been found for these difficulties. Authors limit themselves, therefore, by saying that *the emission of positron by an artificial radioelement does not take place in the extra-nuclear Coulomb's field of the atom, but is caused by the transformation of a proton into a neutron inside the nucleus itself, so that it is to be considered as a nuclear phenomenon.*

Investigations on the nature of the positron

Researches made to determine with precision the nature of the positron consisted chiefly in the measurement of its mass, charge and *e/m* value, as well as its absorption in matter. The *main purpose was to establish the identity of the positron with the electron except for the sign of the charge,* but the experiments on the absorption of positrons were intended also to verify the theoretical prediction of the phenomenon of *annihilation of matter.*

The mass of the positron

Anderson, estimating the amounts of ionization produced by the positron and electron from their respective cloud chamber tracks of same curvature, was able to conclude that the masses of the two particles could not differ by more than 20%.

Chadwick, Blackett and Occhialini have made a more accurate determination of the mass of the positron by measuring the energies of positrons in the pairs produced by the absorption of γ-rays of known energy. The *principle* of the method is as follows:

The phenomenon of pair production is governed by the relation by

$$hv = (m_1 + m_2) c^2 + E_1 + E_2$$

where,
hv is the energy of the incident quantum,
m_1 the mass of the electron.
m_2 the mass of the of the positron,
c the velocity of light,
E_1 and E_2 the kinetic energies of the electron and positron respectively.

Considering the case of maximum kinetic energy (E_{max}) of the positron which corresponds to zero kinetic energy of the electron, the above expression can be written as

$$E_{max} = hv - (m_1 + m_2) c^2$$

Since hv is known, if E_{max} can be estimated experimentally, the value of m_2 can be readily deduced in terms of m_1.

The *experimental procedure* consists in obtaining cloud chamber tracks of pairs, using γ-rays of known energy, say those emitted by ThC" of 2.62 MeV, and a suitable absorber. The energies of a large number of positrons are measured from the track curvatures caused by a known magnetic field in which the cloud chamber is maintained. Drawing the energy distribution curve, the value of E_{max} can be found. Substituting this value in the above relation, the mass of the positron in terms of that of the electron is readily obtained. In this way, it was found that

$$m_2 = (1.02 \pm 0.10) m_1,$$

which proves that within the limits of experimental error *the positron has the same mass as the electron.*

The charge on the positron
Application of the principle of conservation of electron charge to pair production clearly indicated that the charges on the positron and electron must be equal and opposite. Hence no attempt has so far been made to measure *directly* the charge on the positron. Anderson has made indirect estimates of this quantity from studies of cloud chamber tracks.

First of all, from the observed curvature and ionization, he showed that the charge on the positron could not be twice that of the electron.

Secondly, measuring the specific ionization produced by the positron and finding that it did not differ from that of the electron by more than 20%, he argued that, since the specific ionization is proportional to the square of the charge, other conditions being the same, the charge on the positron could not differ from that of the electron by more than 10%.

Other workers, such as Thibaud (1933), Spees and Zahn (1940) have attempted to establish the numerical equality of charge on the positron and electron by measuring their (*e/m*) values with highly refined experimental techniques which will now be considered briefly.

Determination of e/m of the positron
Thibaud was the first to compare the specific charges (*e/m*) of the positron and electron by an ingenious modification of the magnetic spectrograph, known as the **trochoid method,** which permits a *very good concentration of light charged particles* under study.

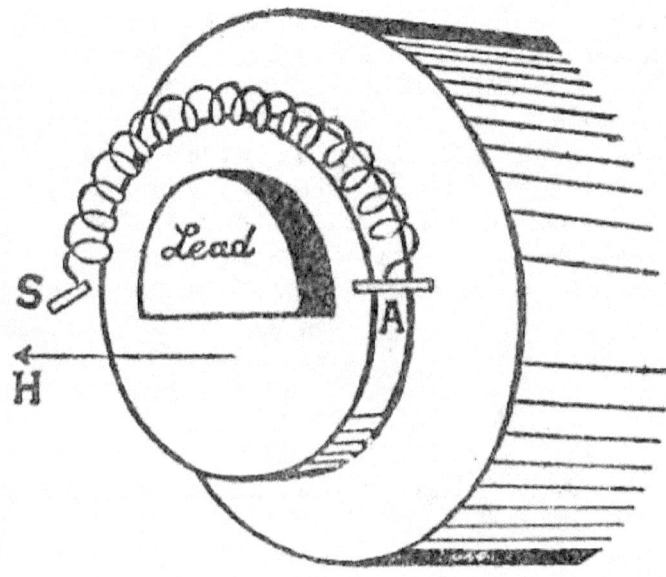

Fig. 318. Trochoid method—only one pole is shown. (Thibaud)

The *principle* of the method, illustrated in Fig. 318, consists in utilizing a peculiar effect produced on the path of light charged particles, like the electrons and positrons, by a strong nonuniform magnetic field. The source S of the particles is placed between the pole-pieces of a powerful electromagnet capable of developing a field of 10,000 to 20,000 gauss, *not at the center but at the outskirts of the pole-pieces,* where the field is bound to be non-uniform due to the tendency of the lines of force to spread out into the surrounding space. The particles issuing from S will be projecting in all directions in the plane of symmetry of the pole-pieces. Each of these particles will describe, under the influence of the marginal non-uniform field, a *little ring, not completely closed, but drawn out a little* so that its path will present the form of a spiral, as shown in Fig. 318, a *trochoid* contained between two close concentric circles. All the particles that have traversed this rather complicated path find themselves concentrated at a definite point A, situated diametrically opposite to the source. It is at this point A that the different devices used for detecting and studying the particles, such as photographic plate, electronic counters, etc., are introduced. It is possible to obtain a very convenient nonhomogeneous radial field by a careful adjustment of the inter-space between the pole-pieces. By merely changing the direction of the current feeding the electromagnet the field can be reserved at will, so that one or the other of the two kinds of electrons can be made to describe the trochoid and concentrate themselves at the point of detection. This elegant and effective arrangement, first devised by Thibaud, was soon adopted by Curie-Joliot, Fermi and others in the analysis of light particles met with in induced radioactivity.

In the *experimental determination of the elm* values for the positron and electron, a semi-circular glass tube connected to a vacuum pump (shown, in the adjacent. photo) was arranged to occupy the variable marginal region between pole-pieces, 24 cms in diameter, of a powerful electromagnet which gave a field varying from 11,000 to 18,000 gauss as the interspace width was changed from 7 to 2 cms. The glass tube contained at one end the source S of positrons or electrons, which was a thin tube containing radon that emitted γ-rays, enclosed in a thin sheet of lead (about 2 to 3 mm. thick) acting as the absorber of γ-rays and producing the electron-positron pairs. The detector A consisted of a photographic plate, below which were arranged two grids raised to a high potential of 12,000 volts. This arrangement was protected from direct radiations from the source by interposing a lead block, 10 to 15 cms thick. The electrostatic field between the grids shifted the trace made on the photographic plate *by* the charged particles and the displacement of the trace enabled one to measure the *e/m* of the particle under test, by the use of the relation

$$d = (L \, l \, X/v^2).(e \, / \, m),$$

where,
d was the displacement of the trace under the influence of the electric field of intensity X acting along a length *l* of path of the particle,
L the distance of the plate from the center of the field and
e and v the mass, charge and velocity respectively of the particle.

Reversing the field direction, the electrons and the positrons could be concentrated separately at the detector and hence their e/m could be separately determined.

Thibaud found that *the e/m of a positron was equal to that of the electron within 10%.*

More recently, Spees and Zahn have employed a **modification of the original Bucherer method**, in which the positions of the source and the detector were interchanged. An electronic tube counter replaced the photographic film. Using the artificial radioelement Cu^{64}, emitter of both positrons and electrons, as source, they were able to compare directly the *e/m* of the two particles by reversing the deflecting electric and magnetic fields. Their result showed that *the value of e/m was the same for both the particles within 2 %* which in consequence argued to the numerical equality of charge of the two kinds of electrons.

Absorption of positrons in matter
The study of the absorption of positrons in matter was first undertaken with a view to establishing the identity of the positron with the electron. Thus Thibaud, using the

trochoid method and introducing metallic absorbers at the point of concentration of the positrons or electrons in the apparatus, was able to show that *positrons behaved in the same way as electrons of the same speed, as regards penetration in matter.*

But the absorption of positron in matter is of much greater interest and importance from another point of view, viz., the possibility of checking the theoretical prediction of the phenomenon of *annihilation of matter* and of acquiring thereby an insight into the *ultimate fate of the positrons.*

Annihilation of matter

The positrons, penetrating into the interatomic regions of matter which swarm with negative electrons cannot live there *free* for long. For, it is an established experimental fact that the positive electrons play no part, unlike the negative electrons; in the phenomenon of conductivity, gaseous, metallic or electrolytic. On the other hand, positrons and electrons, existing side by side, would necessarily rush together and destroy each other, unless a *negative proton* be postulated, which might bind the positron into a new type of hydrogen atom, in a manner analogous to the positive proton binding the electron into the ordinary hydrogen atom. For, in all atoms, including that of hydrogen, the mutual destruction of positive and negative electricity is avoided through the usual quantum conditions holding within the atom.

According to Einstein's relativistic law, such a fusion must give rise to radiant energy, since the disappearing potential energy of two separated attracting systems, corresponding to m in the relation $W=mc^2$, must of necessity appear in radiant form. Further, as the rest-masses of the two particles that have disappeared are equivalent to 1 MeV approximately, the total energy content of the resulting radiation must be about 1 MeV. It is assumed that the kinetic energies of the particles at fusion are negligibly small Thus we arrive at the phenomenon which is usually referred, to as the annihilation of matter or *dematerialization.*

Applying the laws of conservation of energy and of momentum to the process; *two possible cases* arise concerning the nature and energy content of the radiation that results from the union of positron with electron:

(*i*) *if the combination takes place in free space, removed from any third particle,* if the positron unites with a free or loosely bound electron, *two photons of equal energy, about* 1/2 MeV *each, will be emitted in opposite directions,* since only thus can both energy and momentum be conserved.

(*ii*) *If, however, the combination takes place near a third particle, i.e.,* if the positron unites with electron tightly bound to the nucleus, a *single photon, of about 1 MeV will be emitted,* since the third particle can take up the momentum and energy required for their conservation, and, being sufficiently heavy, as would be the nucleus of any atom, it would take up only a negligible amount of energy.

Theoretically then, one would expect both types of photons, some of 1/2 MeV and some of 1 MeV. But, since positrons, traversing matter would lose practically all of their kinetic energy before being annihilated and hence would be unable to penetrate the interior of atoms, the first type of annihilation where two photons, each of 1/2 MeV energy are emitted would be the most probable process that could be observed, while the other single photon emission process would be much less probable.

These theoretical predictions have been well confirmed in the positron- absorption experiments carried out by several experimenters. Chao was the first in 1930, to observe the expected emission of 1/2 MeV photons in his experiments on the absorption of the 2.62 MeV γ-rays of ThC" in lead. Gray and Tarrant, in 1932, performing similar experiments, were able to detect two secondary γ-radiations of energies nearly 1/2 MeV and 1 MeV among the scattered rays. They also placed in evidence the following special features of the observed γ-rays:

(*i*) their energy did not depend upon either the energy of the primary γ-rays or the material used as absorber;

(*ii*) they could not be identified with any of the characteristic γ-rays emitted by the absorber;

(*iii*) their wavelengths were independent of the angle of scattering and hence they cold not be attributed to Compton scattering;

(*iv*) they were observed only when the energy of the incident γ-ray was greater than 1.5 to 2 MeV. All these characteristics of the observed secondary γ-radiations can be easily explained, as Blackett and Occhialini first pointed out, only if it be admitted that these rays rise from the annihilation in lead of positrons which are produced in the lead itself by the primary γ-rays. The chief drawback of these initial experiments was that the estimate of the energies of the secondary radiations could not be made with great accuracy.

More direct and convincing proofs of the existence of the annihilation radiation were soon obtained. Thibaud and Joliot, in 1933, working independently with the trochoid method and using aluminum bombarded by the α-rays of polonium as source of positron-electron pairs and a photographic plate or G-M counter as detector, were able to observe the γ-rays emitted in the annihilation process when a stream of positrons was directed on a metal absorber such as Pt or Pb. By a suitable change of direction of the magnetic field, either the positrons alone or the electrons alone were made to strike the absorber. When the positrons fell on the metal plate, there emerged from the point of concentration γ-rays of much greater intensity than when, by reversing the field, electrons were made to fall on the absorber. By interposing absorption sheets between the emitted radiation and the detector, the energy of the radiation was found to be about 1/2 MeV. Joliot measured also the number of photons corresponding to the annihilation of a single positron and found it close to two in agreement with theory.

Methods of precision of analyzing the annihilation radiation became possible with artificial radioelements emitting positrons of great intensity. Lauritsen and Crane (1934), employing the positrons emitted by N^{13} obtained by bombarding a carbon plate with 1 MeV deuterons, were able to place in clear evidence that γ-rays of energy 1/2 MeV were really generated by the disappearance of positrons, at the rate of two per positron.

Richardson and Kurie (1936-1938) have observed the γ-rays from a number of artificial radioelements such as N^{13}, V^{46}, Cu^{64}, etc. They found that while the positron emitters all gave rise to γ-rays of 1/2 MeV energy, the electron emitters did not emit this γ-ray. No γ-ray of 1 MeV was found by them, which proved that this type of annihilation must be extremely rare.

Positronium

More recent researches have placed in evidence a *three-photon annihilation* and the consequent existence of a new fundamental particle, named *positronium*. This particle is a bound positron-electron system which is formed under certain conditions, before the annihilation process takes place, it is analogous to the hydrogen atom, except that it has no heavy particle as nucleus. In it, the positron and electron revolve around each other with equal speed.

The theory of positronium is much more complicated than that of atomic hydrogen. For, since both the particles in positronium move with high velocity, their motion must be considered relativistically and this relativistic system cannot be simplified by motion about a common center of mass, unlike in the case of the hydrogen atom. Two forms, the para- and *ortho-*, are possible in positronium. In para-positronium the spins of the two particles are anti-parallel, which renders the system very unstable against annihilation. Hence para-positronium would quickly decay into two photons of energy of 1/2 MeV each, after a very short mean life of only 10^{-10} sec. The two-photon annihilation, first observed (as described in the previous section), is due to the para-positronium decay. In ortho-positronium, on the other hand, the spins of the positron and electron are parallel. This state of affairs prevents annihilation into two quanta, since angular momentum could not be conserved. Ortho-positronium should, therefore, rather decay into three quanta to conserve angular momentum and this only after a much longer mean life, being much more stable against annihilation. The calculated value of the mean life of ortho-positronium is 1.4×10^{-7} sec. Further, to conserve linear momentum, the three photons must be emitted in a plane, and if they have equal energies of 1/3 MeV, they will be ejected from the source at equally spaced angle of $120°$. According to theory, the ratio of occurrence of the three-photon annihilation process as compared to the two photon annihilation process is about 1 to 370.

The three-photon annihilation was first observed by J. A. Rich (1951) with three scintillation counters in coincidence placed symmetrically in a plane about a Cu^{64} positron source, and absorber. Do Beneditti and Siegel (1952-1954), with a similar arrangement, setting the three-scintillation counters at various angles in a plane containing the positron source, have shown that the total annihilation energy of 1 MeV may divide into three photons in various ways conserving momentum and having the energies demanded by theory. M. Deutsch (1951) has investigated by a direct method the formation of positronium as an intermediate step in pair annihilation. Using Na^{22} as the positron source and N_2 as the absorber, he has measured the mean life of ortho-positronium and found it to be about 1.5×10^{-7} sec in good agreement with the theoretical value.

The formation of positronium is found much more probable in some absorbers than in others. For instance, the three-photon emission observed with "freon" gas as absorber is about nine times as frequent as when solid absorbers are used. Addition of a small amount of another gas to the one used as absorber has been shown to reduce the three-photon emission process, evidently due to the conversion of ortho-positronium into para-positronium. Likewise, the states of positronium are found to be influenced by a magnetic field, which might naturally increase the probability of an ortho-para conversion.

Fate of the positrons
These experiments fairly well establish that

the positrons can live in a free state only until they have lost the greater portion of their kinetic energy, when they disappear uniting with electrons. The mean life of a positron depends upon the number of electrons near it.

The concentration of the latter in matter being very great, the *mean life of positrons should be very small.* Calculation shows that it is of the order of 10^{-7} to 10^{-10} sec. This extremely short life of the positrons explains why they have escaped detection for a long time, although cosmic ray studies indicate that they are nearly as numerous as the electrons in nature, at least at sea level. They can be observed only in a cloud chamber or with a counter, almost immediately after being formed,

Another consequence of the very short life of the positron is that *matter does not contain positrons in a free state.* Certain authors have put forward the hypothesis that all nuclei are ultimately constituted with neutrons and positrons. But this has not been favorably received by the great majority of scientists on account of a number of serious objections raised against it, as *we* shall see later, when dealing with nuclear structure,

It was for some time thought that the positron might form an important constituent of the universe, as it was supposed that many of the incident .cosmic ray particles were positrons. Since, however, more recent researches have shown that most of the positrons observed in cosmic rays are of secondary origin, it can no longer be said that the universe might contain very many positrons. Thus it appears that, whether within matter or out in the universe, the positron has no independent existence. When it does make its appearance, it is for a very short time, since it soon disappears along with the nearest electron, producing radiation according to the laws of conservation of-energy and of momentum.

10- THE MESOTRON

The mesotron is supposed to be a fundamental particle (like the electron, positron and proton) carrying a single elementary charge and possessing a mass intermediate between that of the electron and that of the proton. It was first discovered, in 1937, in cosmic ray researches. Already two years before the actual discovery, it was predicted by Yukawa in his theoretical study of the nature of *intra-nuclear forces*, which fact indicates its importance in nuclear physics.

Discovery

The credit of the discovery is usually given to Neddermeyer and Anderson as well as to Street and Stevenson, although the former were the first to surmise the existence of the new particle. It is interesting to note the striking similarity between the discoveries of the positron and mesotron in so far as both were theoretically predicted some time before being experimentally detected and both were discovered by the same persons during the same cosmic ray work. But, in contrast to the discovery of the positron, which came about almost through the evidence of a single cloud chamber picture, the discovery of the mesotron was the result of a vast amount of experimental and theoretical work.

The *experimentally observed facts* which constituted the starting point in the discovery of the mesotron were the following:

(1) The existence of the two main components is cosmic radiation *soft* and *hard,* the latter forming about 75% of the total radiation at sea level.

(2) The hard component consisted of high speed particles positively and negatively charged in approximately equal numbers.

(3) These particles often left isolated single tracks in a cloud chamber, the appearance of which was very much like those produced by electrons as illustrated in the adjacent photo containing the tracks of two highly penetrating cosmic ray particles which traverse in one case two lead plates each 1 cm. thick without any appreciable change of curvature, thereby indicating that they lose very little energy in the absorber.

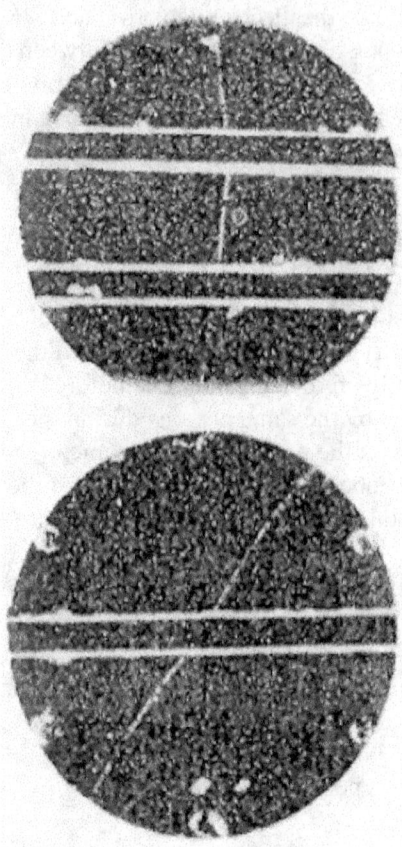

Tracks of highly penetrating particles of cosmic rays.

(4) The most important characteristic of these particles was, therefore, that they were unusually penetrating, nearly 50% of them passing through 100 cms of lead, while the particles of the softer component, presumably electronic in nature, were completely absorbed in about 10 to 15 cms of lead.

(5) A statistical study of the loss of energy of the two types of particles in going through solid absorbers showed that different mechanisms were followed in the two cases. For the particles of the soft component, the loss of energy was found to be proportional to the incident energy of the particle as well as to the square of the atomic number of the absorber, while with the particles of the hard component the loss of energy exhibited no such simple relation with the incident energy and was found to be approximately proportional to the atomic number of the absorber.

Indirect evidence
An attempt to interpret adequately these experimental data in the light of the *existing theories* concerning the loss of energy by fast charged particles passing through matter gave a *strong* though *indirect* evidence to the effect that the highly penetrating particles of cosmic rays were neither electrons nor protons but were of a new type possessing mass

intermediate between those of the electron and proton, though carrying a single elementary charge, either plus or minus.

The theories relevant to the present problem are those which deal with the loss of energy by fast charged particles due to two types of interaction with matter, known as *ionizing collision* and *radiative collision,* whose main and distinctive features may be briefly stated as follows:

Ionizing collision

Ionization takes to inelastic collision between a fast charged the material through which it passes, with a real transfer of energy from the former to the latter. The mechanism of the phenomenon may be visualized as follows:

As the fast moving charged particle passes near an atom, an electric force develops between it and the orbital electrons of the atom, resulting in a transfer of energy to the electrons which in consequence free themselves from the parent atom and ionization follows. The ejected electrons are known as *secondary elections,* and may sometimes be sufficiently energetic to ionize other atoms. Such energetic electrons are sometimes called *"knock-on" electrons,* as illustrated in the adjacent photo, where a very high energy cosmic ray particle of about 10^{10} eV produces in the gas of a cloud chamber a "knock-on" electron curved into a circular arc under the influence of a magnetic field of 10,000 gauss, with an estimated energy of 6 MeV.

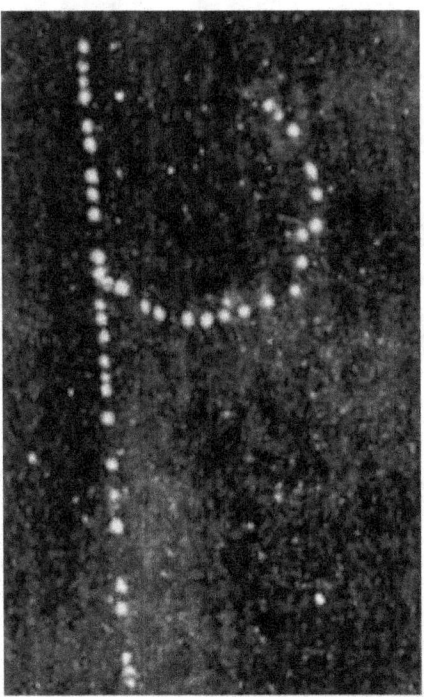

Production of a "knock on" electron by a very energetic cosmic ray particle. (10^{10} eV) (J.G. Wilson)

Approximate theoretical formulae for the loss of energy of a fast charged particle in any material due to ionization have been derived on relativistic wave mechanical principles by Bethe, Heitler and others. According to thorn, the energy loss due to ionization depends on three factors, *viz.,* the magnitude of the charge carried by the particle, its speed and the nature of matter through which it is moving, as follows:

(i) *The loss of energy per unit length of path is proportional to the square of the charge of the particle.* Hence, for instance, a particle with two-fold charge, like the α-particle, will ionize four times as strongly as a singly charged particle, like the proton. For the same reason, the ionization will be the same for the positron and electron.

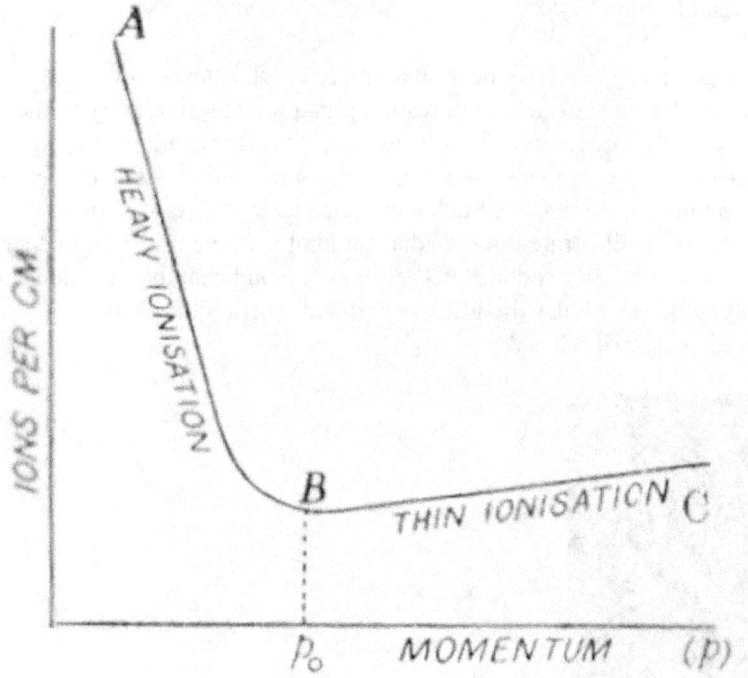

Fig. 319. Relation between energy loss by ionization and momentum of the ionizing particle.

(ii) The dependence of *the loss of energy on the speed of the particle is somewhat complicated* and is best studied by means of a graph predicted by theory. It is shown in Fig. 319, where the number of ions formed per unit length is plotted, not against momentum, since at high speeds the mass of a particle is no longer constant according to the theory of relativity and the quantity actually measured is not velocity but momentum. Two parts AB and BC can be distinguished in the curve. Along AB, the ionization decreases rapidly as momentum increases up to a certain limit p_o, which corresponds to the *rest-mass* of the particle or to a speed approximately that of light. In this region there will be *heavy ionization* and consequent rapid loss of energy. In the part BC, as momentum increases, the ionization increases very slowly and continuously due to increasing contraction of the electric field of the particle towards a plane perpendicular to it motion and the resulting increase of sharpness of the impulse given to the electrons. This is the region of *thin ionization.* Although ionization does not depend on the mass of the ionizing particle, the transition point p_o is very important, since it depends on the rest-mass of the particle and hence offers a means of measuring the mass of the particle under study. It is easily seen that since heavy ionization means rapid loss of energy, *particles which heavily ionize are always near the end of their career as fast particles and will soon, be stopped,* while particles which thinly ionize last longer. Furthermore, since at a given high velocity, kinetic energy and rest-mass are proportional to each other, even in relativistic mechanics, *particles of equal charge but different masses should produce more or less the same ionization at equal velocities.* Thus, for instance, a fast proton which is 1,840 times as heavy as an electron should produce the same ionization and hence the same kind of track in a cloud chamber as the electron moving at the same high velocity and hence with an energy 1/1840 times as great as that of the proton. According to theory, a proton track should not be clearly distinguishable from an electron track of the same speed, beyond 500 MeV about. *But two particles having the same momentum but different masses will ionize to perceptibly different degrees, since their velocities are different.*

(iii) Concerning the nature of the material traversed, theory indicates that *the energy loss is roughly proportional to the number of electrons in unit volume of the material.* Since the number of electrons (Z) in each atom is approximately proportional to the atomic weight (A), the *energy loss is roughly proportional to mass.* Hence the energy loss of a fast Charged particle in one cm of lead will be about the same as in 100 meters of air.

Radiative collision is a phenomenon of nuclear absorption, in which a fast charged particle passing close to an atomic nucleus loses energy by *radiation.* The mechanism involved is best understood by a reference to the production of X-rays. When high speed electrons strike the target of an X-ray tube, due to their sudden stoppage, two types of radiation are emitted, viz.,

- Continuous X-rays arise due to the *deflection* of the electrons by the strong fields surrounding the nuclei of the atoms of the target with the consequent emission of a pulse of radiation.

- Characteristic X-rays are caused by changes in the internal energy of the atoms of the target *ionized* by the electrons.

As the energy of the bombarding electrons is increased to high values, *e.g.,* several million volts, the energy that goes into the continuous spectrum becomes all in all, while that represented by the characteristic X-rays is negligible. In a similar manner, when an energetic charged particle passes close to an atomic nucleus, its interaction with the strong field of the nucleus makes it suffer large decelerations with the consequent transformation of its energy into an electromagnetic radiation, which is usually called *"impulse radiation"* or *"bremsstrahlung"* (i.e., radiation which arises from deceleration). The process is therefore named *bremsstrahlung process* or *radiative collision.*

Bethe and Heitler worked out, in 1934, the theory of this type of interaction of fast charged particles with matter on the basis of quantum electrodynamics and derived expression for the loss of energy involved, which led to the following conclusions:

(*i*) *The energy loss by radiation per unit length of path is inversely proportional to the square of the mass of the particle.* Hence the process *assumes a great importance only in the case of electrons,* its probability rapidly decreasing with heavier particles, for which the energy loss by radiative collision is negligible compared to the loss by ionization.

(*ii*) *The rate of energy loss by radiation is nearly proportional to the energy of the particle,* subject to certain restrictions. For, as the energy of the particle decreases, there comes a stage, when the energy loss by radiation becomes' equal to and then less than that by ionization (Fig. 320).

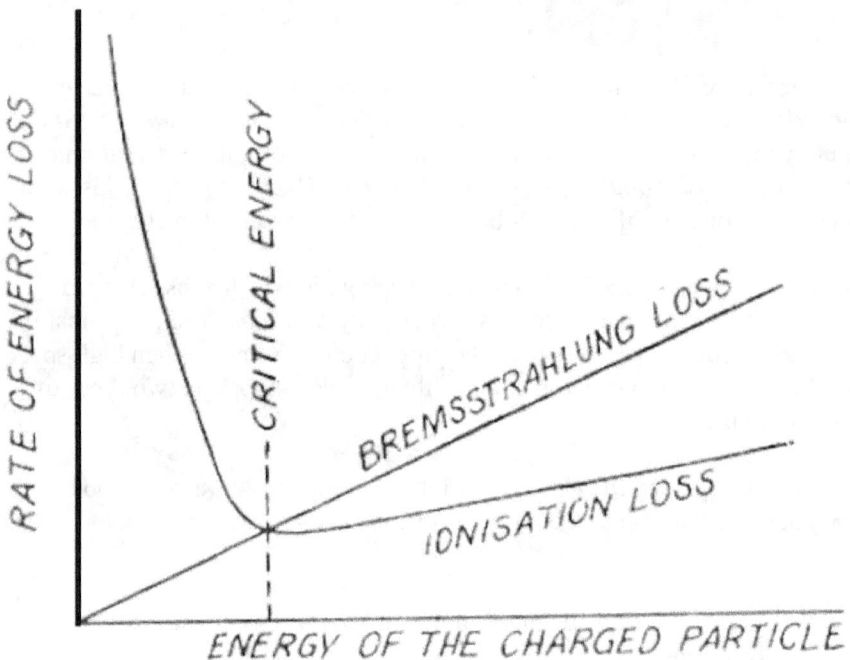

Fig. 320. Critical energy in collision radiation process.

The energy at which the two types of loss become equal is known as the *'critical energy'* of the radiative collision process. Above the critical energy, radiation loss alone takes place, while at lower energies, ionization loss becomes important. It is to be noted also that the particles does not radiate all its kinetic energy at a single collision. Furthermore, the radiation is actually emitted in discrete photons, not in a continuous fashion as in classical theory. At each collision, one photon with energy between zero and the whole energy of the particle will be emitted. According to theory, the probability of emission of photons is inversely proportional to the energy of the photon and hence photons of small energy are very likely to be produced.

(iii) The energy loss by radiation is roughly proportional to the square of the atomic number of the material traversed. This is the celebrated Z^2 law which differs from the Z law, met with in the ionizing collision process, but is similar to that governing the phenomena of materialization of radiant energy with pair production and of the actually observed absorption of soft component of cosmic rays. According to the theoretical Z^2 law, the loss of energy by radiation is very much greater in heavy matter such as lead, than in light matter as water or air. Likewise, the loss of energy due to radiative collision is much greater than that due to ionizing collision for fast electrons, other conditions being the same.

A very interesting and important point about this theory of radiative collision is that although it was originally expected to become invalid at high electron energies of the order of 137 $m_0c^2 \approx$ 70 MeV, experiments conducted with normal electrons of very much higher energies of the order of 500 MeV, met with in cosmic ray showers and bursts, showed that there was no *evidence whatever for the breakdown of the theory* even at those ranges. Nay more, observations of the shape of the altitude-ionization curve in the upper atmosphere made by Millikan and his co-workers (1937) pointed to the *possible validity of the theory even up to energies of 10,000 MeV.*

These theoretical investigations on the loss of energy suffered by fast charged particles in their passage through matter and especially the revealing piece of information concerning the validity of the wave mechanical theory of radiative collision up to very high energies gave the first clue as to the real nature of the highly penetrating particles of the hard component, viz., that they were neither electrons nor protons, whatever be their energy:

Not electrons

The general appearance of their tracks and the existence of positives and negatives among them might, at first sight, appear to suggest that they should be electronic in nature. But there were several serious objections against such an. assumption.

First of all, the approximate energy that an electron should possess in order to have the high penetration of the particle under study, when theoretically calculated, led to values very much larger than any of the energies associated with cosmic ray particles.

Secondly the high penetrating power of the particles suggested that Bethe and Heitler's theory of collision radiation involving rapid absorption as expected by the Z^2 law failed in their case; this, however, could not be due to their high energies, since theory was found valid right up to very high energies. Hence the true cause for the failure must be sought elsewhere.

The *low* probability of radiative collision process for *heavy* particles might well account for the breakdown of the theory. In fact, if the particles were heavy enough not to be suddenly stopped in their interaction with atomic nuclei, their energy would not be largely transformed into impulse radiation and their loss of energy would not be as important as demanded by the radiative collision, but could be explained on the basis of ionizing collisions alone. Hence it followed that the *highly penetrating cosmic ray particles must be heavier than the electrons.* Since, however, when fast, they showed an ionization much like that of the electron, the theory of ionizing collisions, according to which the amount of ionization depends upon the square of the charge carried by the ionizing particle, demanded that *they should have not multiple but single charge, like electrons.*

Not protons

For, in the first place, the small number of slow protons actually observed in cosmic rays at sea level and below (about 10%) could hardly represent all the fast particles constituting the hard component. Secondly, the specific ionization of the particles was found too low to admit their identification with protons and suggested that *they were definitely lighter than protons.*

Thus cosmic ray studies, both from the experimental and theoretical points of view, led to the conclusion that the highly penetrating cosmic ray particles forming the hard component were most probably a new type of particles with a single charge as the electron and proton and a mass intermediate between the masses of the electron and proton.

Direct evidence
Anderson and Neddermeyer, in 1936-1937, by further careful investigations made on the cloud chamber tracks of the penetrating cosmic ray particles were able to confirm the above conclusion and place in clear evidence the existence of the new particle. Their general method of procedure in the analysis of tracks, which were surmised to have been made by the new particle, was to show that the observed range combined with the observed curvature as well as the actual amount of ionization could not be adequately accounted for, if the proton or the electron had been responsible for the tracks, whereas they fitted well a singly charged particle of intermediate mass. Several among the cloud chamber photographs taken, in 1936, at Pike's Peak, subjected to such tests, gave *direct evidence* to the existence of the new particle.

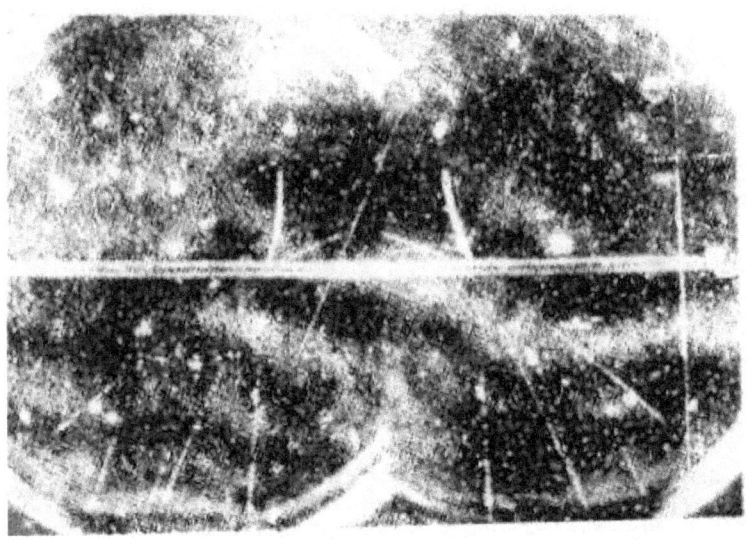

The first photograph which gave evidence to the existence of the mesotron. (Anderson & Neddermeyer)

As an illustration, the first photograph, in which a track of such a particle was recognized, is reproduced here. Out of the six ionizing particles ejected from the same point of the lead plate, probably produced in a nuclear disintegration by a nonionizing cosmic ray, one is far more strongly ionizing than all the others. This one, ejected nearly vertically upwards, has a range of 4 cms nearly; its measured radius of curvature is 7 cms about; the applied magnetic field being 7,900 gauss, its $H\rho$ is approximately 5.5×10^4 gauss-cm. Now the energy of a proton having the observed range of 4 cms would be 1.5 MeV. But a proton of this energy would have a value of $H\rho = 1.7 \times 10^5$ gauss-cm., or a radius of curvature of 20 cms in the magnetic field used. This is about three times the actual value. Hence, if the observed curvature were produced entirely by the applied magnetic field: it would be necessary to conclude that this track must have been left by a massive particle with and *e/m* value much greater than that of a proton, which meant a particle having a mass smaller than that of the proton, but still far larger than that of the electron. Since the track analyzed had a positive curvature, the particle must, be positively charged.

Street and Stevenson, in 1937, in order to make sure that such a particle of intermediate mass really existed, devised a cloud chamber which would record only those penetrating particles which were nearing the ends of their ranges and thereby produced heavy ionization. (It is only in this region of heavy ionization that the ionization is at all characteristic of the mass of the particle, so that analysis of the amount of ionization produced near the end of the range gives clear indications about the mass of the ionizing particles).

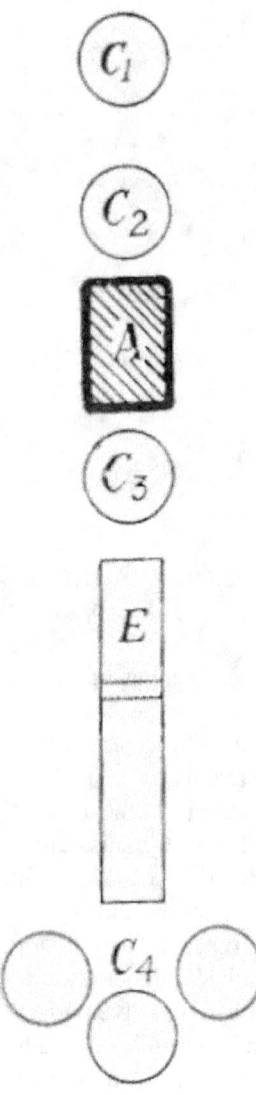

Fig. 321. Apparatus of Street and Stevenson.

The arrangement, of apparatus used by them is schematically shown in Fig. 321. The counter-controlled cloud chamber E was disposed between the counters C_1, C_2, C_3 and C_4 in such a way that it expanded only when a particle passed through the counters C_1, C_2, C_3 but not through C_4. Thus only those particles which stopped before reaching C_4, *i.e.*, particles nearing the end of their ranges, were photographed. A lead block A, 11 cms, in height, was interposed between C, and C_3, in order to filter out the particles of the soft component and allow only the penetrating particles to pass vertically through the chamber.. The expansion of the chamber was further automatically delayed about one second after the passage of the particle, so that the ions formed could diffuse sufficiently and in consequence the droplets condensed on them could be sufficiently separated along the resulting broad track. This permitted with the use of a microscope an easy and fairly correct counting of the number of ions formed per unit length of path, from which a decent estimate of the specific ionization of the particle could be made.

Street and Stevenson were able to obtain photographs of the broad tracks of the penetrating particles almost at the end of their ranges, which were like those shown in the picture (a) on page 1081. Analysis of one of them has been made as follows:

The curvature of the track gives a value of $H\rho = 9.6 \times 10^4$ gauss-cm., while its length a range for the particle about 7 cms. The specific ionization of the particle is approximately six times that of a normal electron. If the particle has entered the chamber from above, as seemed likely, the sign of its charge is negative. The high value of the amount of ionization produced by the particle as compared with that of the electron shows that the particle is much heavier than an electron. On the other hand, the track could not have been made by a proton, since a proton of the observed value of $H\rho$ would have an energy of only 0.45 MeV and hence a range of only 1 cm., *i.e.*, at least 7 times less than the observed value. The, inevitable conclusion is that the particle responsible for the track is neither an electron nor a proton but a new one, of intermediate mass. Furthermore, the heavy ionization can be adequately explained, if the particle has a mass 130 ± 30 times the electronic mass.

Anderson and Neddermeyer, in 1938, using a counter-controlled cloud chamber, where one of the two "triggering" counters was placed in the middle of the chamber instead of its usual place below the chamber, in order to increase the probability of observing particles nearing the end of their ranges, obtained a remarkable photograph. It gives probably the *most direct and the most convincing evidence* for the existence of the new particle. The registered track shows that a particle of 10 MeV energy arriving at the counter placed the chamber passes through it and emerges from it with a much reduced energy of only 0.21 MeV. The residual range of, the particle after it comes out of the counter is 1.5 cms in standard air. This particle which is positively charged cannot have the mass either of an electron or a proton for the following reasons:

The specific ionization of the particle *above the counter is* too great to ascribe it to an electron of the observed curvature. Likewise the specific ionization *below the counter* is definitely greater than that of an electron having the curvature shown, since the radius of curvature, which is about 3 cms, would correspond to an energy of 7 MeV if an electron were involved and such an electron would produce a much thinner track and further would have a range of at least 30 meters instead of the actually observed 1.5 cms.

Moreover, if the particle had electronic mass and emerged from the counter with a velocity such that its specific ionization could correspond to that seen in the photo, its residual range would be only 0.05 cm. instead of the observed 1.5 cms Hence, to have the observed range with the observed curvature, the particle should be considerably more massive than an electron.

The particle cannot be a proton, either. For the curvature of the track *above the counter* would correspond to that of a proton of 1.4 MeV, capable of producing a specific ionization of 7,000 ion pairs/cm., that is at least 30 times greater than what is actually obtained. Similarly, *below the counter,* the curvature would correspond to that of a proton of only 25,000 eV, equivalent to a range of only 0.02 cm. The conclusion which is thus unavoidably forced upon us is that the track under study is due to a new particle of mass intermediate between the proton and electron masses.

(a)   **(b)**

Tracks of (a) a meson (b) an electron under identical conditions. (Corson & Brode)

Other workers such as Nishina, Takeuchi and Ichimiya, Corson and Brode, Ruhling and Crane and Ehrenfest have obtained similar proofs for the existence of this new particle. Corson and Brode, 1938, using the "delayed expansion technique" were able to study the relative ionizations produced under identical conditions by normal cosmic ray electrons and penetrating cosmic ray particles, as illustrated in the above photo and confirm the evidence for the actual existence of the new particle which according to their estimate must weigh at least 185 times as much as a normal electron.

It may therefore be said that a new particle either positively or negatively charged with a mass intermediate between those of the electron and of the proton has been definitely discovered.

Of the several names that have been successively proposed to designate this particle, such as *yukon; heavy electron, baryton, mesotron,* etc., the last-mentioned is now generally accepted, with a frequent shortening into *meson.*

Researches on the meson

The study of the meson is of the greatest interest both as an experimentally observed particle and as a possible realization of Yukawa's theoretical prediction. Since its discovery, intense researches have been and are still being made to study its nature and properties and quite a good amount of data have already been gathered. These findings, however, defy easy co-ordination and sometimes are even in apparent conflict. Chiefly, with the recent discovery of mesons of different masses, current ideas on the subject are in a very rapid state of flux, with a consequent indecision about not only the identity of the cosmic ray mesons with Yukawa's theoretical particles but also the validity of the proposed theory of nuclear forces. We shall now briefly describe the main features of the investigations so far made and the results obtained.

Mass of the meson

The discovery of the meson, as a new particle, cannot be considered as fully established, unless its actual mass is also measured. Hence it is that most of the workers who have studied the evidence for the existence of the meson have attempted to determine its mass as well. In the initial stages of research it was not possible to measure the mass of the meson by a *direct* method, like the one used in the case of electrons and protons for two reasons, *viz.*,

(*i*) no convenient localized source of mesons could be had and

(*ii*) the available cosmic ray mesons move at so high a speed that they could hardly be affected by even the strongest applied electric and magnetic fields.

Hence *indirect* methods which depend on the cloud chamber photographs of meson tracks have been used. The curvature and range of these tracks, the specific ionization and its rate of change along the path of the particle can he directly measured. Since these quantities are connected to the charge, mass and velocity of the particle by simple and well-known relations, they provide means of evaluating the mass of the meson.

Experimental determination of the mass of the meson

In practice the following methods have been used:

(a) Momentum-ionization method

The momentum is determined from the curvature of the track; the specific ionization is measured by counting the droplets along the path of the particle. Although the specific ionization in a given medium for a particle of fixed charge and heavy compared to the electron is independent of the mass of the particle, it depends on its velocity. On the other hand, since momentum is a, function of mass and velocity, a relation between momentum and specific ionization can be readily established from which the mass of the particle can be deduced. The method is subject to an inherent difficulty, in so far as a precise determination of *momentum* requires *sharp* tracks, while, an accurate measurement of *specific ionization* demands *diffuse* tracks.

(b) *Momentum-range method*

The momentum and range of the meson are readily determined from the measured curvature and length respectively of the track in the cloud chamber. Since the range is a function of velocity, it can be connected to the momentum by a simple formula from which the mass of the meson can be evaluated. The chief drawback of tills method is that particles slow enough to stop in the chamber are subject to large scattering effects with the result that the momentum cannot be determined with much precision.

(*c*) *Momentum loss method*

Using tracks that have traversed a metal plate across the cloud chamber, the momentum of the particle before and after passing through the plate can be measured from the curvature of the track above and below the absorber and the loss of momentum $(\rho_1 - \rho_2)$ can he calculated. The difference in range $(R_1 - R_2)$, readily obtained from the thickness, density and nature of the absorber used, is equated to $(\rho_1 - \rho_2)$ and the equation solved for the mass of the particle.

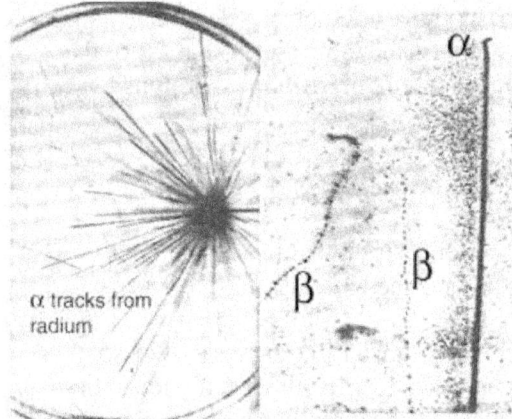

Thickness of tracks of α- and β-particles

(*d*) *Elastic collision method*

Based on the ejection of a "knock-on" electron due to a close collision of a meson with an electron, the method is capable of leading to an accurate measurement of the meson mass by a straightforward application of the simple laws of conservation of momentum and of energy.

With a cloud chamber photograph of the type given here, the initial momentum of the meson is known from the curvature of its track; the amount of energy transferred to the secondary electron can be estimated from the radius of curvature of the electron track and the intensity of the applied magnetic field; the angle between the initial trajectories of the meson and the "knock-on" electron can be measured directly from the photo.

With these quantities, the mass of the meson can be deduced using the formula appropriate to the impact phenomenon involved. The great advantage of this method is that it enables the mass to be determined over wide ranges of meson velocities, even when high, while the previous methods are limited to low speed mesons corresponding to the *heavily ionizing region.*

Results

Using these methods, several investigators have estimated the mass of the meson. Corson and Brode have employed the first two methods, Wheeler and Landenberg the third and Leprince Ringuet and his associates the fourth. The mass of the meson is usually expressed in terms of the electronic mass (m_e). Widely divergent values ranging from 30 m_c to 1,000 m_c, have been obtained; the value close to 200 m_e, having been cross-checked by the different methods, was first generally considered as representing fairly well the meson mass.

But in the course of the researches on the meson mass, it has often been remarked that the values obtained by individual experimenters departed from the most probable value (200 *me*) by far more than the experimental errors permitted. As a result, authors have been suggesting that the meson may not have one single characteristic mass, but may take on, with an equivalent energy change, any mass over a rather wide range.

Bethe, on the other hand, in 1946, has tried to explain away the discrepancies in the experimental values on the basis of the spurious curvature introduced by multiple scattering of the mesons in the gas of the cloud chamber. But, more recently, fresh proofs have been gathered for the existence of mesons of different masses.

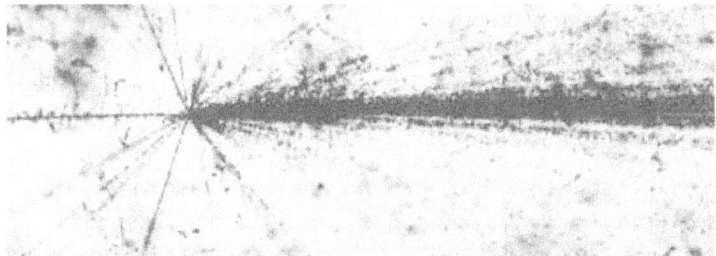

A collision between a high-energy cosmic ray particle and an atom in a photographic emulsion, as viewed through a microscope. Credit: NASA, Dr. David P. Stern.
The cosmic ray photograph was taken from the website "The Exploration of the Earth's Magnetosphere" at URL http://www-spof.gsfc.nasa.gov/Education/index.html

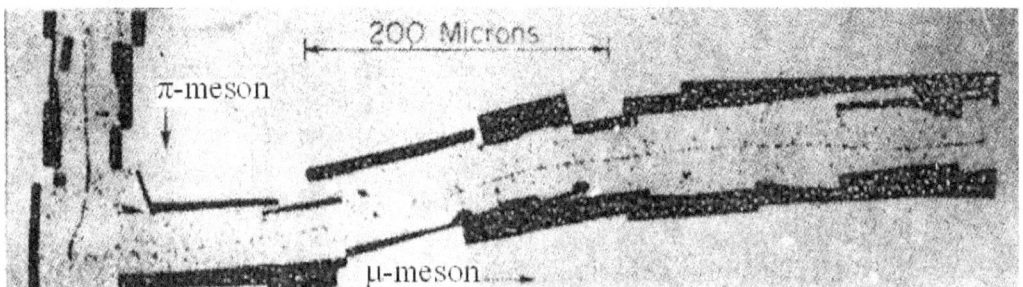

Decay of a π-meson into a μ-meson observed in a photographic emulsion.

In May 1947, Powell, Occhialini and Lattes at Bristol University, England, studying the tracks obtained in photographic emulsions exposed to cosmic rays at high altitudes, found that some meson-like particles decayed with the formation of a second meson rather less massive than the first. One such record, a mosaic of a large number of micro-photographs, is reproduced here. They were able to estimate that the first meson had a mass of about 320 m_e and the second one, a product of decay, was almost certainly the first discovered meson of mass about 190 m_e, while simultaneously a neutral meson of mass about 130 m_e was formed. The first and heaviest is now known as the *π-meson,* while the second and lighter one, the *μ-Meson.* Both types are found to be either positively or negatively charged and appear singly in pairs and in showers.

Confirmation of the Bristol discovery has been obtained by other workers. Thus, in October 1947, Anderson, studying cosmic rays with the help of a large aircraft, has reported a photograph, in which the momentum of the decay electron from a 190 m_e meson is found to be different from what is to be expected in the normal mode of decay involving a neutrino. This unusual feature has been interpreted as probably due to the appearance of the neutral 130 m_e meson, already found by the Bristol workers.

A more exciting corroboration, which at the same time marks a great advance in fundamental particles physics, is *the artificial production of π-mesons in the laboratory.* At the beginning of 1948, Gardiner and Lattes, bombarding different targets, such as carbon, beryllium, copper and uranium, with 380 MeV alpha particles from the 184-inch Berkeley cyclotron, were able to produce mesons from the disrupted nuclei of the targets. The ejected mesons were deflected by the magnetic field of the cyclotron and received on a stack of photographic emulsions placed across the expected path. The exposed stack, when carefully developed, exhibited hundreds of meson, tracks. A 30 secs bombardment with the 184-inch cyclotron gave 100 times as many mesons as those obtained in nearly 2 months of cosmic ray observations. These artificially produced mesons were naturally not so energetic as those produced by cosmic rays; they were about 4 MeV. But wider the controlled conditions of their production, it was possible to determine their masses quite exactly from the measurements of range and Hρ, the latter being given by the geometry of the apparatus and the orientation of tracks in the plate. Practically all the mesons knocked out of the targets had a mass of about 313 m_e agreeing well with the heavy π-mesons of 320 m_e, discovered in the cosmic ray photographs. This successfully carried out experiment has initiated a new era in the study of mesons by the high intensity of mesons it has made available (about 10^6 times that of the cosmic ray meson intensity at sea level) and also by the opportunity it offers for the determination of the sign of the particle from the sense of curvature in the magnetic field and of the momentum from the amount of curvature.. Further researches in neutron-proton interaction using high energy particles from the more powerful proton synchrotrons established

(i) the mass of the charged π-mesons as 273 m_e and
(*ii*) the existence of a neutral **π** -meson of mass 264 m_e and
(*iii*) the mass of the charged μ-mesons as 207 m_e.

There exists also sure experimental evidence for *mesons still heavier than the π-mesons, having a mass of about 1000 m_e.* As early as 1944, Leprince Ringuet, studying the elastic collisions between slop meson-like particles and stationary electrons in the cloud chamber, obtained a single record of a particle about 1,000 times heavier than an electron.

In the summer of 1947, Rochester and Butler of Manchester University, while studying the "penetrating cosmic ray showers", were able to obtain cloud chamber photographs with forked tracks, which could only be explained by the decay of a particle about 1000 m_e which was christened the "τ-meson". Since then, several other particles whose masses are more or less the same as that of the τ-meson have been discovered and called K-mesons.

Rochester and Butler discovered in 1948 also another particle with mass greater than that of the proton (2190 m_e) which decays into a proton and a negative π-meson, with an energy release Q of about 37 MeV. It was called V_1^o particle.

From work with photographic emulsions, still other types of very heavy particles, of mass about 2,500 m_e, which decay into a neutral particle (neutron) and a π-meson, with an energy release Q ≈130 MeV have been found recently and called J^+- particles.

These two last categories seem to be very much more closely related to nucleons than to mesons — some sort of *excited nucleons.* They are now called *hyperons or* Y-particles.

Some authors, like J. G. Wilson, have tried to account for the occurrence of distinct and different sorts of mesons on the basis that the meson field may demand various alternative types of waves, such as transverse, longitudinal, heat conduction, etc., unlike the electromagnetic field restricted to the transverse type alone.

Mean life of the meson
The question of a mean life-time for the meson, which implies that the particle is unstable and decays first arose in an attempt to identify the cosmic ray mesons with the particles predicted theoretically by Yukawa. The Yukawa particle was supposed to be unstable: after a very short life, estimated at 1/2 microsecond, it disintegrated spontaneously in accordance with the laws of radioactivity into an electron and a neutrino. Hence it was highly interesting as a test case to see whether the mesons decayed in the manner predicted by theory.

During recent years, ample experimental evidence have been gathered to show that the mesons are definitely radioactive. Williams, Roberts and Evans (1940) and Shutt, De Benedetti and Johnson (1942) have been able to obtain a few cloud chamber photographs which give *direct* and *spectacular* evidence for the disintegration of the meson. The photo obtained by the last-named workers with a high pressure (70 atmospheres) cloud chamber is reproduced here.

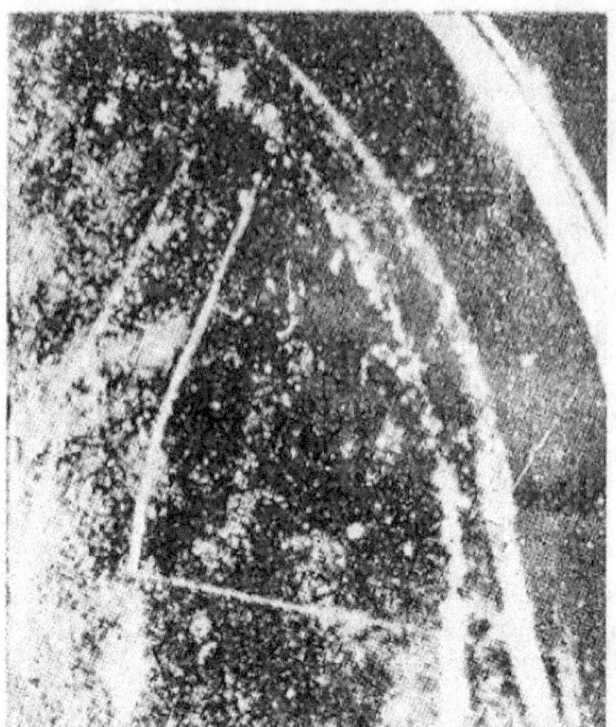

Tracks of a decaying meson and of the electron emitted by it.

It is seen that a meson enters the chamber from top, gradually slows down as indicated by the increase in ionization and scattering and finally stops at a point, from which a lightly ionizing particle of high energy, presumably an of the electron, flies off to the right at an angle of 85° with the direction of the meson-track. The final kinetic energy of the decaying meson estimated from the density of ionization is less than 1 MeV, while the energy of the emitted electron appears to be greater than 30 MeV. The simultaneous emission of a neutrino would also be involved to satisfy the law of conservation of energy and momentum.

Experimental determination of the mean life of the meson
The following three methods have been used to measure the mean life of the meson.

(a) *Method of seasonal change of cosmic ray intensity*
It has already been said that the slight variation in cosmic ray intensity with the season is due to the spontaneous decay of the mesons formed in the upper atmosphere. A rise of temperature means that the air layers in which mesons are formed move upwards away from the earth, so that the mesons, which are the chief carriers of cosmic ray effects downward, have a greater distance to travel and more of them perish on the way.

Blackett, the author of this explanation, was able, from a study of this effect, to estimate the mean life of a meson, when at rest, as 3.4×10^{-6} sec.

(b) *Absorption method*

It has been known for several years that the penetrating components of cosmic rays in traversing the atmosphere are absorbed more rapidly than can be accounted for by the mass of air they pass through, The most logical explanation of this fact is that some of the mesons constituting the hard component disappear by disintegration during the time required to pass through the atmosphere. Hence the absorption of these particles *depends not only upon the quantity of matter they traverse, but also upon the distance they travel in passing through that matter.*

From this it follows that the absorption of mesons in the atmosphere should be much greater than that in an equivalent layer of more condensed material, such as metal plates, since the distance to be traveled in the latter is quite negligible (a layer of lead; for example, is about 1/10,000 of the thickness of the equivalent air layer) and consequently there will be no comparable loss of mesons by decay. Measurements made upon the relative absorption of the mesons in the atmosphere and in equivalent metal absorbers should, therefore, decide whether mesons .really decay, and even permit the estimation of the mean life-time.

Several workers, such as Rossi, Hilberry and Hoag (1939), Neher and Stever (1940) and Rossi and Hall (1941) have utilized the above-mentioned facts in measuring the mean life of the meson. Using a vertical counter telescope with 12.7 cms, of lead between the counters, in order to filter out the soft component, Rossi, Hilberry and Hoag measured the absorption of mesons in air by making observations of the intensity at different altitudes. The absorption in carbon was also measured at each altitude by placing graphite layers of different thickness above the counters. It was found that the mass absorption in air was greater than that in carbon. For instance, an air layer of 82 gms./cm^2 reduced the meson intensity more than twice as much as did a carbon layer of 84 gms/cm^2. The greater apparent absorption in air is presumably due to the decay of the mesons in the greater distance and hence the longer time they take in traversing the air than the solid absorber. From the experimental data

the mean life-time of meson could be deduced and it was found to be about 2 x 10^{-6} sec. with some indication that this quantity might not be constant but depended upon the energy of the particle.

(c) *Disintegration electron-method*
Based on the detection of the electrons given out by the decaying mesons, this method has been successfully employed by Rasetti (1941) and by Nereson and Rossi (1943) for measuring the mean life of the meson.

The *principle* of the method is as follows:

When mesons pass through heavy solid absorbing material, losing energy by ionization, they soon reach the end of their path, come to rest and then decay. Hence if the time interval between the instant at which the meson comes to rest in the absorber and that at which the electron is emitted can be measured in a sufficiently great number of cases, the mean life can be estimated, assuming an exponential decay. This method appears to be the *best,* since the proper time of decay is *directly* measured.

The *experimental arrangement* used by Nereson and Rossi is diagrammatically shown in Fig. 322. A beam of mesons selected by a four-fold counter system represented by 1, 2, 3 and 4 with interposed lead blocks is allowed to pass through a brass block absorber B where the mesons decay and electrons are emitted. The selector circuit S works with the counter sets 1, 2, 3 in coincidence and 4 in anti-coincidence, *i.e.,* it responds only when all the first three act while the fourth does not. The counter of set 3 surrounding the brass block serves also the purpose of detecting the decay electrons.

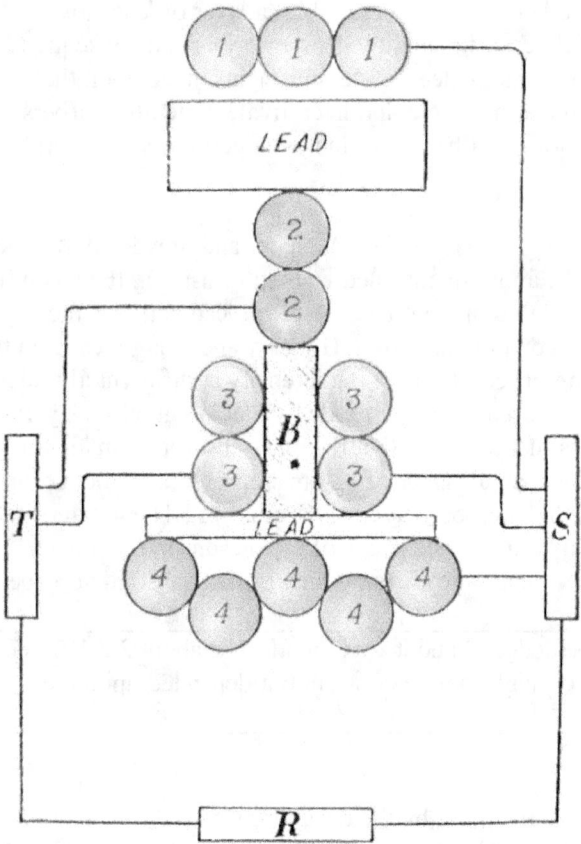

Fig. 322. Apparatus for the study of meson decay.

With this device, the selector will evidently react only when a meson has stopped in B, since otherwise, *i.e., if* the meson .has passed through 4 also, S will not respond. This meson which has come to rest decays with the emission of an electron that activates one of the counters in set 3.

To measure the mean life, counter sets 2 and 3 are also connected to a time circuit T in such a way. that, when a coincidence in set 2 is followed after a certain interval of time *to* by a pulse from set 3, T gives rise to a corresponding pulse whose magnitude is proportional to t_o. The recorder R in the final stage mechanically marks the magnitude of the pulse received from T, whenever a pulse is simultaneously received from S, indicating thereby that the event looked for has really taken place. From the experimental data, it is possible to obtain the decay curve of mesons by plotting the number of mesons decaying after the lapse of a certain time t_o against t_o. It is found to be exponential in form, as in radioactive decay. The mean life deduced from the decay curve is (2.5 ± 0.07) x 10^{-6} sec.

Recently, the *liquid scintillation counter technique* has been employed to measure the mean life of mesons artificially produced with the very high energy accelerators. In one form of apparatus an aluminum absorber is used, placed between two scintillation counters in coincidence. A third scintillation counter placed below the first two and connected in 'anti-coincidence' with them ensures that the events to be recorded take place where they are detectable, since mesons which pass through all the three counters will not be recorded. The entry of the π-meson into the first counter causes a pulse to be recorded, the emission of a μ-meson gives rise to a third pulse. These pulses are timed electronically and hence the process of meson decay can be studied. From the experimental data, it is found that

a π-meson decays into a μ-meson in about 2 x 10^{-8} sec. and the μ-meson in turn decays into an electron in about 2 x 10^{-6} sec.

Comparison with theory

These experiments leave no doubt that the *meson is radioactive* as predicted by theory. The mean life of the μ-meson is 2 microseconds, which, though of the order given by theory, is found to be four times greater. Yukawa, attempting to overhaul the theory in view of this experimental finding, was led to a still more curious result, viz., the correct theoretical decay time of the Yukawa particle is not 1/2 microsecond but only 1/100 microsecond. Hence the μ-meson may not after all be the Yukawa meson as was originally supposed. The recent discovery of different kinds of mesons with their roughly estimated mean lives, as 10^{-10} sec. for 2,190 m_e meson, 10^{-9} sec. for 1,000 m_e meson and 10^{-8} sec. for 300 m_e meson, seems to confirm this conclusion.

NOTE:

Classification of the fundamental particles

With the discovery of so many different kinds of mesons, a certain amount of confusion has been inevitable. Various authors have called the same particle by different names or have attached different meanings to the same symbol. Sometimes the meaning of a symbol has changed in course of time. It has therefore been found necessary to agree upon some classification of the fundamental particles and the International Cosmic Ray Congress held under the auspices of the University of Toulouse, France, in July 1953, has proposed the following suggestions:

(*i*) *The fundamental particles are to be subdivided into the following three main, groups, according to their mass*

(a) *L-mesons*

Light mesons which include π-mesons, μ-mesons and any other light mesons which may be discovered. The masses of these mesons are meson

$\mu^{\pm}$-meson — 207 m_e, life-time 2.2 x10^{-6} ,
$\pi^{\pm}$- meson — 273 m_e, life-time 2.5 x10^{-8} ,
π^{o}- meson — 264 m_e, life-time ≈ 10^{-15} sec.

(*b*) *K-mesons*

Heavy mesons — all particles heavier than mesons and lighter than protons. These comprise of:

$\tau \rightarrow 3\pi$ (considered certain); mass = 966 m_e, life-time = 1.2 x10^{-8} sec.

$\kappa \rightarrow \mu + 2$ neutral particles (considered very probable; nature of neutral particles unknown); mass = 963 m_e, life-time = 1.2 x 10^{-8} sec.

$\chi \rightarrow \pi + 1$ neutral particle (considered as probable, mass = 963 m_e, life-time = 1.2 x 10^{-8} sec.

θ^{o} (neutral heavy meson) previously known as v^{o}, V_2^{o}, V_4^{o}, characterized by the decay scheme
$\theta^{o} \rightarrow \pi^{\pm} + (\pi^{\pm}$ or $\mu^{\pm})$
If it turns out (as suggested by some results) that there are particles with this decay scheme and several Q-values, they could be designated by different subscripts; mass = 965 m_e, life-time =1.3 x 10^{-10} sec.

(*c*) *Y -particles. Hyperons*

These comprise all particles with mass heavier than that of the proton. They comprise of:

A^{o} — previously known as V_1^{o} and (now known as *neutral hyperon*) is characterized by the decay scheme
$A^{o} \rightarrow \rho + \pi^{-}$
If there are particles with this decay scheme and different Q-values they could be designated by different subscripts; mass = 2,182 m_e, life-time = 2.8 x 10^{-10} sec.

$\Sigma^{\pm}$ — with the possible decay scheme:
$\Sigma^{\pm} \rightarrow n^{o} + \pi^{\pm}$
$\Sigma^{\pm} \rightarrow p + \pi^{o}$; mass = 2,327 m_e, life-time ≈10^{-10} sec.

(*ii*) *Based on the empirical features of the decay process, a further phenomenological classification is made as:*

(*a*) *V -event* — previously known as V-particles: decay *in flight* of K-meson or Y-particle; subdivided into V^{o}-event (decay of a neutral particle) and $v^{\pm}$ event (decay of a charged particle).

(*b*) *S-event* — previously known as S-particles: decay *at rest* of a charged K-meson or Y-particle.

Origin of mesons

The fact that *mesons* spontaneously decay with a very short mean life clearly indicates that they *cannot have come from far away and hence cannot themselves be the primary cosmic rays* which certainly originate very far away from the earth as established by the geomagnetic effects.

The mesons must, therefore, be of *secondary origin,* formed close to the earth. Moreover; since there is experimental evidence that the mesons are not produced at sea level, they must be formed *in the upper atmosphere.* Study of the distribution of mesons in altitude not only confirms the above conclusion, but also suggests that the *process of meson creation is almost completed in the top one-tenth of the atmosphere* (in weight).

The most interesting and important problem concerning the origin of mesons is the *mechanism of meson formation* which is bound to throw light on the nature of the primary cosmic rays. It has been tackled both from the experimental and theoretical standpoints, which strongly tend to the conclusion that the *primary radiation responsible for the penetrating component observed at the surface of the earth consists largely of protons.*

Experimental study

Meson production by non-ionizing radiation in cosmic rays

Many workers have obtained evidence that *at high altitudes mesons are generated to a small extent in solid absorbers by non-ionizing radiation.* Thus, for instance, Schein and V. C. Wilson (1939), employing an arrangement of counters in coincidence capable of detecting mesons and an interposed thin lead block in which the mesons were generated and which could be disposed in two different positions in order to obtain the amount of mesons produced by nonionizing radiation alone, made observations in airplane flights and showed that mesons were produced by non-ionizing radiation at high altitudes.

The generating agent might however be either photons or neutrons. To decide this point, Schein, Jesse and Wollan, in 1940, conducted a series of experiments with a new arrangement of coincidence counters with interposed lead blocks and showed that the production of mesons increased in proportion to the intensity of the soft component of the cosmic rays, which was known to contain an appreciable number of photons of high energy. They therefore concluded that the *generating radiation was presumably photonic in* nature. They also found that:

(i) the meson-production was negligible up to an altitude of 40 cms.Hg beyond which it increased rapidly,
(*ii*) the mesons were produced either singly or in narrow pairs to a small extent,
(*iii*) the energy of photons must be greater than 200 MeV for meson production and

(*iv*) the mesons produced were of low energy.

To obtain confirmatory evidence about the photonic nature of generating radiation, Tabin, in 1944, carried out an experiment in which the absorption of the generator of mesons in different substances was studied, in order to see whether the absorption laws for photons were obeyed. A four-stage coincidence counter set, known as the *master-coincidence,* was employed to select cases in which only mesons were present, natural or generated. A large cap of closely shaped counters protected the generator which was a block of paraffin, iron or lead, from the undesirable action of ionizing radiation. The thickness of these blocks could be varied.

Observations were made at high altitude mountain stations (10,000 to 14,000 ft.). It was found that saturation for meson production in lead occurred for a thickness equal to 2 to 3 cms, beyond which the number of mesons decreased, while with paraffin, no saturation in the production of mesons could be reached even at a thickness of 57.2 cms

This result indicated that the generating agent was really photon, since it should take about a meter of paraffin to absorb the photons, whereas all photons could be absorbed in a few cms of lead, any additional thickness merely acting as a filter of the already generated mesons.

In this connection, it may be noted that evidence for the production of neutral π-mesons by high energy X-rays from a synchrocyclotron was first obtained by the use of a scintillation counter technique. These neutral mesons decay into two quanta of energy about 100 MeV and, although the primary particles cannot be detected, the two γ-ray quanta produce scintillations by which their presence can be recorded. By a suitable arrangement of the scintillation counters it has been shown that these quanta are emitted in pairs and that no other ionizing particles are simultaneously emitted. Measurements of the angular distribution provide data from which the approximate mass of the initial particle can be deduced. This has been shown to be about the same as the mass of the π-meson.

Meson production by ionizing radiation in cosmic rays

All the mesons present in cosmic rays cannot be explained as produced by photons in the soft component for the following reasons:

(1) Above an altitude corresponding to 8 cms of Hg the soft component falls off rapidly, while the hard component increases up to the highest altitudes reached, about 2 cms of Hg.
(2) A great number of mesons has a much greater energy than those generated by photons. These facts suggest the *formation of mesons by ionizing particles chiefly.* Furthermore, in view of the steep rise of meson intensity from zero at the top of the atmosphere to a maximum value at about 2 cms of Hg level, a new *mechanism by which several mesons are generated at a time might be expected.*

To verify these indications, experiments were conducted by Janossy and Rochestor of Manchester University *at sea level* (1942) and by Schein, Iona and Tahiti in America (1943) *at high altitudes.* The *basic principle* involved in these researches is to produce mesons in a material of low atomic number such as paraffin, where the undesirable electronic showers are not likely to be produced, and to detect them under large thickness of heavy metal, such as lead, which will absorb any electrons that may still linger about, chiefly of the "knock-on" type. Elaborate coincidence counter arrangements are used for the detection of mesons. The great amount of lead to be used is an obstacle against reaching the highest altitudes of maximum meson formation, while the sea level observations labor under the paucity of the generating particles.

The *results* obtained by these experiments may be summarized as follows:

(i) At sea level penetrating showers are very rare, but at high altitudes half of the penetrating particles are accompanied by mesons showers.

(*ii*) Practically all the meson showers are found to be excited by ionizing particles. A transition curve for these showers clearly indicates that the mechanism involved is quite different from the Z^2 process of the electron showers.

(*iii*) At sea level, 30 per cent of penetrating showers are excited by non-ionizing agency, which shows that neutrons are not ruled out as generating agency. Assuming that what reaches the earth from outside is entirely a proton stream, it is still conceivable that, after some protons have made meson-forming collisions, there will also be neutrons present and that at sea level the two agents, protons and neutrons, will probably be available in comparable numbers, but most of them will be secondaries.

Theoretical study

Yukawa's theory
Yukawa, in 1935, in order to explain adequately the 'short range attractive force' between nuclear particles (protons and neutrons), suggested the existence of "heavy quanta", in analogy with the photons that appear in the electromagnetic field associated with the 'long range Coulomb's inverse square force' between charged particles. Just as charged particles interact electromagnetically through the agency of ejected and absorbed photons, so may the nuclear particles interact through ejected and absorbed "quanta" of some sort.

Following up this analogy and speculating on the correct mathematical equations that would represent the specifically nuclear force, he was led to conclude that the quanta involved were not photon-like particles which have no mass, but were rather particles which had a definite rest-mass about 200 times that of the electron, and carried a single elementary charge, positive or negative. These particles were called *' mesons'* and the field of force *'meson field',* about which we shall deal more in detail later. For the present, we are interested in the fact that according to Yukawa's theory, *mesons are generated in close collisions between the nuclear particles, analogous to the formation of photons.*

Hamilton-Heitler-Peng theory of meson formation in cosmic rays

On the basis of Yukawa's meson theory, Hamilton, Heitler and Peng, in 1943, proposed the hypothesis that the *primary cosmic ray particles are protons which interact with the atoms of air in the uppermost part of the atmosphere to form penetrating mesons with a considerable multiplicity.*

Considering the two types of action characterizing the meson field, viz., interaction of electromagnetic field of moving particles with meson field of stationary particles, represented by

$h\nu + proton \rightarrow neutron + meson$ (+)
or
$h\nu + neutron \rightarrow proton + meson$ (-)

and

(ii) interaction of meson field of moving particles with meson field of stationary particles, represented by

$proton + proton \rightarrow proton + neutron + meson$ (+)
or
$neutron + neutron \rightarrow neutron + proton + meson$ (-)

they calculated the respective cross-sections for meson production in the two cases and showed that

(a) for energies met with in cosmic rays the second process has a much larger cross-section than the first,
(b) one single proton entering the earth's atmosphere will lead in the end to several mesons (in the first effective collision of this proton with a nuclear particle, the proton will change into a neutron with the emission of a positive meson; the neutron, in turn, makes a similar collision and changes into a proton with the emission of a negative meson; the process is repeated until several mesons are formed with the dissipation of the energy of the primary particle) and
(c) for incident protons of energy greater than 3×10^9 eV, the probability of the process is so high that practically all the meson production takes place within about 2 cms of Hg of the top of the atmosphere.

Thus protons constitute the primary cosmic radiation, which produces the mesons, forming the hard component, within a very small region at the top of the atmosphere.

The mesons are then supposed to be the source of the soft component, by knock-on or decay process. The formation of mesons by protons may be considered to be similar to the production of 'bremsstrahlung', in which mesons are scattered from the meson field.

An interesting feature of the meson-forming collision, first pointed out by Janossy, is that *in close collisions* a *single primary particle can produce several mesons from a single complex nucleus,* since the first particle passing through the nucleus is likely to have several meson-producing collisions with more than one of the nuclear particles. This also might explain the rapid formation of practically all the mesons in the very top of the atmosphere.

Experimental tests so far made touch only the broadest outlines of the theory. The production of several penetrating particles by one primary particle has been demonstrated not only by such investigations on multiple meson production as those of Schein and collaborators, but also by cloud chamber pictures of *'penetrating -showers'* obtained by Powell (1941), Haven (1944) and Fretter (1947), of which we shall speak in the section on cosmic ray showers. Since according to theory, a good number of fast protons should be expected close to the top of the atmosphere, experiments with cloud chambers to determine the number of high energy protons at great altitudes are very desirable. Likewise, one should expect to find high energy neutrons at those altitudes, which however, cannot be observed until they are slowed down by sonic means.

Decay of mesons
We have above treated the problem of mean life as though both positive and negative mesons always behaved in exactly the same way. This is not quite true. For, experiments conducted on the decay electrons show that the number of these electrons is not quite as great as would be expected. For, instance, Rasetti found that the number of disintegration electrons was slightly less than half the number of mesons that disappeared in the absorber, which indicated that about *one decay electron was produced for every two mesons absorbed.* Cloud chamber photographs of decaying mesons, in their turn, prove that decay electrons are produced only in the case of positive mesons, but not with negative mesons. So far, no photograph has been obtained, which contains the track of a decaying negative meson producing an electron.

A very probable explanation of this difference of behavior of mesons according to the sign of their charge is as follows:

When a negative meson has come almost to rest, the strong electrical attraction between it and atomic nuclei become important, and there is a high probability that before it has had time to decay, the negative meson will have been attracted to and captured by a nucleus which subsequently might explode.

Cases of such disintegrations by slow meson capture have been observed, in 1947, by Perkins and by Powell in photographic emulsions exposed for several days at high altitudes, as illustrated in the adjacent photo, where a nucleus disintegrates by capturing a. slow meson presumably negative, and emits a proton and an α-particle and two other, probably heavier, nuclear fragments.

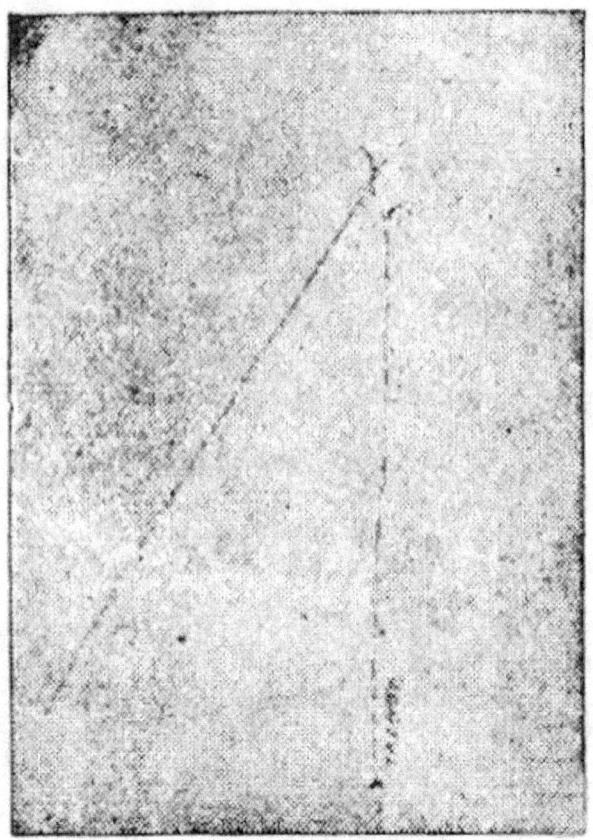

Disintegration produced by a meson. (Powell)

In the light of these findings and other data, the following *provisional scheme for the decay of mesons* has been suggested:

(a) π- *mesons,* positive or negative, are formed in nuclear disintegrations, while *μ-mesons* of either sign are produced by the decay of π-mesons.

(b) *The positive π-meson d*ecays spontaneously with a mean life 10^{-8} sec. into a *positive μ-meson* and a *neutral meson,* regardless of the atomic number of the absorber in which it is stopped.

(c) *The negative π-meson,* on the other hand, may *either undergo spontaneous decay,* with the same mean life as the positive π-meson into a *negative μ-meson,* and a *neutral meson, or suffer capture by a nucleus,* followed by the disintegration of the nucleus— "star" effect. The former process is less likely and the latter more likely to occur with increasing atomic number of absorber.

(d) *The positive μ*-meson decays spontaneously with a mean life of 2×10^{-6} sec. into a *positron* and probably a *neutral meson* and a *neutrino,* regardless of the atomic number of the absorber, in which it is stopped.

(*e*) *The negative μ-meson* may undergo *either spontaneous decay* of mean life 2 x 10⁻⁶ sec. into an *electron* and probably a *neutral meson* and *a neutrino, or be captured by a nucleus,* followed by a process as yet undetermined, but apparently not a *"star"* effect. Decay is less likely and capture more likely to occur with increase in the atomic number of absorbers.

(*f*) The nuclear interaction of μ-meson being, however, smaller than that of the π-meson, in light absorbers most μ-mesons will decay, while most π-mesons will be captured.

It may to be noted that neutrino is required to reconcile the decay of π-and μ-mesons with the conservation laws, as in the case of β-radioactivity. As an example, consider the decay of the μ-meson. The electron, which is the only visible product, can have an energy between zero and 55 MeV. This shows that there must be at least two other decay products, since if there were only one, the decay electron would always have the same energy. The maximum energy which the electron can take also shows that the combined mass of the neutral products is very small. The only known particle having these properties is the neutrino, and accordingly the decay scheme of the μ-meson is written as

$$\mu^+ \rightarrow e^{\pm} + \nu + \nu$$

It might be argued that an alternative to the neutrinos would have been two photons (i.e., γ-rays). This possibility has been investigated experimentally by searching for electron pairs caused by the materialization of the γ-rays. None, however, has been found.

Energy spectrum of mesons
The energy spectrum of mesons has been studied by two methods chiefly, the counter-controlled cloud chamber and the coincidence counters. The former has been employed by Anderson and Neddermeyer (1934), Leprince Ringuet and Crussard (1937), Blackett and Brode (1937), Jones (1939), Hughes (1940), and J.G. Wilson (1946). The latter method has been utilized by Rossi (1931) and by Benardini and collaborators (1945) with an auxiliary iron-core technique which separates particles of opposite signs. To extend the investigation of the spectrum to high values of energy, very thick absorbers such as deep lakes, mines etc., have been used by Clay (1937), Ehmert (1939) and others. Recently, the liquid scintillation counters have also been used in this type of research. The energy spectrum has been analyzed both at sea level and at high mountain altitudes (15,000 ft.).

The results obtained from these researches may be summarized as follows:

At sea level the differential meson spectrum increases from zero intensity to a maximum at about 10^9 eV, falls off approximately as the inverse square of energy between 2 to 3 x 10^9 eV and 10 to 15 x 10^9 eV and roughly as the inverse cube from 15 x 10^9 eV to 60 x 10^9 eV. This means that the mesons have energies ranging from 200 to 15,000 MeV, the majority of them having an energy in the whereabouts of 1,000 MeV (corresponding to the maximum in the energy spectrum curve). There is also a small number having energies up to 69 x 10^9 eV and more. At high altitudes, there is a much larger number of low-energy mesons, the peak intensity occurring at about 100 to 200 MeV. This emphasizes the fact that *a large number of low energy mesons fail to reach sea level, all of which are subject to decay, at least half of them, giving rise to decay electrons, constituting the soft component in the lower parts of the atmosphere.*

Comparing these results with those obtained for *electrons* by Lombardo and Hazen, it is seen that the electrons have much lower energies, ranging with decreasing intensity from 10 MeV up to 500 MeV.

11. COSMIC RAY SHOWERS

'Showers' is the name given to a very interesting cosmic ray phenomenon, in which the *rays arrive in groups that appear to have been produced simultaneously by some common cause, i.e., without becoming associated merely by chance.*

Several years before the phenomenon itself was formally discovered, its effects had been noticed in the occasional sudden *bursts* of ionization in cosmic ray ionization chambers at all altitudes.

Discovery
Rossi, in 1932, with his coincidence counter technique and Blackett and Occhialini, in 1933, with their counter-controlled cloud chamber, were the first to place in *clear* and *direct evidence* the phenomenon of cosmic ray showers.

Evidence obtained with coincidence counters

Rossi discovered the simultaneous emission of groups of particles from matter as secondary products of cosmic rays, using an arrangement of three coincidence counters in a triangular pattern with lead plate above the counters, as shown in Fig. 323.

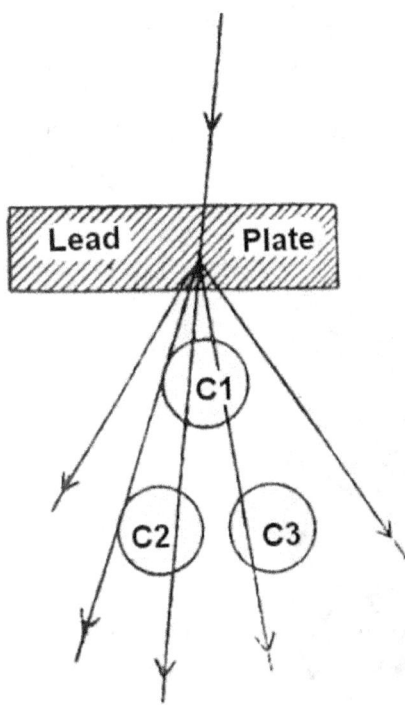

Fig. 323. Rossi coincidence counter arrangement for the study of cosmic ray showers.

A coincidence discharge of the three counters can be produced only by the simultaneous passage of at least two particles including the incident one. Rossi observed that an appreciable number of coincidence counts was registered and these counts became still more numerous if a lead plate about 1 cm. thick was placed above the counters, From this result he naturally inferred that the three counters were affected by two or more secondary particles produced simultaneously by a single cosmic ray as it penetrated the lead plate. Thus the production of showers in the lead plate was detected.

Discovery of skewers with cloud chamber

Cloud chamber photographs of cosmic ray tracks occasionally revealed **a** group of two or more tracks. which appeared to have been produced simultaneously. Often the group seemed to have a common focus lying in the walls of the chamber or even outside it, from which the tracks diverged. Skobelzyn was the first, in 1927, to notice this peculiar feature of the cosmic ray tracks.

Blackett and Occhialini with their counter-controlled chamber situated in a magnetic field were the first, in 1938, to obtain good photographs of such groups and demonstrate that the tracks in a given group were bent in opposite directions in about equal proportions thereby indicating that each group consisted of approximately equal number of positive and negative particles. They further showed from the measurement of energies and the amount of ionization involved in the tracks that the constituent particles were positrons and electrons. The groups of such tracks were named by them *'cosmic* ray *showers.'*

Anderson and Neddermeyer, in America about the same time, obtained photographs of these 'showers' and confirmed the discovery of Blackett and Occhialini. In the adjacent picture we have reproduced one of their photos, where a direct ocular evidence is had of **a** shower which, consisting of three positrons and three electrons, originates at the wall of the chamber. They extended very much the study of these showers by introducing a lead plate across the chamber, as we shall see presently.

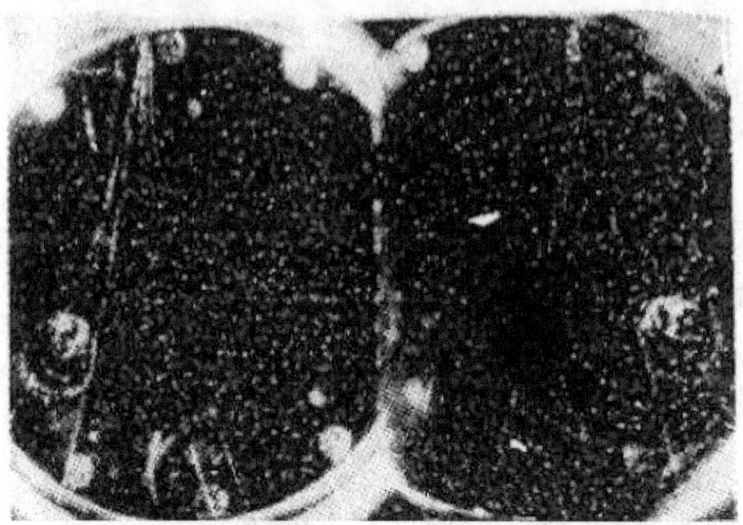

Stereoscopic cloud chamber photo of a shower originating at the wall of the chamber (Anderson & Neddermeyer)

Researches on cosmic ray showers

From the theoretical point of view, a fairly impressive theory, known as the 'cascade process of showers' has been worked out in great detail. On the experimental side, however, the study of the showers has not advanced very much, chiefly as regards the data required to check the proposed theory. We shall first, give a brief summary of the experimental investigations so far made and the results obtained and then pas on to the theory of showers.

Experimental study of showers

The different characteristics of the shower phenomenon have been studied with three different techniques, viz.,

(*i*) *the cloud-chamber,*
(*ii*) *shower counting array of coincidence counters* and
(*iii*) *bursts* in cosmic ray ionization chamber.

The following points have been investigated:

(a) Shower-producing agents

The counter-controlled cloud chamber with a lead plate across it has been extremely useful in elucidating this point. The showers seem to diverge form points ill the lead plate or in dense material close to the cloud chamber, sometimes from the very wall of the chamber, but rarely form a point the gas inside the chamber. This indicates *that the chance of shower production depends upon the amount of matter there is at any point and that solid matter, probably the heavier the better, is particularly conducive to shower formation.*

But heavy matter alone is not enough for producing showers; a radiation that can effectively interact with matter is required. Cloud chamber photographs make it clear that *showers maybe induced ether by an ionizing particle or a non-ionizing photon,* as illustrated in the following two photos:

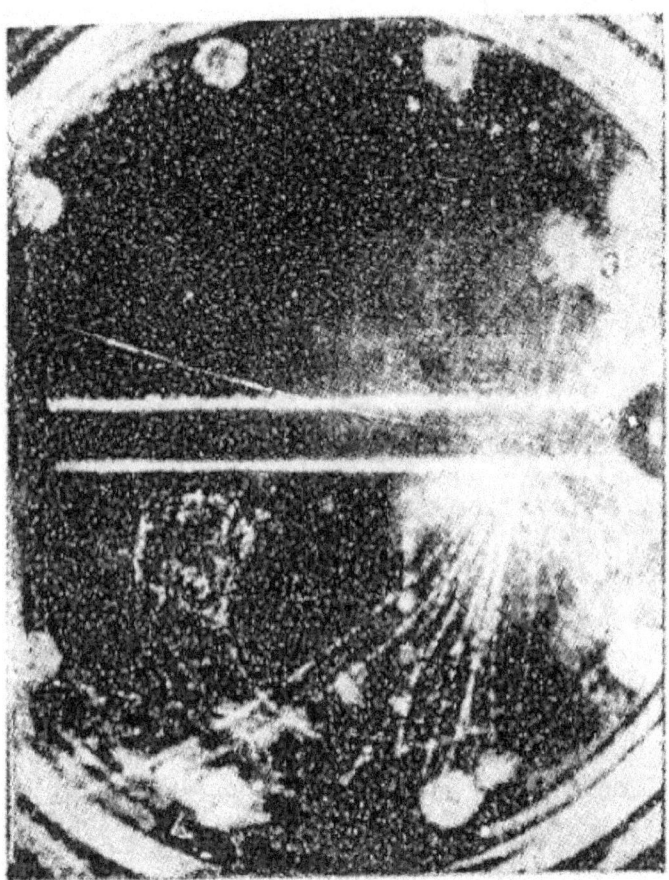

Shower produced by an ionizing particle (Anderson)

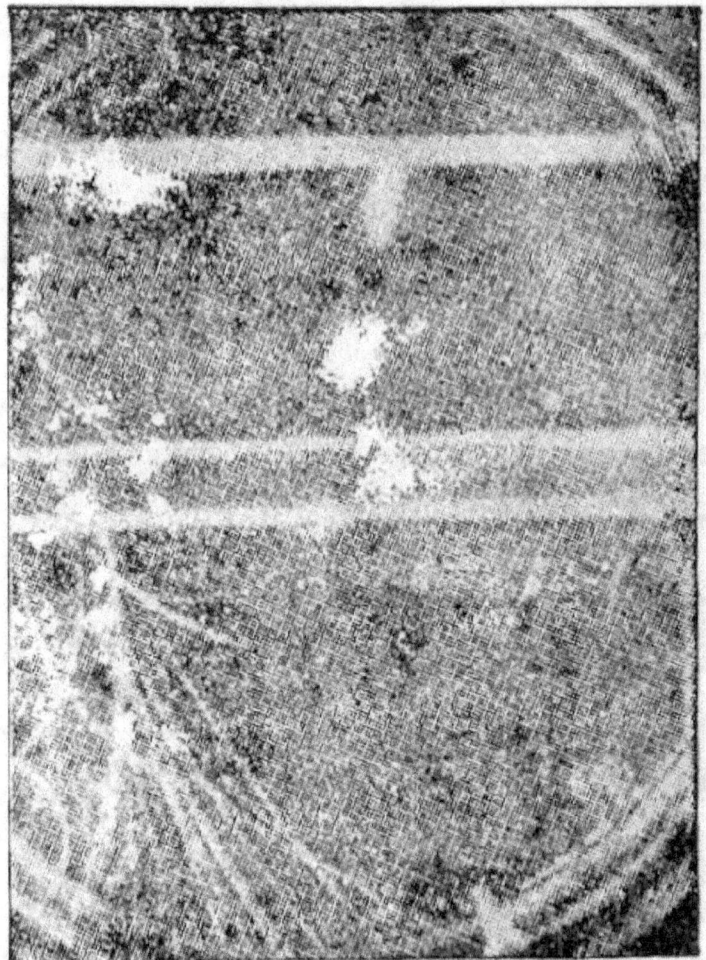
Shower produced by a non-ionizing photon. (Anderson)

In the first one an ionizing particle passing through the lead plate ends at the point of origin of the shower below the plate, while in the second a shower emerges from the plate with no apparent track of a particle entering into the plate. *There is some evidence for the two types of radiation taking roughly equal shares in the production of showers.*

Even within a single shower, sometimes the particles originate not from one point but rather from a narrow region. Furthermore, a shower passing through the lead plate produces other showers radiating from different foci. These observations indicate that *showers do not happen suddenly in a single elementary process, but they grow as a result of* many *successive elementary processes* taking place within a short distance in the substance.

(b) Shower particles
It can be easily established from the shower tracks that a *shower is made up of an approximately equal number of positrons and electrons.* There are also strong evidences to admit that a *shower contains photons as well*, though these leave no tracks in the

302

chamber. Thus, for instance, when a shower reaches a metal plate and more particles are formed in the plate, small showers appear to come out where nothing visible went in. Granting that something connected with the shower must have gone into the plate to produce this effect, one can conclude that there must be non-ionizing particles, i.e., photons associated with the shower which can do this. It belongs to theory to suggest how these non-ionizing links are able to transfer energy from shower to shower. As far as the shower photographs are concerned, one has no way of deciding this point or of judging how many photons may have been present in the shower. There are indications, however, from other types of measurements, that *the number of photons present in a large shower is comparable to that of ionizing particles, perhaps even slightly in excess.*

Anderson and Neddermeyer in 1936, using a counter-controlled cloud chamber, with a lead plate 3.5 mm. thick across it, placed in a magnetic field of 7,900 gauss, were able to obtain very interesting data about the *number of ionizing particles contained in showers.* At Pasadena (sea level), out of 2,684 photographs of cosmic ray tracks, 383 or 14% contained showers. On Pike's Peak (4,300 meters altitude), out of 1775 photographs, 752 or 42%, contained showers. In both places *two-particle showers were most frequent;* next came those containing 6 to 10 particles; among the photographs taken at Pike's Peak, some half a dozen showed about 100 particles each as the one in the photo below; one particular photo showed more than 300 tracks of electrons and positrons. The number of particles in a shower often increased, as the shower traversed the lead sheet.

Two associated showers, one originating from above the chamber and the other in the lead plate. (Anderson & Neddermeyer)

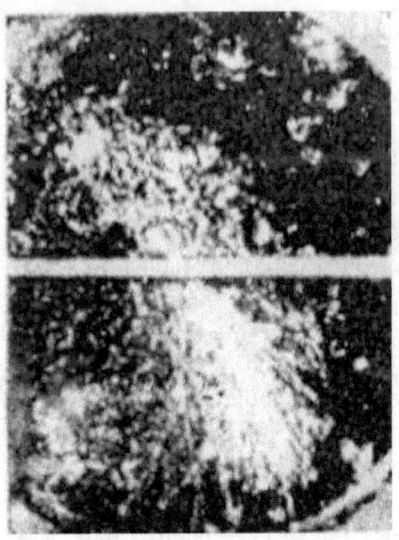

A complex shower with more than 100 tracks obtained at Pike's Peak.(Anderson)

The increased frequency of occurrence of showers at higher altitudes is a point worthy of note. At Pasadena, only 34 photographs per hour contained showers, while at Pike's Peak, 120 per hour. Chiefly the larger showers were noted to increase at a higher altitude. This observation points out that *the shower particles form the chief constituents of the soft component of cosmic rays.*

From a statistical study of the curvature of the shower tracks, it was found that the *energies of the shower particles* (both *positive and negative*) *ranged from 1 MeV to 500 MeV, 5 to 20 MeV being the most frequently occurring energy.* The total energy in the biggest shower of 300 particles was estimated to be greater than 15,000 MeV.

When shower particles passing through a plate produce other showers, the particles of the newly produced ones are found to be *less* energetic, as seen from the greater curvature of their tracks in the magnetic field. This suggests that *shower formation, we are dealing with a process of continuous dissipation of the energy available at the start, without any replenishment in the intermediate stages;* hence the more particles there are in a shower; the less energy each will have and the quicker the shower will die out.

Most of the particles constituting a shower appear to leave the point of origin within a cone of 40° and less. The maximum number of particles diverges at still less angles, about 20°, which, however, increase slightly with the thickness of the metal plate used for producing the shower. This indicates that *a large fraction of the momentum of any given shower particle lies along the direction of the shower producing particle.* This would be expected if the shower particles received most of their energy directly from the primary shower-producing ray, which fact is bound to throw light on the mechanism of shower formation.

Shower particles seem to give rise to further showers far more readily in a metal plate than the normal single particles, which might mean that the two kinds of particles are different in nature. As a matter of fact they are; the shower particles are mostly electrons belonging to the soft component of cosmic rays, while the bulk of single particles are mesons constituting the hard component.

(c) The Rossi curve

Rossi with his simple shower counting array of counters (Fig. 323) was able to obtain very important results concerning the shower phenomenon. By varying the thickness of the lead plate above the triangle of counters and measuring the rate of coincidence counts for different thicknesses he studied *the number of showers as a function of the thickness of the shower producing material.* When the experimental data were plotted, *i.e.,* the counting rate against the thickness of the lead plate, a curve of the form shown in Fig. 324 was obtained. It is sometimes called *shower production curve,* but more often as *"Rossi curve",* since it is concerned with the transition of cosmic rays from passage through air to passage through solid matter.

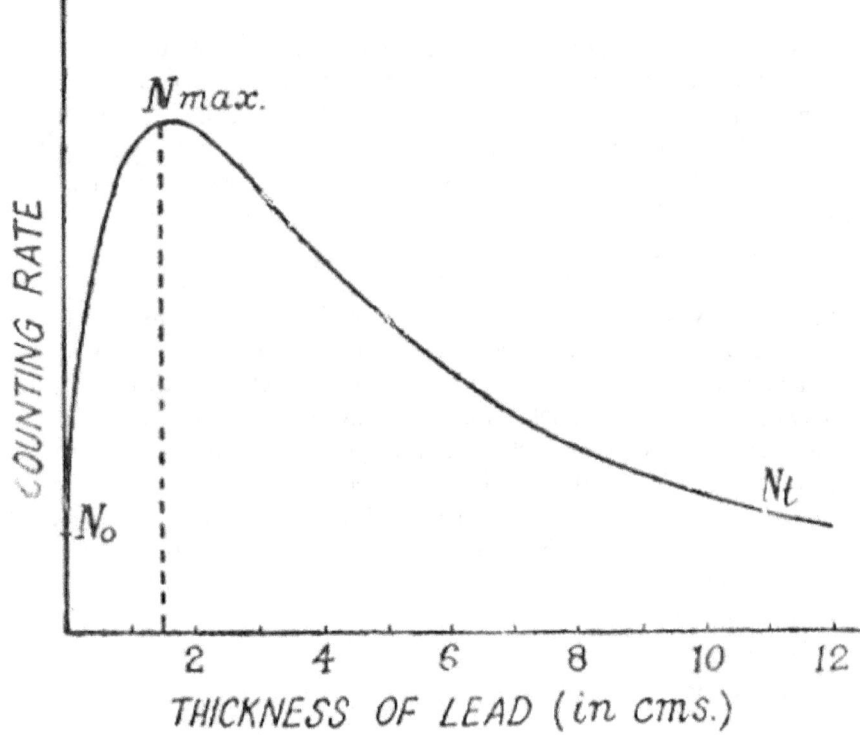

`Fig. 324. Rossi curve.

Such a curve represents the showers produced by primary particles of all energies and of a particular type of angular divergence depending on the geometrical disposition of the coincidence counters. The counting rate N_o, for zero thickness of lead, *i.e.,* when there is no lead, gives the rate at which showers coming from outside set off the coincidence counters. The rate of counting increases rapidly as thickness of lead is increased, until a.

305

maximum (N_{max}) is reached at a thickness of about 1.5 cms This implies that with increasing thickness there are either more showers or the showers have more particles originating in the lead, which by the increase in its size, helps the shower-producing agency to do its work efficiently.

With greater thickness of lead, the shower counting rate decreases, at first sharply and then very slowly. The general decrease is due to the fact that the shower producing agency is absorbed by the increasing thickness of lead beyond the optimum size, so that fewer showers will be generated. If the lead is thick enough so that only a few of the shower particles can get through it, the chance that they will stop rather than give rise to more particles becomes great. At this stage, the shower has passed its maximum size and is dying.

The initial rapid decrease of the curve up to 10 cms of lead followed by a very slow continuous decrease indicates that the *shower-producing agency, be it particle or photon, consists roughly of two parts*: a highly absorbable *"soft"* component and a very penetrating *"hard"* component.

For great thickness of lead, the counting rate does not fall to zero or even to its initial value N_o, but settles down at a rather higher rate N_l, which decreases very slowly. Evidently, a few new shower-producing particles are being formed continuously in a thick lead layer by some very penetrating primary agency. Since the value of μ/ρ in this region agrees well with that obtained for the absorption in lead of cosmic ray particles filtered through 25 cms of lead, the penetrating particles responsible for the showers at this stage would certainly represent the hard component.

The nature of this new shower-producing agency can be understood by considering the two ways in which mesons of the hard component give rise to fast electrons moving in more or less the same direction as the original mesons. These electrons may be produced either by the decay of mesons or as 'knock-on' electrons from very close ionizing collisions. At sea level, there are more electrons of this type present than those of the soft type, majority of which fail to reach there. In open air, decay electrons and their shower children ate the most abundant, but under thick layers of solid absorber, practically all the electrons are 'knock-on' electrons or their shower progeny which are mainly responsible for the tail of the curve. There are indications, however, for the existence of a new type of penetrating showers in the 'tail' region, which we shall consider when we deal with different kinds of showers.

The Rossi curve throws light also on the 'soft' component of cosmic rays. In order to explain the occurrence of a maximum in the curve at less than 2 cms of lead one must suppose that the showier particles themselves are easily absorbed. Rossi has tested this hypothesis directly by measuring the penetrating power of the shower particles with an absorbing screen placed horizontally between the top and bottom counters.

It is found that a considerable fraction of the shower particles are absorbed by a few centimeters of lead. Furthermore, other observations have shown that the absorption coefficient of the shower particles agrees fairly well with that of the soft component of cosmic rays. It has also been found that the frequency of occurrence of showers in a given material increases with altitude, but much faster than the total cosmic ray intensity in the light of these results, it is now generally held that *a great majority of showers are produced by the soft component.*

(d) Shower production in different materials
In the Rossi arrangement for shower production, replacing the lead above the counters by other material, such as aluminum, copper, iron, etc. and repeating the experiment, the efficiency of shower production in different substances can be studied. Drawing the Rossi curve for each of the metals used, the thickness corresponding to maximum efficiency of shower production (N_{max}) can be estimated. The results obtained show not only that lead is much more effective than any of the other metals, weight for weight, but also that

the shower efficiency is directly proportional to the square of the atomic number Z of the absorber used.

Thus we arrive at a very important property of showers, *viz., their production is a Z^2 process,* like the pair production and radiative collision processes. This gives a valuable clue to the mechanism of shower formation, which is utilized in formulating an adequate theory of the phenomenon. The Z^2 law enables one also to establish the identity of shower particles with the soft component.

Note. The Rossi curves can be drawn also from data obtained with a cloud chamber which is placed in a magnetic field and in which can be inserted plates of various materials of varying thicknesses. Investigations of this type have been made (in 1946) by Nassar and Hazen, but the data obtained are few and incomplete.

(e) Different types of showers
According to the different modes of production, some of which still remain obscure, showers are classified under different categories:

(i) *Electronic cascade showers*
Those are very frequently met with in cloud chambers, especially at high altitudes, as illustrated by several photos already given. They are called 'cascade' showers, because they can be adequately interpreted by a theory known as the 'cascade process' which involves the multiplicative transformations of electrons into photons by means of pair production. This mechanism of shower production will be dealt with in greater detail below. These showers consist of positrons and electrons as well as photons and form a good portion of the soft component of the cosmic rays. The primary particle responsible for the whole family of these showers is very probably an electron, hence the name 'electronic' showers. They are produced in air of the atmosphere, as well as in localized material objects. In the former ease they are known as *air showers* and in the latter *local showers.*

(ii) *Auger showers*

Pierre Auger, the French scientist, was the first, in 1938, to discover the existence of *'very extensive air showers'*, whose breadth may be considerable, up to a quarter of a mile. Using shower counting arrays of large counters spread over a distance of 10 meters apart from each other, he was able to obtain coincidence counts. About 60 such counts per hour were recorded at sea level, while 600 per hour at Jungfrau station (3,500 meters-altitude). At the latter place coincidences were found even up to distances of 300 meters, indicating the spread of an extensive shower originating high up in the atmosphere. Such showers are known as *Auger showers*.

Experimental data indicate that the *total number of particles in such a shower might well exceed a million and that the primary particle which is responsible for the entire shower must have* an *energy of about 10^{15} eV.* Other still bigger showers have been detected by large arrays of counters; the biggest that has been observed must have developed from an initial particle with an energy of the order of 10^{17} eV. It may be noted that large volume liquid Cerenkov counters have been recently used in the study of such showers.

Large air showers are accompanied by bursts of visible light due to Cerenkov radiation generated by the high speed positrons and electrons.

That the light is really Cerenkov radiation and not that arising from the ionization produced by the cosmic rays is established by the polarization of the light. An extensive shower containing millions of particles spread over a large area is bound to be densest in the center where there may be hundreds of particles per sq. foot, while thinning down rapidly towards the outlying part with only one particle per sq. yard.

It is possible to get some idea of the central dense portion of an extensive shower by the use of cloud chambers. Thus, for instance, in 1939, Lovell and J.G. Wilson, operating two cloud chambers, some 20 ft. apart, simultaneously by a counter system which detected air showers, obtained photographs of parts of a large air shower in the two chambers, as the one illustrated in the photo on the next page. The particles are seen to produce nearly parallel tracks in the chamber. If a cloud chamber occupies the outskirts of the shower, it will just receive one or two particles and will therefore give no indication of the extensive shower to which the particles belong. In the shower phenomena observed in a cloud chamber with a. lead plate across, one does not strictly see the beginning of showers in the plate, but rather the transition of showers from air to lead and then back again to air. The one or two particles of the large air-shower reaching the lead plate become sources of smaller local showers which emerging into air show signs of growing sideways to a size more suitable for a shower in air.

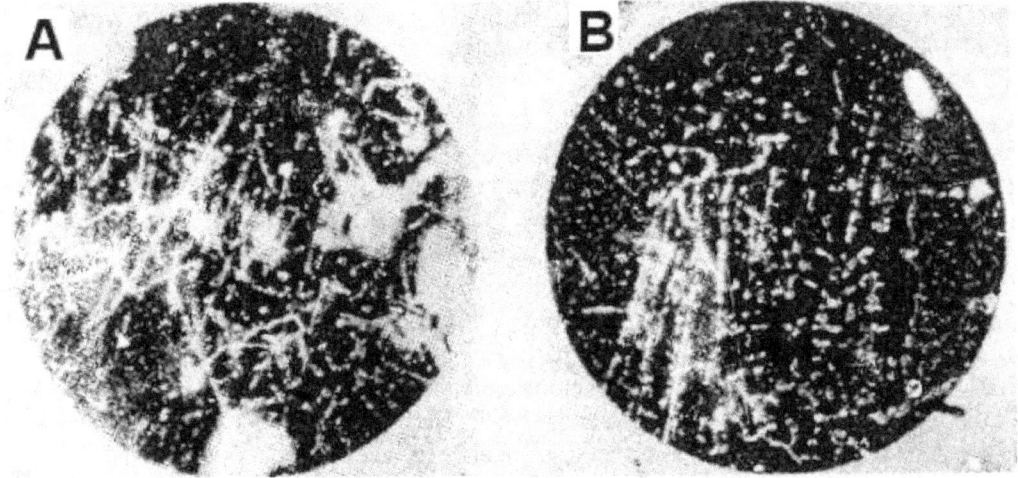

An extensive air shower passing through a pair of cloud chambers A and B, 18 ft. apart (J. G. Wilson)

Several investigations have been made on these extensive air showers using counter systems called *hodoscopes* or ionization chambers, but the interpretation of data obtained is not easy due to variations in the density of the particles across the shower, different sizes of individual showers, etc.

Most of the particles in these showers are electrons. There is, however, some experimental evidence to believe that *most air showers contain a small fraction (25%) of penetrating particles.*

Auger, in 1947, has tried to establish their presence by measuring the absorption of the extensive air showers and even to identify them with a new kind of meson, the *lambda meson,* singly charged, with rest-mass of 3 to 10 times that of the electron. Other investigators have tried to produce evidence for the presence of ordinary mesons in air showers. Further researches have shown that air showers have often a narrow central core of μ-mesons. These μ-mesons are the decay products of charged $\pi^{\pm}$-mesons created in the same nuclear explosion that produced the neutral π°-mesons whose decay γ-rays initiated the shower. Additional evidence that the decay of the π°-meson into two γ-rays is the primary event in the initiation of the shower is given by cloud-chamber photographs of complex showers that have double cores of electron-positron cascades. It is therefore possible that Auger showers may be, at least partly, different in nature from those ordinarily seen in cloud chambers.

(iii) *Penetrating showers*
The tail of the Rossi curve at great thicknesses of absorber, and chiefly the presence of a second slight maximum in the curve at a thickness of 18 or 20 cms of lead indicated the possibility of the existence of penetrating showers. Although the question of a second maximum was controverted for a time as some observers could not find it, Brussard and Graves, in 1941, were able to obtain definite evidence for its existence at about 17 cms,

of lead through a cloud chamber study of the showers produced by thicknesses near this second maximum.

The existence of penetrating showers was established first with the cloud chamber. The first direct and convincing evidence for a *pair* of penetrating particles was obtained, in 1939, by Braddick and Hensby who, working in a tunnel 30 meters under the London clay, photographed cloud chamber tracks of a pair that passed through 15 cms of lead above the chamber and a lead plate 2.5 cms thick inside the chamber.

By 1944, Rochester at Manchester was able to show photographs of 18 separate showers of more than two particles in which the particles had penetrated 53 cms of lead. Powell (1941), Wollan (1941), Hazen (1943) and Daudin (1944) obtained cloud chamber photos of these penetrating showers at mountain altitudes. In 1948, Fretter at California built a large cloud chamber with 16 lead plates, 1/2 inch thick each, which, operating at an altitude of 10;000 feet, photographed about four penetrating showers daily, as the one reproduced here.

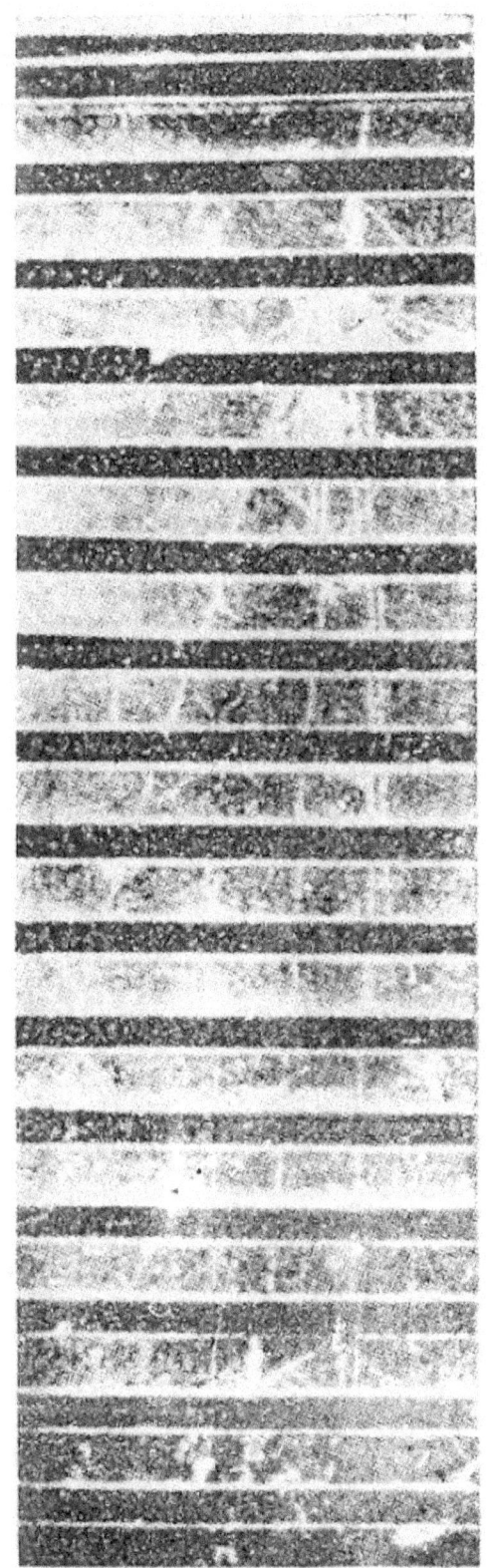

A penetrating shower made up of four particles which pass through many lend plates. At the top of the showers is a burst of electrons (Fretter)

It is seen from the photo that the particles of a penetrating shower are capable of traversing several lead plates without deviation or multiplication. They also tend to diverge less widely than do those of electronic showers.

From these facts it has been concluded that the particles are more massive than electrons and consequently named *meson showers.*

Janossy and Ingleby, using the *counter technique,* were able, in 1941, to show definitely that showers of penetrating particles were found under lead absorbers, which could not be either ordinary electronic cascade showers or knock-on showers. In 1913, Janossy and Rochester, making a more informative counter investigation, found that *penetrating showers were produced by both ionizing and non-ionizing radiation in the proportion of about 2: 1 and that the non-ionizing radiation was not photonic in nature.* They have speculated on the nature of these penetrating shower-producing agents and remarked that *protons and neutrons might be the ionizing and non-ionizing radiations respectively.* Janossy, with a slight improvement of the Hamilton-Heitler-Peng meson formation theory, has suggested that if a fast proton collides with a complex nucleus containing several protons and neutrons, the result is a small shower of mesons forming the penetrating shower. More recent investigations by Broadbent and Janossy and by Rochester, in 1947, tend to show that the phenomenon of penetrating showers might be even more complicated than that expressed by the above hypothesis.

(iv) *Non-collimated type of showers*
Sometimes a relatively rare type of shower is obtained in cloud chamber photographs, as the one shown here.

A non-collimated type of shower

In such a shower, the particles do not appear to be collimated at all, unlike the more common collimated type; they diverge from the place of origin in all directions, some of them even backward. The total number of particles involved is seldom large. Some of them possess very great transverse momenta, while the particles of the ordinary showers have a preponderance of momentum in the forward direction. These facts indicate that this rare type of non-collimated shower is something quite different from the normal collimated type, involving quite a different mechanism of production. They are to be assimilated to the "star" effect, which results from a nuclear disintegration produced in some way by the cosmic rays.

(f) Bursts
Hoffman was the first, in 1927, to observe cosmic ray bursts. Studying the ionization caused by cosmic rays in a high pressure ionization chamber, he noticed that occasionally a large deflection of the electroscope occurred, indicating momentary excessive

ionization. The cosmic ray phenomenon responsible for such excessive ionization is called a *"burst"*. Since 1932, the phenomenon has been extensively studied. Steinke and Schindler (1932) demonstrated the production of bursts in a sheet of lead above the ionization chamber. Subsequent observations have established definitely that *bursts are identical with showers*. Both vary in much the same manner with altitude, latitude and thickness of material in which the effect is produced. Bursts thus constitute an alternative method of studying showers.

Bursts are found to vary greatly in size, their frequency decreasing with increasing size. Their frequency increases with altitude, more rapidly for large bursts than for small. Schein and Gill, in 1939, using an ionization chamber 36 cms in diameter filled with argon at 50 atmospheres, observed at an altitude of 3,350 meters, a burst of 10^9 ion pairs, which they ascribed to some 10,000 cosmic ray particles, having a total energy of 10^{12} eV. Lewis and Kingshill, in 1946, had investigated extensive air showers by coincidence bursts in two thin-walled, unshielded ionization chambers. They measured at two different altitudes, 3,100 meters and 190 meters, the frequency of bursts of ionization occurring simultaneously in the two chambers as a function of burst size and separation between the chambers. The frequency was found to increase with altitude. At a given altitude, frequency decreased with increase of size and separation of the chambers. In the largest showers observed, at least 6,000 particles per sq. meter over an area of 100 sq. meters were involved, so that the total number of particles must have been near a million. A few bursts, as in the case of air showers, must have resulted from nuclear disintegrations caused by the cosmic rays.

Theoretical study of showers
The normal type of well collimated showers, which includes the extensive air showers and bursts; where the constituent particles are electrons and photons, has been satisfactorily explained by what is ordinarily known as the 'cascade' theory. Special showers of the penetrating and non-collimated types have not yet received such a complete theoretical interpretation.

Homi Jehangir Bhabha (30 October 1909 – 24 January 1966)

The cascade theory of showers

The theoretical explanation of showers as a *cascade process* of elementary happenings was suggested *qualitatively* by several workers. In 1937, a *quantitative* treatment was given independently by Bhabha and Heitler in England and Carlson and Oppenheimer in America. The cascade theory involves two other theories, *viz.,* that of *pair production* and *radiative collision* which have already been treated. A high energy electron moving through matter loses a large amount of its energy by radiative collision with an atomic nucleus and hence in the form of a photon which will be emitted according to the laws of conservation of momentum in the forward direction. The photon thus generated moves with the speed of light and after travelling a short distance is completely absorbed by interaction with the electric field of an atomic nucleus resulting in pair production. The positron-electron pair thus created sharing between them the energy of the lost photon, tend to be shot off in the forward direct. Each of the pair will, in its turn, produce, by nuclear collisions, new impulse photons of slightly lower energy and these produce new pairs of still lower energy. Thus a *cascade shower* of electrons and photons is built up, as represented diagrammatically in Fig. 325, by alternate transformation of mass into energy and of energy into mass, a single particle starting the cascade of events that results within a minute fraction of a second in a large group of more or less equal numbers of positrons and electrons.

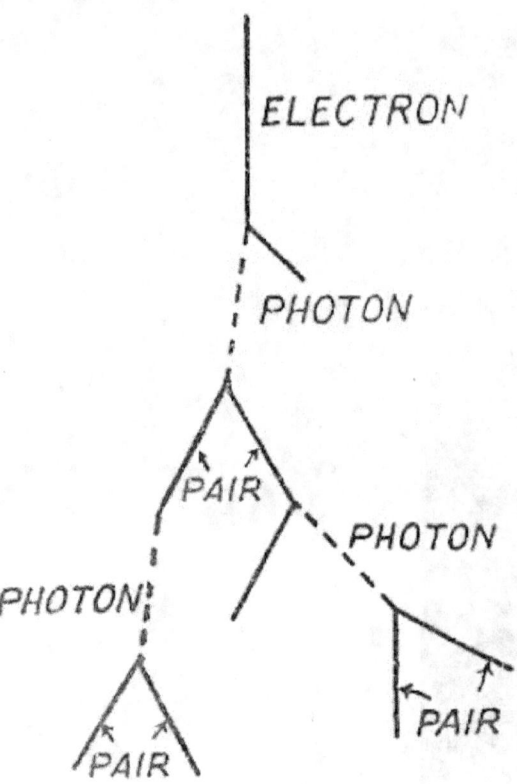

Fig. 325. Cascade process of shower formation.

The multiplication will continue until the initial energy becomes divided between a large number of pairs and the individual energies of the particles fall below the *'critical energy'*, when photon emission and pair production can no longer occur.

The energy of the shower is ultimately dissipated through Compton and photoelectric effects produced by the photons and through ionization of atoms by the particles. As these slow down to rest, the positive electrons in them will disappear by uniting with any negative electrons (not necessarily those belonging to the shower) by the process of annihilation of matter. Thus the number of particles and photons in a shower increases to a maximum and then decreases.

In the *quantitative treatment* of the phenomenon it is necessary, therefore, to take into account the probabilities of the two elementary processes, viz., radiative collision and pair production. Due to the *straggling* effect involved in such a process and the variation of energy of individual particles and photons over wide limits, it has not been possible to give a rigorous mathematical theory of the problem. The approximate solution leads to the following conclusions:

(i) The *general course of the shower* is the same whether it is initiated by a particle or by a photon. At the high energies involved, it can be shown that the particles and photons should move almost in the same forward direction, so that the whole group will proceed in close array.

(ii) The *number of photons* in a shower should be nearly double the number of particles.

(iii) The *critical energy for shower formation depends on the density of the material in which the shower is produced,* being lower for heavy substances than for light. Thus, for example, in water the shower production will cease when the energy has degenerated to 115 MeV, while in lead the process goes on down to energies of about 7 MeV.

(iv) *Showers occurring in different materials differ chiefly in the spatial scale of the phenomenon.* The amount of space occupied by a shower will depend on two quantities: the distance a fast electron is likely to travel before it gives rise to an energetic photon and secondly, the distance which this photon is likely to go before it is absorbed by pair production. It happens that these two quantities are almost the same. But as the two processes depend on Z^2, the distance covered by a single step of the cascade multiplication, known as the *'cascade unit',* varies greatly from substance to substance; for lead it is 0.4 cm., for iron 1.5 cms, for aluminum 8 cms., for water 34 cms and for standard air 275 meters.

(v) *The number of particles in a shower at its maximum size is approximately given by the energy of the particle that started the shower divided by the critical energy,* since the shower stops growing when the electrons in it have an energy equal to the value of the critical energy. The number may attain any value if the initial energy is sufficiently great; certainly showers with millions of electrons occur, as we have seen. A 10^{10} eV electron in passing through lead will activate a maximum shower at about 3 cms, *i.e.,* in 6 to 7 cascade units from the upper surface producing about 100 particles and about 150 photons. But if the electron carries 10^{15} eV energy it will produce a shower of 10 million particles, the maximum number being reached in 8 cms of lead. By contrast, a shower initiated by the same 10^{15} eV electron in water would not reach its maximum until 800 cms had been penetrated.

(vi) *The full range of the shower is greater and varies slightly with the distance corresponding to the maximum size of the shower.* Only exceptionally energetic showers extend beyond about 20 cascade units, so that all cascade showers can be effectively filtered by the use of 10 to 15 cms of lead. For the same reason, the vast majority of extensive air showers, starting high up in the atmosphere could hardly reach the ground level, since the whole atmosphere corresponds to about 26 cascade units.

(vii) *The lateral spread* of a shower does not depend much on its size, for the sideways motion always takes place in the last step or two, when the particles and photons are of the lowest energy and hence subject to *scattering. In any material the lateral .spread is only about one cascade unit.* As a result, a shower coming out of a lead plate, appears all to come from a point, since the cascade unit in lead is so small; but in the atmosphere, the lateral spread will extend to several hundreds of meters corresponding to the cascade units in it. Furthermore since the scattering of shower particles is inversely proportional to the energy, the highly energetic shower particles form the core of the shower and the less energetic particles are pushed off to the sides. Due to the same cause, the density of a shower at a given level should be greatest in the center and should de reease rapidly towards the sides.

These predictions of the cascade theory are in fair agreement with the results obtained in the experimental study of the ordinary showers, both of the local type in different absorbers and of the extensive air type. Perhaps *the best and the most direct proof of the essential correctness of the theory* comes from cloud chamber .photographs of showers produced in successive lead plates within the chamber.

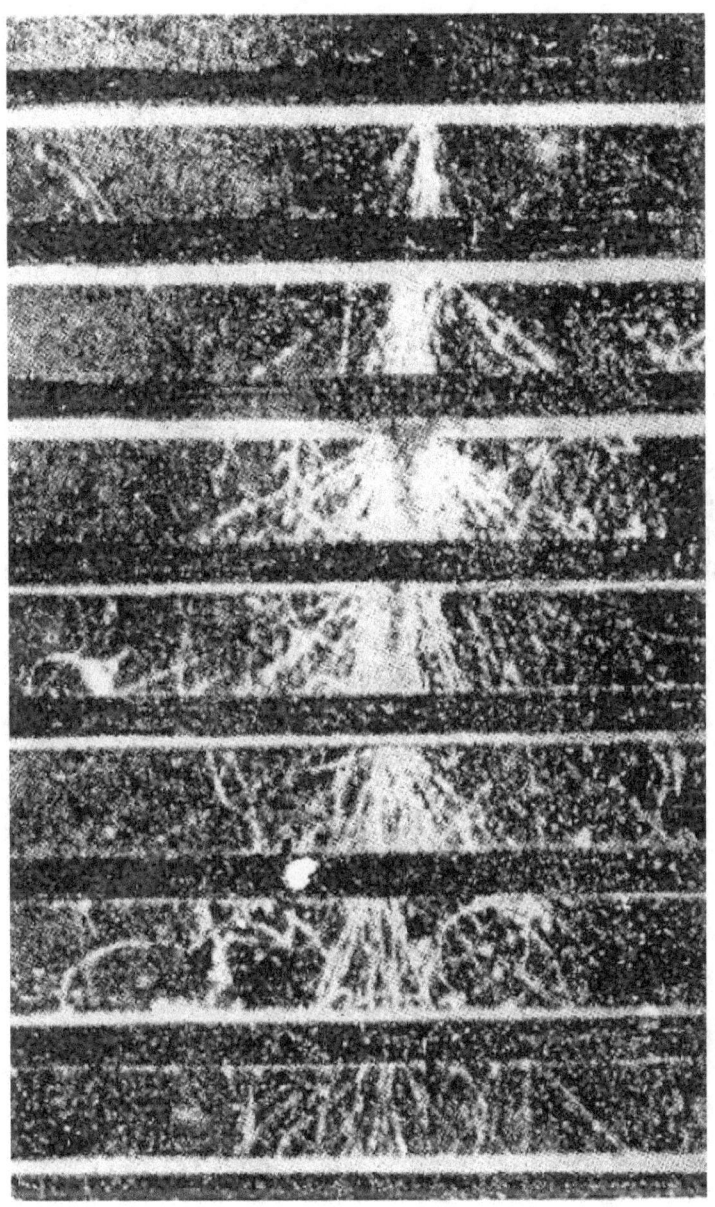

Growth of a cascade shower.

In the adjacent photograph it is seen that a single shower-producing particle coming from above gives rise to a shower consisting of a few particles as it penetrates the first top plate. The shower then grows to larger sizes in the successive plates, until the energies of the shower particles become so small that the probability of further photon production and pair creation decreases sufficiently that no further increase in the size of the shower occurs.

Nassar and Hazen, in 1946, using a counter-controlled cloud chamber with 4 lead plates 0.7 cm. thick each, at distances, 5.3 cms apart and a magnetic field of 1,100 gauss, were able to confirm roughly some of the particular theoretical predictions regarding the number aparticles present at various distances along the path of the shower, the distribution of enrgy among the particles at maximum size, etc.

Theory of penetrating showers
Showers of the penetrating type **are primarily** caused by a peculiar type of explosion of atomic nuclei, which gives rise not only to nuclear fragments, protons and neutrons, but also to a certain number of high energy π-mesons both charged and neutral. The decay of neutral π^o-mesons produces photons which, in turn, produce cascade showers of the electron-photon type. Protons, neutrons and charged π-mesons pass through matter and collide with atomic nuclei causing another explosion with the emission of new nuclear fragments, but this time of lower energy. And finally if the birth of a penetrating shower takes place in the air, the charged π-mesons quickly decay into μ-mesons, which do not produce any new nuclear explosions but pass through big thicknesses of matter without further multiplication and appreciable loss of energy. As the process is thus intimately connected with profound changes of atomic nuclei, the penetrating showers are known as "nuclear" showers to distinguish them from the pure electronic cascade showers. The general scheme of production of a nuclear shower is represented diagrammatically in Fig. 326.

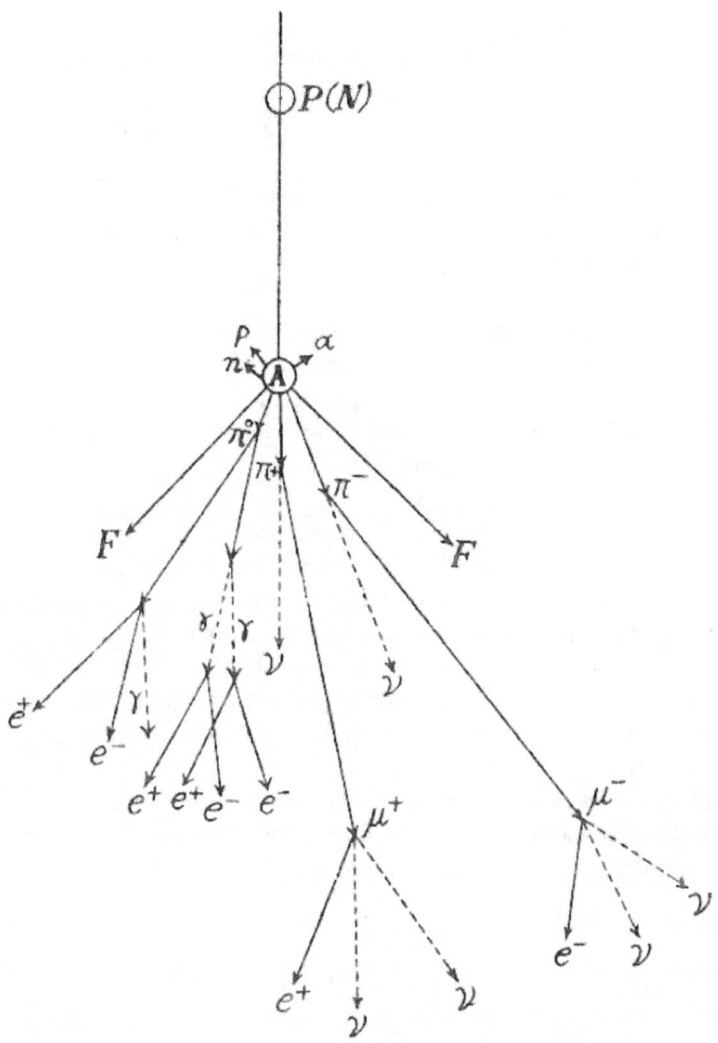

Fig. 326. Process of nuclear shower formation.

A proton (P) or neutron (N) of high energy acting on an atomic nucleus A of air disrupts it with the ejection of nuclear fragments (F), alpha particles (α), protons (p), neutrons (n) and charged and neutral π-mesons (π^+, π^-, π°). The charged π-mesons decay into charged μ-mesons which, in turn, decay into electrons (e^+, e^-) and neutrinos (ν). The neutral π°-meson decays into an electron pair either directly ($\pi^\circ \rightarrow e^\pm + \gamma$) or indirectly ($\pi^\circ \rightarrow \gamma + \gamma$), each of the γ-rays producing electrons by pair production. The electrons thus formed will still have high enough energy to produce cascade showers of the electron-photon type.

As regards the nature of the primary particles that initiate the nuclear showers, numerous experiments with photographic emulsions show that at mountain altitudes they may be both charged and uncharged particles, while in the stratosphere they are mostly charged particles. Experiments conducted under thick layers of lead show that these particles cannot be electrons or photons, since they are too readily absorbed in lead. On the other hand, the sharp increase in number of nuclear showers with altitude shows that the particles responsible are absorbed in the atmosphere much more readily than ordinary it-mesons. They cannot be π-mesons either, since the latter decay in air very quickly. We are therefore left with the conclusion that the primary particles responsible for nuclear showers are protons and neutrons.

Further, since the primary cosmic ray entering the top of the atmosphere has been shown to be made up of protons chiefly, the ultimate originator of nuclear showers must be traced to very high energy protons and not neutrons.

The electrons and photons generated at the tail end of the nuclear shower are fully capable of initiating an extensive air shower through subsequent pair production and radiative collision. This method of conceiving the formation of air showers explains different types of anomalies in the behavior of extensive air showers, which cannot be understood from the viewpoint of the ordinary cascade theory.

Proof of the essential correctness of the theory proposed is had photographs of both cloud chamber and nuclear emulsion. In the cloud chamber photo shown earlier, three types of particles may be distinguished, viz., electrons that produce cascade multiplication in lead, mesons that pass through many lead plates and heavy nuclear fragments. A microphotograph of the formation of a penetrating shower in a highly sensitive nuclear emulsion plate is reproduced here. The shower arises at the point A due to the collision of a very high energy proton (p) from above and an atomic nucleus. Besides the several nuclear fragments produced in the nuclear explosion, represented by the thick tracks forming a "star" there are other fast particles, indicated by numbers, 1, 2, 3, 4...8, that form the shower; they are π-mesons and high energy protons. One of these particles (5) produces farther down a new nuclear explosion at the point B with the ejection of two heavy fragments and three fast particles (a, b, c).

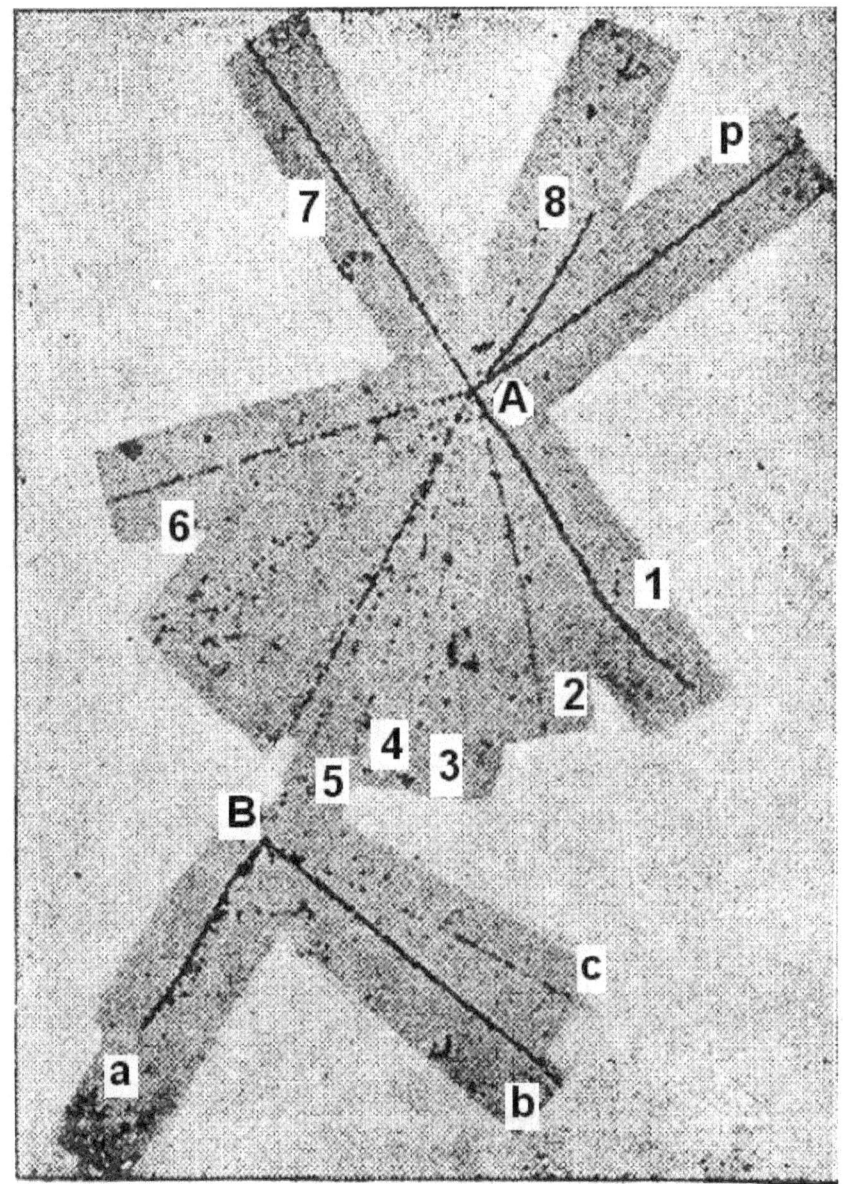

Formation of a penetrating shower.

The most direct proof of the theory has been obtained by studying the nuclear showers with counter arrays forming a horoscope, as shown in Fig. 327.

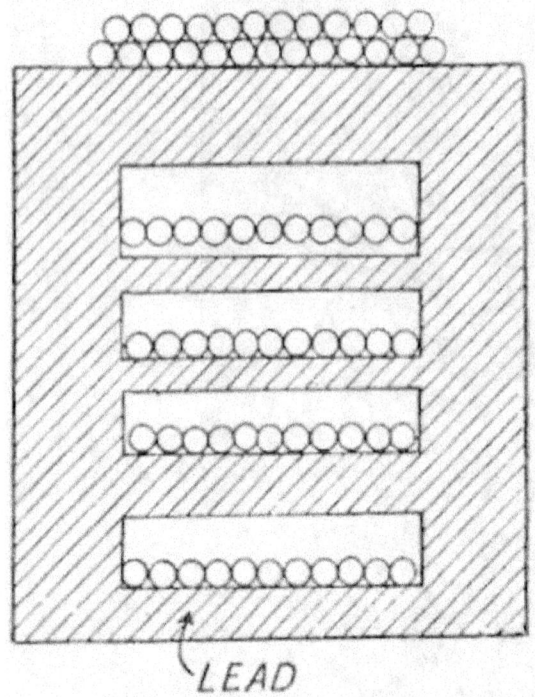

LEAD

Fig. 327. Hodoscope for studying nuclear showers.

The apparatus consists of several rows of counters separated by layers of lead of different thickness. Through a special electronic circuit, which responds only when a nuclear shower is generated in the hodoscope, neon lamps connected with the counters glow in such a way as to give a vivid picture of the birth and development of a shower as it passes through matter, closely following the trajectories of all charged particles that pass through the corresponding counters and lead. Thus it is possible to observe many cases when particles, both charged and neutral, generated in a nuclear shower, are able in their turn to produce similar type of showers in matter. It has thus been shown that the penetrating showers can be somewhat cascade-like, though this nuclear cascade differs sharply from the cascade multiplication of electronic shower, which is determined by only electromagnetic processes in the vicinity of atomic nuclei and is in no way related to transformations of the nuclei themselves.

Despite this difference, there is a close relationship between the two types of cascade processes. For, first of all, it is found that the nuclear shower is pretty frequently accompanied by an extensive air shower. Secondly, particles capable of producing nuclear processes within thick layers of lead have been found in extensive air showers. Both these facts indicate that an extensive air shower in the air need not necessarily be caused by a very high energy electron or photon; it can as well be a proton of the same high energy producing in the upper layers of the earth's atmosphere a nuclear shower of great power. The electrons and photons which form the end products of this shower can initiate an extensive air shower, as already indicated.

It is of interest to note that Klein, in 1944, has proposed quite a different mechanism for the production of extensive showers. According to him, the primary component of cosmic rays are interstellar dust particles consisting of many atoms of *"inverted matter"*, *i.e.*, atoms built up of negative protons and positive electrons. Such dust particles, containing about 10^7 atoms each, as soon as they reach the earth's atmosphere, are annihilated, their entire rest-mass being transformed into a number of photons of the order of 10^7, each of an energy of 10^9 eV, which photons at once give rise to large cascade showers. As all the annihilations would take place simultaneously at the top of the atmosphere, the, net result as observed in the lower part of the atmosphere would be the same as if the showers had been produced by a *single* high energy particle. This hypothesis has the merit of not assuming that each air shower is produced from just *one* primary particle of extremely high energy of the order of 10^{16} eV, for which there is no direct experimental proof.

12. HEAVY PARTICLE COMPONENTS OF COSMIC RADIATION

A. **Primary cosmic rays**
It has been remarked that the *primary* cosmic radiation, arriving at the top of the atmosphere from outer space, is constituted largely with *protons*. This view is supported by the following facts:

(*i*) Inasmuch as hydrogen is the most abundant element in the universe, the most numerous particles encountered in free space should be protons;

(*ii*) The study of the geomagnetic effects on cosmic rays shows that the primary cosmic rays should consist of positively charged particles like protons;

(*iii*) The recent finding about meson-formation in the upper atmosphere by positive primaries strongly suggests that these are protons.

Experiments conducted in 1948 by Frier, Lofgren, Ney, F. Oppenheimer, Bradt and Peters at Minnesota and by Schein at Chicago with thick emulsion photographic plates and cloud chambers carried by sounding balloons to altitudes of 80,000 to 100,000 feet, hence near to the top of the atmosphere, have however led to the very interesting discovery of *heavy nuclei* among *the primary cosmic rays*. The masses of these heavy nuclei range from about 4, that of helium, to about 56, that of iron, as judged from the heavy tracks of ionization left on the plates. The primary cosmic rays contain only a very small fraction of the heavy nuclei. Estimates based on the data available at present indicate that the 21,500 primaries falling per minute on every sq. foot of the top of the atmosphere at $\lambda = 41°$ include about 80 heavy nuclei, the remainder being protons.

Measurements on the *charge spectrum* of the primary radiation (*i.e.*, the relative number of different nuclei) show that

the intensities decrease with increasing charge and no certain cases of nuclei with charge greater than 26 have hitherto been observed. Further, the nuclei of even charge are found to be about three times as numerous as those of odd charge.

It has been suggested that this is a result of the fact that the evenly charged nuclei have a large number of stable isotopes. If the nuclei are produced as a monotonic function of mass, then the resulting β-decays of the unstable nuclei will produce just such a charge distribution. A mass distribution of the above type might be expected if the majority of the primary nuclei were secondary particles from nuclear reactions in inter-stellar space. When the charge distribution of the primary cosmic radiation is compared with the abundances of the elements in the universe, some degree of similarity is found, *e.g.,* hydrogen is the most abundant; helium comes next; lithium, beryllium and boron, which are known to be very scarce in the universe, are found only in a small proportion in cosmic rays also. A more detailed study of the charge distribution however shows that the abundances of nuclei of charge greater than 10 are all greater in cosmic rays than those observed in the universe. These findings may be useful in framing a theory about the origin of the primary radiation.

B. Secondary cosmic rays

As for the *secondary* cosmic radiation reaching sea level through the atmosphere, there are evidences to show that it also contains, in addition to electrons and mesons, more massive components, such as *protons, neutrons* and the so-called *"star-particles",* most of them being secondaries of course. Although these heavy particles constitute only a small fraction of the total radiation, yet many of them have large energies and cause powerful nuclear effects, so that their study is bound to help greatly in understanding the fundamental nuclear processes. The investigations made to detect and analyze them will now be briefly stated.

The proton component

Since 1933, several investigators such as Anderson and Neddermeyer, Blackett and Occhialini, Street and Stevenson and others have suggested the existence of a small percentage of protons in cosmic rays, in order to account for the heavily ionizing, tracks found in the cloud chamber pictures. More recently, direct attempts have been made to detect the proton component. Thus, Korff, in 1941, using specially arranged counters which were carried aloft by balloons, detected about 4 protons per sec. per cubic meter at an altitude of 10,000 feet. Powell, in 1946, using a *random operated* cloud chamber at an altitude of 10,000 feet, was able, from 22,400 photographs taken, to identify as protons 4% of the rays that were observed to penetrate at least 1 cm of lead. In spite of the paucity and indefiniteness of experimental data, there seems however to be a strong indication for the existence of a proton component in cosmic rays, about 0.1% at sea level and about 3% at an altitude of 10,000 feet. This component shows no preferred direction and is probably not primary radiation, but rather a secondary produced by nuclear disintegrations in the atmosphere.

The neutron component

The neutrons form a more important component in cosmic rays than protons. Rochester and Janossy found that about a third of the penetrating showers are produced by a non-ionizing radiation, presumably neutrons, as we have already remarked. Free neutrons in the atmosphere have been detected and analyzed by the photographic emulsion, proportional counter and ionization chamber methods:

Photographic emulsion method

Schopper (1930), Locher (1933- 36), Blau and Wambacher (1937) and Heitler, Powell and Fertel (1939) have employed this method. Several specially prepared photographic plates covered with paraffin, carbon and lead were sent up in balloons to detect the neutrons.

A five-fold increase in the number of heavy tracks was noticed for the plates covered with paraffin compared with the plates covered with carbon and lead.

Most of the tracks were attributed to recoil protons excited in the paraffin by neutrons in cosmic rays. It was concluded that at a depth of 1/2 meter water equivalent, an appreciable number of neutrons was found.

Proportional counter method

Korff and his collaborators in America (1939-1946) have used proportional counters with thick walls (to exclude electrons and mesons) filled with boron trifluoride (BF_3 gas) to detect the neutrons through the α-particles emitted in the nuclear reaction of neutron with boron. A sleeve of boron carbide was employed, in order to ensure that the neutrons came from outside the walls of the instrument. Likewise, a shield of cadmium was used to sort out the slow neutrons from the fast ones. When the counter is surrounded by the cadmium shield alone, only fast neutrons are allowed entry into the counter. On the other hand, when the sleeve of boron carbide alone is slipped over the counter, since boron absorbs both slow and fast neutrons up to 1 MeV, no cosmic ray neutron below that energy level is likely to get through. By comparing the counts obtained with the different shields on, it is possible to determine not only the total number of neutrons in cosmic rays, but also to separate them into slow and fast ones.

Korff and Hamermasch (1946) devised a mechanism that automatically changed the shields at two-minute intervals, leaving the counter unshielded for an equal period. In the stratosphere exploration by sounding balloons, extending up to 100,000 feet, the counts were transmitted to the ground station by means of short wave radio. Their observations showed that the neutrons were more numerous at high altitudes, with a steady decrease in intensity downwards.

Ionization chamber method

Montgomery has used an ionization chamber filled with BF_3, in conjunction with a linear amplifier, to detect the neutron intensity. Von Halban and his associates (1939) utilized the radioactivity induced by neutrons in bromine in their experiment in which a chamber containing ethyl bromide was carried in an aircraft to a height of 30,000 feet.

These investigations have led to the following conclusions:

(i) The total number of neutrons in cosmic rays is considerable, possibly of the same order of magnitude as that of electrons or photons;
(ii) Their total energy is probably much less than that of the soft rays;
(iii) Their intensity and rate of production increase with altitude;
(iv) Most of them are released in nuclear disintegrations (stars).

C. Cosmic ray stars

The phenomena known as "stars" *were first discovered in cloud chamber photographs,* where occasionally star-like set of tracks of several heavily ionizing particles radiating from a common origin was obtained.

Extensive researches were undertaken by Hazen (1944) and Powell (1946) to study these cosmic ray stars employing *cloud chambers* containing several lead plates. The photo below illustrates one of the few cases of a star formed in the gas of the cloud chamber obtained by Powell.

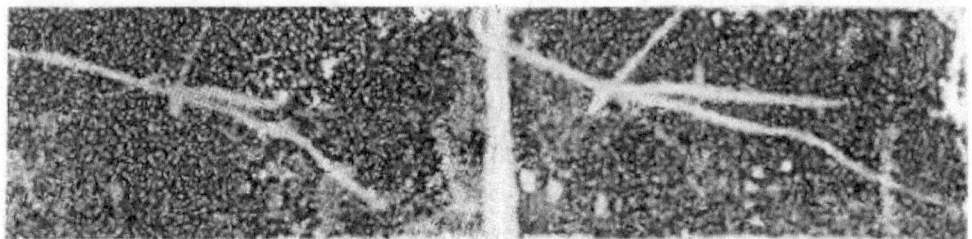

A cosmic ray star formed in the gas of a cloud chamber (Powell)

The star contains five particles and is probably excited by a non-ionizing radiation, although it may be caused by the fast heavy particle making the thin track coming obliquely from below. The cloud chamber, though capable of furnishing excellent qualitative and quantitative data, has the drawback of producing only a few stars over a long period of operation.

The photographic emulsion method has the great advantage that many more stars may be obtained in a reasonably short period of observation and with less trouble, since the plates have merely to be exposed without any complicated operational difficulties, unlike in the case of the cloud chamber. Blau and Wambacher (1937) were the first to find cosmic ray stars in photographic emulsion plates exposed for five months at an altitude of 2,300 meters. Since then several others, such as Shapiro (1941), Powell and Occhialini (1946), Leprince Ringuet and his co-workers (1947), have employed this method in the study of the stars. We have already given a photo of a "star" obtained by Occhialini and Powell. Here we reproduce another magnificent one obtained by Leprince Ringuet.

A cosmic ray star in photographic emulsion (Leprince Ringuet)

In this star there are 34 visible tracks, chiefly protons and - α-particles. The total kinetic energy of all the emitted particles exceed 1,000 MeV. The longest track within the emulsion is that of a 50 MeV proton, extending to about 3,500 μ. The actual stars obtained in photographic emulsion are evidently much smaller than those produced in the cloud chamber; the photo given here is evidently a very large magnification of the original star.

The *ionization chamber technique* has also been utilized for the study of stars, by Bridge, B. Rossi and Williams (1947). Analysis of the shape of the pulse as a function of time permits fairly definite identification. The ionization pulses in an unshielded chamber appear to be caused not so much by air showers as by heavily ionizing particles from cosmic ray stars.

From these different investigations of cosmic ray stars, the following *results* have been obtained:

Experimental

The usual number of particles in a star is from two to ten, averaging about four. The mean energy of an individual star particle is of the order of 10 MeV. Most of the observed tracks refer to protons, although some of them may very well be mesons; α-particles also occur, occasionally even heavier nuclear fragments. Neutrons also form a component of the star particles, probably as numerous as the protons. As regards the direction of the particles, although the tracks appear to be distributed at random there is some evidence for a preponderance of downward direction.

With increasing altitude, the number of particles per star increases very slightly; but the energy of the particles increases rapidly.

The frequency of occurrence of stars increases to a great extent with altitude, (of the order of fifty in going from sea level to an altitude of 15,000 ft.)-in somewhat the same manner as several components of cosmic rays, which fact proves definitely that cosmic rays are responsible for these stars.

Theoretical

The stars are presumed to be the outcome of nuclear disintegrations that are produced in some way by cosmic rays, similar to those occurring in natural radioactive materials. As regards the particular component of cosmic rays which is responsible for star production, some authors have suggested the *high energy gamma rays of the soft component;* but this has been contradicted by experimental observations, such as, stars not being associated with the soft radiation, the increase of the average energy of stars with altitude, etc:

Others have expressed the opinion that they *are, due to the capture of slow mesons.* Although there exists some experimental evidence for such a process, it cannot be the main source of stars since the altitude variations of stars and of slow mesons are strikingly different. Powell has proposed that *high energy neutrons are the producers* of *stars,* which seems to be the most satisfactory explanation, since most of the experimental data can be understood fairly well on this hypothesis. The primary rays (protons) colliding with air nuclei produce, in addition to mesons, several high energy nucleons (protons and neutrons). These secondary nucleons may in turn produce further nuclear explosions as they penetrate into the atmosphere. The protons, being charged, lose energy by ionization as well, and hence in the lower part of the atmosphere the majority of nucleons ate neutrons. Thus most of the stars observed are caused by these neutrons.

It is found, however, that *Bohr's 'liquid drop evaporation model',* which should be applicable to such a process; does not represent satisfactorily the star phenomenon, which may be due to either meager experimental data or the model being inadequate for high energy collisions. That the process of nuclear disintegration of cosmic rays is really quite complex is indicated by the observation in the photographic emulsion of stars-with forked tracks.

13. NATURE AND COMPOSITION OF COSMIC RADIATION

As a result of the many investigations which we have described in the preceding pages, we are led to the following comprehensive, though not quite complete, picture of the cosmic rays as regards their nature and composition.

In the first place, it becomes necessary to distinguish between the *primaries, i.e.,* those, coming from outer space, that enter the top of the earth's atmosphere and the *secondaries, i.e.,* those found in the space between the top of the atmosphere and the surface of the earth including a certain distance underground.

Primary cosmic rays

The vast majority of the incident particles are *protons.* Recent experiments have shown that there is a very small percentage of *heavy nuclei.* The velocity range of all types of nuclei is approximately the same.

At any given velocity, there are 85 helium nuclei and 6 heavier nuclei for every 1,000 protons.

There might be also a *small number of electrons.* Even if these are not really primaries; they certainly come into existence in the first few layers of the atmosphere. These primaries as they reach the top of the atmosphere, have energies ranging between 3×10^9 and 3×10^{10} eV roughly, the others being prevented from reaching the earth by the action of the magnetic field of the sun and the earth. But some of the primaries have possibly much higher energies going up to 10^{17} eV; as deduced from the study of extensive air showers. Within these energy limits, the number of particles N(E) crossing unit area per sec. at a given energy E varies roughly as the inverse square of the energy, $N(E) \approx KE^{-2}$ (K a constant). This means that there is a preponderance of particles of low energy.

Secondary cosmic rays

These are found in the atmosphere, which have been the subject of more *direct* research, are highly complex in composition and nature. They contain *heavy nuclear fragments, α-particles, protons, neutrons, positive* and *negative mesons* and *electrons* as well as *photons,* in different proportions, that vary considerably with the different points of observation, due to the different complicated processes by which they are produced, The sequence of events from the top to the bottom of the atmosphere, as far as our present knowledge permits, appears to be somewhat as follows:

As the primary protons, having energies of the order of 10^{10} eV and a consequent relativistic mass ten times as great as their rest-mass penetrate the atmosphere, they begin to interact with the atoms of oxygen, nitrogen and other gases of the air in a complex series of collisions. These disrupt the nuclei of the atoms that are hit, causing them to explode into heavy fragments, protons, neutrons and mesons-in the so called *"star" effect.* Each of the members of disruption, in its turn, demolishes other nuclei, creating other stars, and the process goes on until the energy of the primary ray is dissipated. At 60,000 feet, less than a third of primary protons survive. But at this altitude an enormous number of secondary particles has already been released, including secondary protons, neutrons, mesons and electrons. As the surviving primaries plunge earthward into the relatively dense lower atmosphere, the frequency of collisions increases and the particles rapidly lose energy. At 14,000 feet, only a comparatively small numbers is sufficiently energetic to be detected, and at sea level the number is still less. Thus ends the life-history of the primary protons.

As regards the other two probable components of the primary radiation, the heavy nuclei undergo a fate similar to that of the protons. But the primary electrons, coming down into the atmosphere, immediately lead to *cascade showers,* thereby causing a sharp increase in the total number of electrons, quite at the top of the atmosphere, as indicated by the maximum in the altitude intensity curves. The increase, however, does not extend very far, since the majority of the primary electrons have only energies 40 to 100 times the critical energy (10^8 eV) in air, and beginning at almost one-seventh of the atmosphere from the top, the number of electrons steadily falls. Some of the higher energy primary electrons give rise to the *extensive air showers* which however will not reach sea level, unless the energy of the initiating particle is of the order of 10^{14} or 10^{15} eV.

The main result of the action of the primary rays on the top layers of the atmosphere is the *production of mesons.* Each primary particle loses only a fraction of its energy in each meson-producing collision and is therefore able to make many collisions and to produce many mesons.

It is also possible that a single primary particle may give rise to a shower of mesons, consisting of five to ten mesons, in a single collision. In both cases, the mesons formed are of comparatively low energy and therefore will come to the end of their career relatively soon. Furthermore, there is evidence that the process of meson production may be a two-stage affair, consisting of generation of a heavy π-meson which decays quickly into a lighter μ-meson. Both types are positively or negatively charged. These mesons lose their energy chiefly by ionization, in which they occasionally produce fast electrons through the so-called "knock-on" process. The final behavior of the meson depends on what kind of charge it carries; if positively charged, it will spontaneously decay producing a positive electron, while if negatively charged it will be captured by a nucleus and eventually absorbed by it, thereby releasing an electron. Sometimes a slow meson can give rise to a star disintegration, thereby producing fresh supply of protons, neutrons, α-particles and heavier nuclei. By these different processes of collision, decay and capture, electrons both positive and negative are created.

These electrons are capable of radiating photons when they decelerate in the electrostatic field of the air nuclei. The generated, photons are absorbed quickly by pair production at energies above 20 MeV in air or 5 MeV in lead and by Compton effect below these values. The combination of radiative collision and pair production often results in the building of *cascade showers,* a self-multiplying process, in which a single electron may set off a train of events to create hundreds and even tens of thousands of pairs of particles in a fraction of a second. At the level in the atmosphere where the cosmic ray intensity is greatest, such showers make up approximately 5/6 of the total radiation.

Fast protons and mesons lose energy in *small* amounts chiefly by ionization when their energies are below 1,000 MeV. On the other hand, fast electrons lose energy in *large* amounts by radiative collision and the consequent shower production, when their energies are above 100 MeV in air or 10 MeV in lead. On account of this fact, the total cosmic radiation divides itself into a penetrating or *hard component,* consisting of mesons, a very small number of protons and possibly some very energetic electrons or photons, and a readily absorbed or *soft component,* consisting of electrons; about an equal number of photons and possibly some very slow mesons or protons.

At the position of the maximum for the total intensity, the hard component is only about a fifth of the soft component. The soft component after passing through its maximum decreases uniformly till at sea level its intensity is less than one per cent of the maximum value. The hard component, on the other hand, decreases continually from the top of the atmosphere until at sea level its intensity is about 6% of the value at great altitudes. The total radiation at sea level is made up of about 25% soft and 75% hard component and has an intensity of a little more than one ionizing particle per minute per sq. cm

Neutrons also are present among the cosmic rays in the atmosphere, and its intensity increases with altitude. The source of these neutrons is the nuclear interactions between primary protons and air atoms; secondary neutrons originate from cosmic ray "stars" which are nuclear disintegrations caused by neutrons in most cases or by the capture of slow mesons.

The important cosmic ray particles entering the, ground are electrons .and mesons. Because the material of the earth is much denser (about 3,000 times) than air, the scale of showers is reduced to this extent. The electrons therefore develop small showers in the first layers and are removed from the beam. The mesons, however, travel very much further, lose energy by ionization much faster than in the atmosphere and finally decay, giving rise to electrons and their rather rudimentary shower children which form a very small portion of the soft component underground. In solid matter, however, knock-on electrons are the main source of the soft component, which therefore comes to be substantially in equilibrium with the meson component. At great depths only mesons of extremely high energy of the order of 10^{12} eV can exist.

14. ORIGIN OF COSMIC RAYS

In spite of the great advances made in understanding what cosmic rays are, there still remain several unsolved problems. Of these the roost fundamental is about *the origin of the primary radiation.* There are two main points to be made clear, viz.

(*i*) *Where do the primary cosmic rays originate?*
and
(*ii*) *How are they produced ?*

These two aspects of the problem are closely related to each other; but, for the sake of clearness, we shall deal with them separately. At the very outset, it may be remarked that no simple solution exists for the problem proposed, though many interesting suggestions have been made.

The general criterion for any hypothesis to be acceptable is that it should be able to account satisfactorily for the following experimentally established special features of the cosmic rays:

(*i*) *Isotropic distribution, i.e.,* cosmic rays are found to approach the earth almost uniformly from all directions.

(*ii*) *Nature of the components of the primary radiation.* It is today believed that the primary cosmic radiation consists mostly of *protons* and some *heavier nuclei.*

(*iii*) *Energy spectrum.* The energy of the primary radiation is limited to a range between 10^9 and 10^{17} eV, possibly extending beyond both figures, but mainly concentrated in the region of 10^{10} eV, the rays with extremely high energy being very rare.

As regards the first problem of the **place of origin,** it is now well established *that the cosmic* rays *originate outside the earth's atmosphere:* but as to the *exact place* of origin, no definite solution has been arrived as yet. Of the different suggestions made, the following are the more important:

(*a*) *Remote interstellar space far beyond our galaxy*
This is proposed to be an effective distance of about 10^{10} light years (a light year = 10^{18} cms). The universe is made of stars and nebulae that form different galaxies, one of which is the Milky Way containing our solar system. The galaxies are distributed at large distances apart with mean spacings of 10^6 light years and are endowed with rotational motion. The intergalactic space is very empty and contains very few stars or diffuse matter. Cosmic rays are supposed to have originated in this space. This opinion was the first to be suggested by several authors, such as Compton, Blackett, etc., in an attempt to account for the isotropic distribution of cosmic radiation. A supposed slight variation of cosmic ray intensity with *sidereal* time was adduced as an additional support for the interstellar origin. If the source of cosmic rays were outside our own galaxy and at rest relative to its center of gravity, then the motion of our galaxy as a whole, combined with earth's rotation would lead to a small diurnal variation with sidereal time. But, more recently, it has been definitely proved that this sidereal time effect is fictitious. Furthermore, it is not easy to account for all the protons and especially the heavy nuclei contained in cosmic rays if their source is placed in the extremely attenuated interstellar space.

(*b*) *Initial explosion of the expanding universe*
In order to account for the existence of protons and heavy nuclei in the primary radiation as well as the high energies involved, Lemaitre, Regener and others have suggested that the cosmic radiations are the dust from the original cosmic explosion that occurred some 3×10^9 years ago, when the whole mass of the universe was concentrated in a single nucleus. There can be no doubt that during such an explosion a fantastically great amount

of radiation was formed, with the highest individual energies now known and possibly much more. Certainly protons and other nuclei were shot out in all directions with all energies and in sufficient numbers. These particles would then have been circulating in the universe for the requisite time (a few times 10^9 years) during which some of them would have disappeared by absorption. But the earth would find itself embedded in this cosmic particle radiation, isotropically distributed. This hypothesis though intriguing is not an impossible one and cosmic rays may indeed be residual radiation left over from the largest nuclear explosion of all time.

(c) Within our own galaxy

Considerations of the main protonic constitution of the primary radiation and an adequate mechanism which could accelerate the protons to the requisite high energies have led Fermi to suggest that the cosmic rays originate in our own galaxy. The diffuse material in our galaxy, photo-ionized by the star nearby, is the source of cosmic rays. Clouds of ionized diffuse matter in rotational motion along with the galaxy can set up fields which can impart to the charged particles in them the required high energies. This theory appears to be an attractive one, which may contain the right solution to the problem, but the chief difficulty is that it fails to explain the presence of heavy nuclei in the primary rays.

(d) Within the "solar system itself

In order to account for the presence of heavy nuclei in the primaries, the apparent large intensity of cosmic rays and the observed influence of the sunspot activity on them, E. Teller and R. D Richtmyer, in 1948, have suggested that cosmic rays originate in the solar system itself, on *or near the sun*. The heavy nuclei would be more readily found along with the protons in the solar system than in interstellar space. These particles (both protons and heavy nuclei) could be accelerated to the high energies of cosmic rays by an *extended* (*wandering*) *Magnetic field* of solar origin, reaching out to the outermost planet. The same field could distribute the particles throughout the solar system with a high degree of uniformity. This theory, which seems to be a revival of Swann's idea proposed already in 1933, though in a different form, has the chief merit of assuming that a negligibly small fraction of the sun's energy would be sufficient to generate the accelerating field of the necessary intensity. Such a field has not been actually observed so far, obviously due to the fact that the intensity is too low to be easily detected. We shall come back to this theory presently.

The second important point to decide concerning the origin of cosmic rays is the way in which the cosmic ray particles acquire their enormous energies. The possible mechanisms of generating particles with cosmic ray energies may be divided into *microscopic* and *macroscopic* processes, according to their occurrence in the atomic and large scales respectively.

The **microscopic processes** have been invoked in the following different ways:

Millikan, already some thirty years ago, had suggested that cosmic rays were *sparks of energy released in the process of the building up of more complex nuclei from hydrogen in interstellar space.* At a later date, the *annihilation of atoms with a complete conversion of their masses into energy* was proposed as a possible source of the high energy carried by the cosmic rays. The annihilation energy of a proton is about 10^9 eV. Heavier nuclei will have proportionally higher annihilation energies, up to about 2×10^{11} eV for the heaviest uranium atom-just the energy range experimentally found for cosmic rays. But, in order that there may be no degradation of energy from the place of complete annihilation to the earth, the process cannot occur in stars or in any portion of the universe where matter is present in great abundance.

More recently, Klein and Arley have proposed the novel hypothesis that *a collision between a piece of 'normal matter' and a piece of 'inverted matter' would lead to a very powerful annihilation explosion and release of energy. Inverted matter* is supposed to consist of *negative protons* and *positive electrons* since *normal matter* consists of *positive protons* and *negative electrons.* Just as positrons and electrons exist and can annihilate themselves when they come together, so also positive and negative protons can annihilate themselves liberating at least the rest energy of the two protons.

The microscopic processes labor under the *common drawback* that they set an upper limit to cosmic ray energies, which is far too low. In particular, against the annihilation hypotheses the following objections could be raised. Up to now, there is no sure experimental evidence whatever of the annihilation of heavy particles. But if it does happen, it would take the energy of millions of nuclei to supply **a** single very energetic cosmic ray particle of 10^{17} eV. Such a large aggregate of nuclei does not fall under our experience, the largest known nucleus, uranium, containing only about 240. There is the further difficulty of explaining how the liberated energy is concentrated in a single particle, unless it be assumed that the nuclear annihilation itself gives rise to fast protons. But then, since it is well known that electron annihilation gives rise to γ-rays, symmetry suggests that annihilation of heavy particles would do likewise. Hence, *no nuclear methods of supplying energy, as far as we understand them, appear to offer much hope for a solution of the problem.*

The **macroscopic processes** refer to mechanisms that are effective over very large distances and impart to particles the very high energies met with in cosmic rays. They must necessarily be some sort of accelerating field, of which there are only three known types, gravitational, electric and magnetic. Because of large distances only small fields are needed. The gravitational field can influence all material particles, while the electric and magnetic fields only charged particles. All the three modes of acceleration have been utilized for explaining the energy of cosmic rays.

Milne proposed that *the cosmic* ray *particles owe their energy to the gravitational field of the universe.* This opinion, however, did not gain much favor with scientists, as it could not adequately explain all the observed peculiarities of cosmic rays, such as their charged nature, isotropic distribution, etc.

C. T. R. Wilson suggested that *the electric field produced by thunderstorms might be responsible for the highly energetic cosmic rays.* But this theory was soon abandoned on account of the many difficulties it raised, viz.

(i) Cosmic rays should then have a *terrestrial atmospheric origin* which would lead to results contrary to observed facts;

(ii) Variation of cosmic ray intensity with altitude and latitude cannot be adequately explained;

(*iii*) The great conductivity of the atmosphere at high altitudes renders impossible all supposition of a radial electric field, sufficiently great and permanently existing;

(iv) as the potential difference between charged clouds is ordinarily of the order of 10^9 volts, the very high energy cosmic rays cannot be produced by them, specially the hard component;

(v) electric fields, in general, even when they are extended over large spaces, which renders them effective accelerators in spite of their small intensity, must probably be abandoned as possible sources of great energy, since space is a good conductor and must be essentially neutral for long-time stability.

Mechanisms implying magnetic fields have been more promising. What is essentially required is a magnetic field which varies with time; it is not necessary that the intensity should be great, since charged particles could be accelerated to very high energies by repeated action of a weak field, similar to the *cyclotron effect.* There are two possible places where one may seek for such varying fields. The first is in the stars and the second in the galactic field. In stars there are two kinds of fields, the general field of the star as a whole and the field associated with sun-spots. The general field arises from the rotation of the galaxies or more locally by double or individual stars. The field chosen will naturally determine also the place of origin of the cosmic rays: *e.g., it* the field of the sun is chosen, then the cosmic rays may be said to originate in the solar system; if the galactic field, galactic origin and if intergalactic field, pervading all space, then really cosmic origin.

Authors have made use of these different magnetic fields to account for the supply of energy to cosmic ray particles.

Swann was the first to point out *that the changing magnetic fields of large 'sun-spots' on giant stars might constitute the cosmic ray generator.* Teller and Richtmyer have recently made this mechanism more local by referring the *required varying magnetic field to the sun.* The prominences in the solar corona with their fiery fountains and streamers extending from the surface of the sun and changing with time, as observed with the help of the solar coronagraph, constitute the accelerating mechanism which produces cosmic rays. The balls of fire and streamers contain ionized gas and hence charged particles which are tied to magnetic lines of force, The most powerful among them actually escape from the neighborhood of the sun and carry their magnetic fields along with them to the orbit of the earth and presumably even further. The magnetic curtain around our planetary system may serve a double purpose; it may transform cosmic rays into an effectively isotropic radiation and at the same time increase the cosmic ray intensity isotropic forcing the same particle to pass repeatedly the orbit of our planet a great number of times.

As the authors themselves admit, their hypothesis does not explain how an extended magnetic field is established throughout our solar system and in what way and for how long cosmic rays are confined in our system. Other difficulties that may be raised are:

(i) the sun could, from the known spot fields; supply the low energy part of the radiation, but not the high energy part;

(ii) the lack of marked dependence on solar time;

(iii) the rapidly falling weak magnetic field originating from the sun cannot adequately explain the observed isotropic distribution;

(*iv*) even assuming that sunspots are very common, one is still left with the difficulty that the total energy of cosmic radiation is so great that the spots would have to emit more energy than is possible, at least in the case of the sun. Calculation shows that the total energy of the cosmic radiation is much greater than the total radiant energy output of not only the sun, but even of all the stars and nebulae in the entire known universe put together.

Fermi has proposed *the magnetic field of our own galaxy as the cosmic ray generator;* as we have already remarked. Since the entire galaxy is in rotational motion, the ionized diffuse matter in it revolving in a curved path will set up a galactic magnetic field. Clouds of charged matter passing by each other or occasionally colliding, can certainly set up fields with time variations, which could generate the requisite high energies. On the assumption that the motion of the clouds is quite random, it can be shown that the net result of the transfer of energy in such collisions is that the cosmic ray particles gain energy exponentially.

On the other hand, a particle can lose its energy again if it is involved in a nuclear collision in intersteller space, which will, on the average, happen about once in 6×10^{7} years.

The equilibrium between these two effects results in an energy spectrum which falls off according to a power law with increasing energy, and by using reasonable quantities for the various parameters which enter into the theory, one may obtain good agreement with the exponential energy spectrum of the primary cosmic- radiation. Furthermore, since the process has the entire rotational energy of the galaxy to draw upon, it can, without depleting itself too fast, supply the total energy; the galactic field can also serve the purpose of the trapping mechanism which keeps the particles circulating within the galaxy.

This hypothesis appears to be a really good one, where known sources are used to account for the total energy. Alfven, the Swedish astrophysicist, in 1943, cited evidence for the existence of varying magnetic fields near double stars, which may reproduce the operation of the cyclotron in accelerating the cosmic ray particles to their high energies. In 1948, Babcock of Mount Wilson Observatory reported on a variable star with a changing magnetic field. The field force of this source rises to + 7,800 gauss, then in a few days time reverses its polarity to - 6,500 gauss. Such a fluctuating field could certainly produce cosmic ray energies.

Intergalactic magnetic fields have also been invoiced for the generation of cosmic ray energies. Such a mechanism requires no trapping process; it implies that cosmic radiation is the preponderant energy of the universe in consonance with Lemaitre's hypothesis that considers the universe as an immense atom which at the initial explosion emitted the cosmic rays by a sort of super-radioactive process.

It is theories such as these involving magnetic fields on the macroscopic scale that will probably help us to understand in time how cosmic ray particles become so energetic.

The problem of the origin of cosmic rays is, however, of greater interest to the cosmologist than to the physicist, since its solution is bound to contribute to the description of the physical universe and its ancient history. For the physicist, what the cosmic rays are and what they can do are of more immediate interest, since through them he can learn a lot about the mysterious nature of fundamental nuclear processes and the laws that function in the realm of extremely high energies.

CHAPTER 16

STRUCTURE AND PROPERTIES OF THE NUCLEUS

In this last chapter a short account of the structure and properties of atomic nuclei, as revealed by the experimental investigations of atomic and nuclear processes described in the course of the book, will be given.

1. STRUCTURE OF NUCLEI

A. Theories of nuclear composition

Rutherford's nuclear theory of the atom shows that the nucleus must have a *very compact structure* unlike the extranuclear family of electrons, although practically the whole mass and the total positive charge of the atom are concentrated in the nucleus. Natural radioactivity, in its turn, clearly points out that the nuclear structure, though compact, must be *very complex,* capable of ejecting different particles like α-particles and electrons as well as electromagnetic radiation in the form of γ-rays. These γ-rays, however, are not in any real sense constituents of nuclei, but are emitted when a nucleus in an 'excited state' returns to the normal, in exactly the same way as visible light or X-rays are emitted in the return of the peripheral electronic system of an atom from a higher energy state to its normal state. The study of the α-ray and γ-ray spectra indicates that *there exists great order in the internal structure of the nucleus,* in spite of the complexity, while the interpretation of the continuous nature of the β-ray spectra introduces a new particle, the *'neutrino',* which, though having practically no mass, plays the important role of a discrete carrier of energy. Researches in artificial transmutation of elements establish that α-particles, protons, neutrons, positrons and electrons enter into the constitution of nuclei in some way or other.

There are reasons to believe that the α-particles are stable sub-units built up of four elementary particles.

Finally cosmic ray studies demonstrate the existence of another fundamental particle, the 'meson', positively or negatively charged, which being unstable decays into an electron, positive or negative. Now the problem is to find *out which of these elementary particles enter into the actual constitution of nuclei* in the most compact and complex but orderly fashion.

Several theories have been proposed which may be called, according to the elementary particles selected for the nuclear constitution, the *proton-electron, proton-neutron, neutron-positron* and *negative proton-neutron* theories. Of these, the *proton-neutron* concept alone has found general acceptance, as possessing a number of distinct advantages over the others.

(a) The proton-electron theory

The proton-electron constitution of the nucleus was the first to be proposed and was in vogue for a time, until the neutron was discovered. The conception arose almost naturally from the following experimentally observed facts:

(i) The discovery of the *'whole number rule'* by mass spectrum analysis *justified the age-long idea that the different nuclei are built up from the same simple element, hydrogen (proton).* The slight discrepancy of the actual mass of a free proton (1.00813) from the whole number was accounted for as either due to the arbitrarily chosen standard (1/16 of the mass of an atom of oxygen) or due more probably to the *'packing effect', i.e.,* according to the laws of nuclear reactions, a small amount of mass disappears when several protons are packed together to form a stable nucleus, as will be explained when dealing with *mass defect.*

(ii) *The electrical neutrality of the atom as a whole argued, in its turn, that electrons also should enter into the constitution of the nucleus.* For, except hydrogen, in all other atoms, it was found that the number of extranuclear negative charges fell short, by very much, of the supposed number of positive charges in the nucleus. For instance, considering the helium atom, since its mass is nearly four times as great as that of hydrogen, it may be legitimately supposed that its nucleus is composed of four protons. The fact that 4 free protons are slightly heavier than a helium nucleus can be explained by the *mass defect* arising from the packing together of the protons, as stated above. Under these conditions, the helium nucleus carries four positive charges. On the other hand, it is experimentally proved with certainty that there are only two electrons outside the nucleus. Then how to account for the electrical neutrality of the atom as a whole? The discrepancy between the positive and negative charges becomes all the more marked as one proceeds to the heavier elements of the periodic table. Thus, carbon atom should have 12 protons and hence 12 positive charges, while it has only 6 peripheral electrons and hence only 6 negative charges; oxygen atom 16 protons but only 8 peripheral electrons. In many elements the number of peripheral electrons amounts only to about half the assumed number of positive charges. A natural explanation of this state of affairs would be to introduce the negative charges required by the conditions of neutrality of the atom into the nucleus itself, which would balance the excess positive charges found there. As the mass of the electron is relatively very small, electrons introduced thus into the nucleus would not greatly affect the total mass of the atom while they would effectively render the atom neutral. Thus, in the case of helium, two electrons would be found inside the nucleus to neutralize the two excess positive charges; carbon would have six electrons inside its nucleus, oxygen eight and so on.

(iii) *The β-ray emission by natural radioactive elements seemed to confirm the existence of electrons in the nucleus,* since the emission of electrons in such cases is essentially a nuclear process, as already shown.

The proton-electron theory, which appeared thus to be sound in many respects, brought, however, in its wake, a number of *serious difficulties,* some of which may be briefly mentioned here:

(A) Supposing the electron to be a simple spherical charge of finite dimensions, it can be shown that the dimensions of the nuclei of atoms, chiefly of the heavier ones, are so small that there is no possibility of introducing several electrons inside the nucleus.

(B) The application of wave mechanics complicates matters still further. The wavelength of an electron is found to be much larger than the nuclear radius, which therefore excludes the possibility of keeping the electrons inside the nucleus.

(C) There are also other difficulties that are connected with nuclear spin, magnetic moment, statistics, etc., which will be pointed out later.

For these reasons, it appeared almost certain that *no electrons can exist inside the nucleus.* The discovery of the neutron came just in time to solve these difficulties by the following radically different. theory of nuclear constitution:

(b) The proton-neutron theory
According to this theory which is generally held to-day, *nuclei are composed of protons and neutrons.* Considering the case of helium again, if, instead, of 4 protons and 2 electrons, there are 2 protons and 2 neutrons inside the nucleus, then also the atom as a whole is neutral even with only 2 extra-nuclear electrons, while the same mass is retained, since the neutron is an electrically neutral particle with nearly the same mass as that of the proton, The sum of the masses of 2 free protons and 2 free neutrons is, of course, somewhat greater than the mass of the helium nucleus but this excess is explained, as before, by the mass defect occurring when the four particles go to form the helium nucleus. In a similar manner, the carbon nucleus would contain 6 protons and 6 neutrons, oxygen nucleus 8 protons and 8 neutrons and so forth.

In general, representing nuclei by the two important numbers that characterize them, *viz., the mass number A* which is the integer nearest to the actual mass of the nucleus and *the atomic number Z* which gives the total number of positive charges in the nucleus, according to this theory, the total number of protons and neutrons is given by A, the number of protons by Z and the number of neutrons by A - Z: the number of extranuclear electrons in the normal neutral atom is evidently Z.

The *β-ray emission or electron, disintegration of radioactive substances,* which appeared to give a good experimental support to the previous theory, is accounted for in this theory, in a very ingenious manner, as follows:

The electron does not pre-exist in the nucleus but is formed just at the instant of emission, caused by the transformation a neutron into a proton:

$n \rightarrow p + e^-$.

The positron emission is likewise due to the converse process, *i.e.,* when a proton transforms itself into a neutron:

$p \rightarrow n + e^{+}$.

As protons and neutrons can be converted into each other in the nucleus, they are regarded as *two alternative states of a single heavy nuclear particle,* to which the name of nucleon has been given.

This theory has the merit of being based on two actually existing fundamental particles, the proton and the neutron, while it dexterously removes the difficulties encountered in introducing formally electrons in the nucleus as in the previous theory. It has the additional advantage of reducing nuclear constitution to a single particle, the nucleon, though of complicated characteristics. This conclusion is very close to Prout's old hypothesis.

Hyper-chemical classification of elements
The proton-neutron theory leads also to a more refined classification of elements than classical physics or chemistry can achieve. In chemistry, a rough classification is made by the easily changeable accidental physical and chemical properties depending upon the peripheral electronic configuration, while here a highly perfected differentiation based on. the more intrinsic nuclear properties, less prone to changes, is obtained, Atoms are classified into three different categories as follows:

(i) *Those that have different mass numbers A, but the same atomic number Z, are called* **isotopes** which having the same physical and chemical properties cannot be distinguished or separated by the ordinary physical or chemical methods. The isotopes represent elements in the strictest sense, while the chemical elements that occur in nature are mixtures of several isotopes. Since the essential individuality of an clement is determined ultimately by the nucleus, we may say that the number of simple elements is very much greater than the 92 elements of the periodic table, and should include about 300 different isotopes that are found in nature. The factor that distinguishes the different isotopes is the relative number of neutrons in them, since Z which gives the number of protons is constant, while A, the total number of particles, varies and hence the number of neutrons, given by (A - Z), changes from isotope to isotope.

(ii) *Atoms that have the same mass number A, but different atomic numbers Z, are called* **isobars** which naturally exhibit different chemical and physical properties. Forty-four pairs of isobars are known, excluding some ten cases for which one of the two members is doubtful. Most of the isobars have even mass number and even atomic number, the last differing by two units, while those of odd mass numbers and odd atomic numbers differing by one unit are rare. It is evident that in isobars both the number of protons and the number of neutrons are different. Isobars with even number of protons and odd number of neutrons occur rather frequently.

(iii) *Atoms which, though having the same mass number A and the some atomic number Z, are distinguished by certain differences in the internal structure of the nucleus, manifested by different decay periods in the case of radioactive nuclei; are called* **isomers.** A number of such atoms has been brought to light in the course of a careful analysis of artificial radioactivity. We have already given an account or the experimental study and theoretical explanation of nuclear isomerism.

It has been already pointed out that the theory of proton-neutron constitution of the nucleus fits in well the theoretical explanation of a great number of nuclear phenomena. We shall meet with other cases, where the essential soundness of this theory is clearly brought out. The real crux of its all-round validity rests, however, upon the correct understanding of the nature of the intranuclear forces that bind together the uncharged neutrons and charged protons into stable nuclei and the experimental verification of the proposed type of force. Attempts have been made with very promising results to unravel the mysterious nature of the specifically nuclear forces by Heisenberg, Majorana, Yukawa and others.

Jean Baptiste Perrin (30 September 1870 – 17 April 1942) France

(c) The neutron-positron theory

This was proposed by Jean Perrin, states that *nuclei are built up with neutrons and positrons alone* as fundamental elementary particles. The idea arose from an attempt to solve a difficulty raised against the otherwise satisfactory proton-neutron theory, viz., which of the two constituent particles, the proton or the neutron, is to be considered the primordial elementary particle?

If the proton were to be favored with such a privilege, in consonance with the time-honored belief, the neutron would be a complex particle built up with a proton and an electron. But, since it is definitely proved that the mass of the neutron is greater than that of the proton, laws of nuclear reaction, which demand that the mass of the resulting stable nucleus should be smaller than the sum of the individual masses of masses of the constituent particles, do not favor such a hypothesis.

If, on the other hand, the neutron were to be the primordial simple particle, as confirmed by the mass of the neutron being greater than that of the proton, then the proton would be a complex particle made up of a neutron and a positron. This might mean that to a great extent the *'inertia of matter is not to be linked with electric charges'.*

Perrin, choosing the second alternative as the more probable of the two, argued to the neutron-positron constitution as follows:

Instead of considering a nucleus of mass number A and atomic number Z as composed of A protons and (A - Z) electrons as in the proton-electron theory, or of Z protons and (A - Z) neutrons as in the proton-neutron theory, it is to be regarded as made up of *A neutrons and Z positrons.*

In this scheme also, the neutrality of the atom as a whole is secured. On account of the Coulomb's repulsion between the positrons, the number of positrons which are united with the A neutrons in the nucleus cannot exceed A/2; if it exceeds, the nucleus becomes less stable with positron radioactivity; if less, then also the nucleus is unstable but with electron radioactivity. Although the exact mechanisms of these processes are not definitely known, it may be surmised that in either case the unstable nucleus, using up its internal energy creates *pairs.*

In the first case, the electrons of the pairs annihilate the excess of positrons in the nucleus, while the positrons are expelled.

In the second case, the positrons of the pairs are retained to make up for their deficiency in the nucleus, while the electrons are emitted.

The stability of the nucleus composed of neutrons and positrons is to be attributed to a special *'close-range' property* of the neutron, viz., *a very great mutual attraction between a neutron and a positron and a similar repulsion between a neutron and an electron.*

Since the capture of a positron by a neutron to form a proton is accompanied by appreciable loss of mass and consequently of potential energy, it follows that there should exist, at close range, a powerful mutual attraction between the neutron and positron. On the other hand, since the positron and electron are counterparts, symmetric by nature, it seems reasonable to suppose that there should exist an equally powerful mutual repulsion between a neutron and an electron at close range. Such a conception in no way goes against the existence of the Coulomb's force of attraction between the nucleus and the planetary electrons at relatively great distances. Considering, for instance, a proton, made up of a neutron and a positron, an electron will be repelled by it at close range, this repulsion, by the neutron probably annulling itself against the Coulomb's attraction due to the positron.

Hence, a proton, and, in general, *any nucleus,* in spite of its positive charge, *cannot incorporate into itself electrons;* but at relatively great distances, where the influence, of the neutron does not reach out, it can still attract the electrons and keep them imprisoned in the extra-nuclear orbits.

As regards the stability of the nucleus, it seems necessary to postulate the simultaneous existence of protons and neutrons in the nucleus. On the further assumption that a proton is composed of a neutron and positron, the neutron devoid of all charge, placed close to a proton, eagerly solicits the positive charge of the proton. When the exchange takes place the proton becomes a neutron, Thus the interaction between a proton and a neutron at close quarters, through the intermediary of the positron, results in the stability of the nucleus, which is not merely *static, i.e.,* not a mere huddling together of the different constituents, but highly *dynamic,* consisting in the active exchange of a positron between the proton and the neutron and thereby implying discontinuous changes in the energy states of the nucleus.

This theory is, in a certain sense, the same as the proton-neutron theory, but it tries to go deeper into the ultimate composition of the nucleus by supposing that a proton is a complex entity made up of a neutron and a positron. It is, however, subject to a number of serious objections. Although the electron is excluded from the constitution of the nucleus, the positron is formally introduced into the nucleus. Considering the. symmetrical identity of the two kinds of electrons, both as regards mass and magnitude of charge, except for the sign, all the difficulties raised against the proton-electron theory seem to be pitted against the neutron-positron theory also. Furthermore, the emission of electrons and positrons from radioactive bodies is explained by a somewhat artificial and complex process of creation of pairs which is an extranuclear phenomenon as we have seen. It is strange that a specifically nuclear phenomenon, as β-ray and positron emission, should depend on an extranuclear action, as pair production.

For these reasons, authors are not inclined to accept this theory and prefer to treat, the question which of the two particles, proton or neutron, is the simple one, as having no physical meaning. For, 'elementary particles' are ultimate indivisible building blocks of matter *a very limited sense only,* since they can change into one another practically without restriction, as long as it is compatible with the laws of conservation of mass, energy, etc. As a matter of fact, transformability is one of the characteristic properties of elementary particles. Thus a proton can change into a neutron and a positron, or a neutron into a proton and an electron. But, just for this very reason, it is meaningless to describe any of them as being composed of some of the others: *e.g.,* a proton composed of a neutron and a positron, or a neutron composed of a proton and an electron. Admitting the proton-neutron model, the protons and neutrons should be considered as two different states of one and the same fundamental heavy particle, the nucleon, subject to changes from one to the other with the emission or absorption of energy and the consequent liberation of negative or positive charge, *i.e.,* of electron or positron.

(d) The negative proton-neutron theory
This was suggested by some authors, as Gamow, Klein and Arley. It speculates on a *nuclear constitution with ordinary neutrons and negative protons.* It is based on the assumption that there is a certain possibility that negative protons exist. One general argument is drawn from considerations of symmetry. With the discovery of the positron, symmetry has been established in the realm of elementary *light* particles. In a similar manner, symmetry among the *heavy particles such* as protons, and neutrons, will be secured if negative protons exist. These would be symmetrical to positive protons with respect to neutrons. It can be shown that the relation between the negative protons and ordinary protons is not, however, quite analogous to that between positrons and electrons in Dirac's theory and that in consequence negative protons are not to b conceived as *Dirac's holes* for ordinary protons. These are rather to be regarded merely as independent particles subject to intertransformation with neutrons involving the emission of electrons or positrons, as the case may be.

When a neutron transforms itself into a negative proton, a positron will be emitted
$(n \rightarrow p^- + e^+)$;
when a negative proton is transformed into a neutron, an electron is given out
$(p^- \rightarrow n + e^-)$;
Hence, a negative proton may be conceived to be formed by a neutron and an electron. Negative protons, like positrons, cannot exist free for long within ordinary material, since they will be immediately attracted and absorbed by the nearest positively charged atomic

nuclei and most probably will be turned into neutrons after entering the nucleus. This shows that negative protons may easily escape detection. If the two particles, proton and negative proton, should originate simultaneously like positrons and electrons, an energy of over 4,000 MeV would be required to create such a pair.

Substances formed with nuclei composed of negative protons and neutrons will not represent ordinary matter that we know, but *"inverted matter"* where the atom will be made up of *negatively charged nucleus surrounded by positive electrons.* These "inverted" atoms will have properties identical with ordinary atoms and there will be no way of differentiating the two, unless they are brought together, when a terrific explosion will take place, with the mutual neutralization of oppositely charged nucleons along with the mutual annihilation of the positrons and electrons.

Gamow thinks that isomeric nuclei could be readily constructed on the negative proton-neutron nuclear model by replacing a pair of neutrons in the nucleus by two protons, one positive and the other negative. Such isomeric nuclei would differ naturally as regards binding energy and spin, and if they were unstable also in their mode of disintegration.

Klein has made use of "inverted matter" to explain the origin of showers in cosmic rays. Klein and Arley have introduced the same idea in dealing with the origin of cosmic rays by nuclear annihilation processes.

Evidence for the negative proton

It should be possible to create a negative proton by a proton-nucleon collision, provided the incident proton is accelerated to at least 5,600 MeV energy.

At this energy, the production of negative protons is not expected to be very copious and their abundance, relative to other particles formed, will probably be far smaller than in many cosmic ray phenomena which have already been extensively investigated. The problem of the conclusive identification of the negative proton may be a difficult one for some time to come. A direct mass and sign measurement might be expected to identify it. However, this would require a high precision method to distinguish it

from known negative heavy mesons of mass greater than half the proton mass. Further, any such direct mass and sign determination could, in itself, always be attributed to a new type of heavy meson. In a cloud chamber photo obtained by E.W. Cowan in 1954, there seems to be evidence for the negative proton, on the assumption that the inverted V track seen there is caused by the rare decay of a neutral hyperon into a π^+-meson and a negative proton. From the sense of deflection in the magnetic field, the right-hand limb of the fork (marked A) is due to a positive particle; further from the nature of the track, it can be well presumed to have been produced by a π^+-meson. The left-hand limb (marked B) is caused by a negative particle whose mass, as determined from ionization density and momentum is $(1,850 \pm 250)$ m_e. Taking into account the errors of measurement, the mass of this particle is nearly equal to that of a proton and does not agree with the mass of any negative particle that has been so far identified. Hence the track B might have been caused by a negative proton.

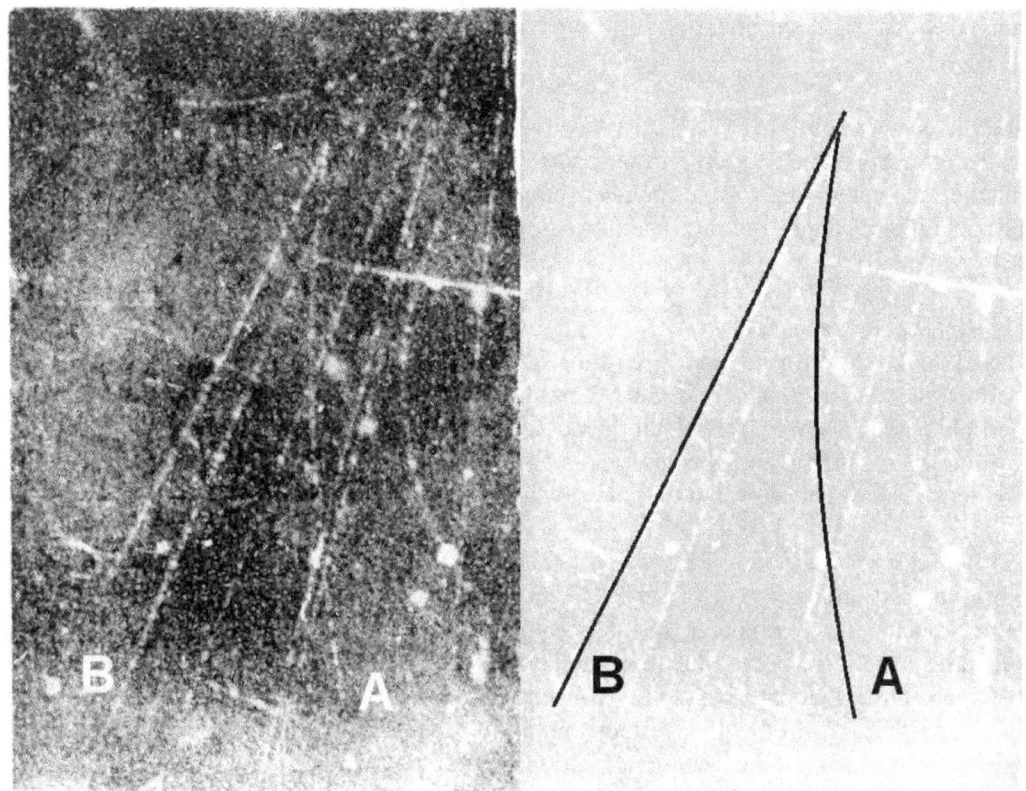

Decay of a Λ° hyperon into a π^+-meson (A) and a possible antiproton (B)-E.W. Cowan.

There is a characteristic feature of the negative proton that should render its identification more simple. On coming to rest in matter, a negative proton should be attracted into the nucleus of an atom and annihilate itself by combining with ordinary proton. This process, liberating as fast mesons (or less probably γ-rays) twice the rest-energy of the incident particle, is, by definition, unique to a negative proton and should be readily observed. One cosmic ray event has been seen in a multi-plate cloud chamber by De Staebler, Bridge, Courant and Rossi (1954), which is consistent with this process and hard to reconcile with any alternative interpretation.

Within the past four years, the use of large blocks of nuclear emulsions and the combination of magnet and multi-plate cloud chambers have both greatly increased the amount of information that may be obtained from the observation of a single particle. It is quite possible that, with the aid of these techniques, conclusive evidence fur the existence of the negative proton may be found in the near future. Moreover, it may become possible to identify the annihilation product of machine-made negative protons by a counter experiment, even if the particle is not detected.

In October 1955, workers at the University of California, Chamberlain, Segré, Wiegand and Ypsilantis, have announced that they have definitely placed in evidence the existence of negative protons or *anti-protons* as they are now known. Bombarding copper with protons of energy 6.2 BeV from the Berkeley bevatron, they have been able to identify mesons of the right specification resulting from the combination of protons and anti-protons.

B. Nuclear Models
In the attempts made to interpret the special properties of nuclei such as stability, spin, magnetic moment, etc., various nuclear models have been proposed, as in the case of the peripheral electronic structure. The more important among these models are the following:

(1) *Alpha-particle model*
Based on the experimental facts that α-particles are ejected by nuclei in disintegrations, both natural and artificial,. Wheeler, Weizsäcker and Fano have proposed the alpha-particle model, according to which the nucleus contains *α-particles,* at least as *substructures,* although they cannot maintain their identity for a very long time inside condensed nuclear matter, but will dissolve into more elementary particles.

(2) *Liquid drop model*
This was first proposed by Bohr and Kalckar, then accepted by Heisenberg, Majorana, Wheeler and others, compares the nucleus to a *liquid drop,* the nucleons corresponding to the molecules of the liquid, due to several points of similarity, such as large interaction between constituent particles, nearly constant density, surface tension effect, etc. This model has been utilized with a certain amount of success in the interpretation of intra-nuclear forces and of nuclear transformations and, in particular, nuclear fission, as we have seen.

(3) *Shell model*
This was proposed by Haxel, Mayer, Feenberg, Nordheim and others, assimilates the nucleus rather to a *gas, made of independent particles* with small interaction, caused by their movement in a common field of force and influencing one another only by means of the requirements of the Pauli principle which excludes identical particles from occupying the same quantum state and thus endows the nucleus with a shell-structure, as in the case of extra-nuclear electron family. This model has been successfully used to explain certain nuclear phenomena such as stability, spin, magnetic moment, etc.

These various models prove insufficient, if an attempt is made to put them on some solid theoretical foundation or to subject them to a really detailed comparison with experimental data. They should be considered as not exclusive of one another, but complementary in scope in the understanding of the extremely complicated behavior of nuclei.

C. Potential barrier
Intimately connected with the nuclear structure is the *electrostatic field* of the positively charged nucleus, ordinarily known as the *potential barrier.* The contents of the nucleus are, so to say, walled in and well defended against interference from outside by this potential barrier which therefore plays an important role in natural and artificial disintegrations.

The special characteristics of the potential barrier have been understood as follows:

The experiments on α-particle scattering showed that up to a very short distance from the center of the nucleus the Coulomb law of inverse squares holds, but that there is a critical distance within which it becomes ineffective. *In the region of the inverse square law,* all positively charged particles, which are directed towards the positive nucleus, will be repelled by the latter the greater being the force of repulsion the closer the approach. On the other hand, the negatively charged electrons, surrounding the nucleus in different peripheral orbits in an atom, will be kept in their proper positions, without flying off at the slightest provocation, by the same Coulomb law of forces, thereby endowing the atom as a whole with a stable structure. The electrons that are closer to the nucleus will experience a greater force of attraction, which explains why more work has to be done to remove an electron close to the nucleus than another in the extremities.

Within the critical distance, the simple inverse square law becomes less important, giving place to a *new and more complicated law* which is essentially characterized by *attractive forces.* The existence of such a non-Coulomb's attractive force is readily inferred from natural radioactivity and artificial transmutation. Given that both the nucleus and the α-particle are positively charged, one cannot understand how the α-particles could remain inside the nucleus for any appreciable time (an experimental fact) unless attractive forces exist between the α-particles and the rest of the nucleus,

Artificial disintegration experiments, in their turn, show that even positively charged particles can be assimilated by the positive nucleus to form a new stable nucleus provided the external particles approach close enough to the nucleus, *i.e.,* within the critical distance. The same attractive force is also implied in the generally accepted proton-neutron constitution of the nucleus. For, otherwise, one cannot explain how the *uncharged neutrons* and *charged protons* can unite together so intimately to form a stable nucleus.

These peculiarities of the potential barrier may be represented graphically by what is known as the *barrier diagram of the nucleus,* shown in Fig. 328.

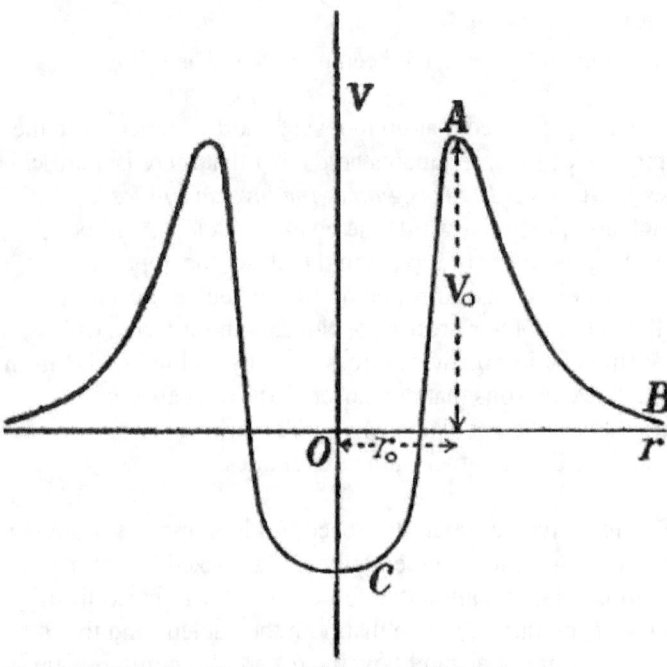

Fig. 328. Barrier diagram of the nucleus.

It is drawn by plotting the electrostatic potential V as a function of the distance r from the center of the nucleus. It is seen that as r diminishes the curve rises rather steeply, reaches a maximum at a certain distance r_0 and then drops down sharply. The portion AB of the curve represents the region where the law of inverse squares is effective so that the electrostatic potential V at any point which is at a distance r *is* given by Ze/r, Where Ze is the nuclear charge. The ordinate at any distance r is a measure of the amount of kinetic energy which a positively charged particle must sacrifice, (*i.e.,* which must be converted into potential energy) in order to come from infinity up to r. If the nucleus were to be a mere point without room for parts or structure, the potential barrier will be of infinite height, since, when $r = 0$, $V = \infty$. In such a case, transmutation of one element into another is inconceivable, since no actual projectile, however powerful, could surmount an infinitely high barrier.

But, in reality, the nucleus is not a mere point charge, playing simply the part of the center of Coulomb force; it has finite dimensions, however small, and is packed with protons and neutrons as constituent particles, as experimental researches have clearly proved. Moreover, transmutations, both natural and artificial, have been actually observed. These facts are represented by the sharp descent of the curve, AC, beyond the critical distance r_0, which indicates that:

(i) The height of the barrier cannot be infinite, but has a finite value corresponding to the maximum in the curve,

(ii) There exist attractive forces in this region and

(iii) These attractive forces become considerable when r becomes smaller than r_0. At the center of the nucleus the intensity of the field must be zero and the potential a minimum. The shape of the curve resembles the crater of a volcano, a deep pit surrounded by high hills on all sides, which has given rise to the name *the crater model of the atom*. The pit itself is known as the *nuclear potential well*.

The quantity V_0 corresponding to the maximum point in the curve is known as the *height of the potential barrier* and the distance r_0 the *radius of the potential barrier* or the *nuclear radius*. These two factors are important, the first in the study of transmutation of elements and the second in the determination of the size or the nucleus. An approximate relation between V_0 and r_0 may be obtained in the following simple manner:

If Ze be the nuclear charge, the part of the curve which represents the Coulomb law may be analytically expressed by $V = Ze / r$. Including the fact that this law is replaced at a certain critical distance by some other law of attractive forces, we modify the above relation as

$V = (Ze / r - 1 / r^n)$
to a first approximation.

The expression for the potential therefore contains not only a term of repulsion $(+ Ze / r)$, but also another of attraction $(- a/ r^n)$, where a is a constant; r^n indicates that the attractive forces become considerable when r becomes smaller than a certain value r_0. In other words, for distances $r < r_0$, the second term becomes all important and positive particles are attracted by the nucleus while for distances $r > r_0$, the first term is the important one, a/ r^n becoming negligibly small and positive particles will be repelled by the nucleus.

Since r_0 corresponds to the maximum value of $V = V_0$, to get an expression connecting V_0 and r_0, the relation for V is differentiated and equated to zero:

$$\frac{dV}{dr} = \left(- \frac{Ze}{r^2} + \frac{na}{r^{n+1}} \right) = 0$$

$$\therefore \quad \frac{na}{r^{n+1}} = \frac{Ze}{r^2} \quad \text{or} \quad \frac{na}{Ze} = \frac{r^{n+1}}{r^2} = r^{n-1}$$

Since in this case $r = r_0$, we may write

$$r_0^{n-1} = \frac{na}{Ze} \quad \text{or} \quad r_0 = \left(\frac{na}{Ze} \right)^{1/(n-1)}$$

To this value of r corresponds the maximum value V_0 of the potential. Hence we may write:

$$V_0 = \frac{Ze}{r_0} - \frac{a}{r_0{}^n}$$

$$= \frac{Ze}{r_0} \left\{ 1 - \frac{a}{(Ze/r_0)r_0{}^n} \right\}$$

$$= \frac{Ze}{r_0} \left\{ 1 - \frac{a}{(Ze)r_0{}^{n-1}} \right\}$$

$$= \frac{Ze}{r_0} \left\{ 1 - \frac{a}{Ze \left(\dfrac{na}{Ze} \right)} \right\}$$

$$= \frac{Ze}{r_0} \left(1 - \frac{1}{n} \right)$$

The value of n is not quite definite; theory attributes to it a value $3 < n < 5$. The height of the potential barrier V_0 varies from element to element, *e.g.*, for gold 10 MeV, aluminum 6 MeV, lithium 2.8 MeV, etc. Likewise r_0 also will vary with different nuclei.

The potential barrier can, therefore, be analytically expressed by the relation

$V = (Ze / r - a / r^n)$
with
$r_0 = (na / Ze)^{1/(n-1)}$
and
$V_0 = (Ze/r_0)(1 - 1/n)$.

Study of α-disintegration by natural radioactive substances and of artificial transmutation of elements has shown that the potential barrier is *"transparent"* to particles whose energy may be far less than the height of the barrier. This cannot be understood on the basis of *classical mechanics* which considers material particles as simply corpuscular in nature. But the new *wave mechanics,* which attributes to every material particle a wave aspect also, argues to the existence of the above-stated transparency of the potential barrier.

2. SIZE OF NUCLEI

Rutherford, as a result of his experiments on α-particle scattering by thin metallic foils, came to the conclusion that the distance of closest approach of the α-particle to the nucleus of the scatterer on the basis of the law of inverse squares was a measure of the size of the nucleus. Since then, a great deal of information has been gathered, of the most diverse character, so that we have now a more precise *idea* of the size of the nucleus.

As there are reasons to believe that *nuclei are very nearly spherical,* we may legitimately think of a *nuclear radius.* Likewise there is evidence for the *density of nuclear matter remaining approximately constant over most of the nuclear volume and then decreasing rapidly to zero,* as shown in Fig. 329.

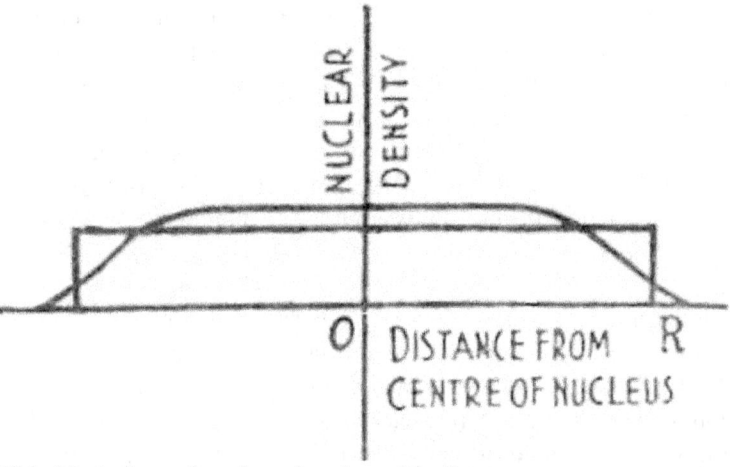

329. Variation of nuclear density with distance.

Hence, to a first approximation, the nucleus may be considered to resemble a billiard ball (though very much smaller) whose radius does not evidently include the outermost limit of the density distribution. It has also been established that the density of nuclear matter varies very little from nucleus to nucleus, except for the very lightest nuclei. This means that *the volume of the nucleus is approximately proportional to the total number of the constituent particles, i.e., to the mass number A;* hence the radius of the nucleus is proportional to $A^{1/3}$. This fact is expressed by the relation

$$R = R_0 A^{1/3} \times 10^{-13} \text{ cm}$$

where R_0 is a constant. It has been the purpose of recent experiments to verify this relation and determine the value of R_0.

Experimental measurement of nuclear radii
When we speak of the size of the nucleus we are dealing with an extremely small quantity, of the order of 10^{-13} cm and its measurement implies great ingenuity and refined technique. Further, it is not to be expected that all the methods of measuring the nuclear radius will yield identical results, For, one method may measure the average radius of the

equivalent nucleus of uniform density, another the radius of the outermost limit of the density distribution. The different methods that have been employed can be divided into two main groups, *viz., nuclear* and *electric.* In the former, the measuring probe is a light nuclear particle such as a proton, a neutron, a deuteron or an α-particle, whereas in the latter, it is an electrically charged non-nuclear particle, such as an electron or a μ-meson.

A. NUCLEAR METHODS

(a) α-disintegration method

Application of wave mechanics to α-disintegration leads to an expression for the mean life of the α-emitter in terms of energy of the emitted α-particles and the effective radius of the nucleus. As the mean life and the energy of the α-particles can be experimentally measured in the case of any given radioactive α-emitter, the radius of the nucleus of that element can be readily evaluated. The nuclear radii thus obtained for elements belonging to the same radioactive series are comprised between 7×10^{-13} and 10×10^{-13} cm, in good agreement with order of magnitude of the values given by other methods. It is also found that the nuclear radius is proportional to the cube root of the atomic weight. The method can be used only for *heavy radioactive elements* whose mass numbers A are greater than 208 and which are α-emitters. Further, it is not clear which particular radius, the average or of the outermost limit of density distribution, is measured in this method.

(ii) Interaction method

The nuclear radii have been determined from observations made on the interaction of light nuclear particles such as α-particles, protons and neutrons with the nuclei of the elements under study. Of the nuclear particles mentioned above, neutrons alone are uncharged. In the interaction between a charged nuclear particle and a nucleus, when the two are at a considerable distance apart, there is an electrostatic repulsion between them, given by Coulomb's law. But, when the particle gets very close to the nucleus, say due to its high speed, it is found that deviations from Coulomb's law set in and finally an effective attractive force comes into play, as indicated by the particle being absorbed by the nucleus, in spite of their like; charges. In the case of neutron, the interaction is evidently nil at large distances, but, at small distances of the order of the nuclear radius, it is 'specifically nuclear'. We shall now consider briefly the researches made on the size of the nucleus, using each one of the above-stated nuclear particles.

(a) α-particle scattering

Rutherford's simple *classical scattering formula* was derived on the assumption that the nucleus of the scatterer was so heavy that its motion during the interaction (considered as a collision) might be neglected. It led to a rough estimate of the radius of the nucleus, which was found to be of the order of 10^{-13} cm

If the condition assumed by Rutherford is not fulfilled, as is to be expected in the scattering caused by *light* elements, the phenomenon becomes complicated. *Two things might happen:*

Inverse square scattering

The collisions might take place outside the critical distance, where Coulomb's law alone holds. This is known as *elastic collision* which would occur for moderate velocities α-particles. In such a case, the scattering will be still governed by the law of inverse squares, hence called *inverse square scattering,* but the final formula will be more complicated, on account of the fact that, instead of the nucleus at rest throughout the impact, a coordinate system, where the center of gravity of the two interacting particles is at rest, must be considered. The general formula holding good for all cases of inverse square scattering is found to be

$$N = \frac{Qnt}{r^2} \left(\frac{Ze^2}{mv_0^3}\right)^2 \operatorname{cosec}^2 \varphi \; \frac{\left[\cot \varphi \pm \sqrt{\operatorname{cosec}^2 \varphi - (m/m')^2}\right]^2}{\sqrt{\operatorname{cosec}^2 \varphi - (m/m')^2}}$$

where the + or - sign is to be taken according as $m < m'$, or $m > m'$, m and m' being the masses of the α-particle and the nucleus respectively.

Considering the nucleus of the light scatterer it can no longer be assumed to remain at rest during impact. It will experience a recoil motion which must be taken into account. As the laws of conservation of energy and of momentum must be true quite independently of the particular laws of force, assuming the nucleus to be at rest before the impact, the *velocity of recoil* of the nucleus can be estimated as follows:

After collision, if the α-particle of mass m and initial velocity v_0 moves with a velocity v in a direction making an angle φ with its initial direction, and the nucleus of mass m', initially at rest, is chased away with a velocity v' in a direction making an angle θ with the same initial direction, then we have

$$\left.\begin{array}{l} mv_0 = mv \cos \varphi + m'v' \cos \theta \\ 0 = mv \sin \varphi - m'v' \sin \theta \end{array}\right\} \quad \text{(conservation of momentum)}$$

and

$$\left.\begin{array}{l} mv_0 = mv \cos \varphi + m'v' \cos \theta \\ 0 = mv \sin \varphi - m'v' \sin \theta \end{array}\right\} \quad \text{(conservation of momentum)}$$

From these relations, v, v', θ and φ can be evaluated. In the case of a *direct head-on collision, i.e.,* when $\theta = \varphi = 0$, the velocity of recoil v' of the struck nucleus can be readily obtained. Among the cases studied for direct head-on collision of α-particles with light nuclei, perhaps the most important is that of hydrogen nuclei or *protons* set in motion by the impact of α-particles, investigated by Rutherford. He found that in the case of such direct collisions the proton acquired a velocity 1.6 times as great as that of the α-particle. If it be assumed that the range of the proton in relation to its initial velocity follows the same experimental law as in the case of the α-particle ($R \propto v^3$), it is to be expected that the range of the proton set in motion by the head-on collision will be about 4 times the range of the α-particle, since 4 is the cube of 1.6 roughly.

This conclusion has been experimentally verified by Marsden and Rutherford. The α-particles emitted by radium were projected into a vessel containing hydrogen gas. The nuclei set in motion, which were identified as protons by electric and magnetic determination of e/m, were made to fall on a ZnS screen and scintillations were observed even when the screen was at a distance of 90 cms, and more. As the α-particle was known to have a range of only 24 cms in hydrogen gas, the observed result proved that the range of the recoil proton nuclei was nearly four times that of the α-particles.

The velocity of recoil of the proton in a direct collision with the α-particle having thus been experimentally determined, consideration of the electric charges at play showed that the centers of repulsion must have approached to within a distance of 1.73×10^{-13} cm This, then, would be the sum of the radii of the two colliding particles.

Blackett and others have employed the cloud chamber to study the same phenomenon of collision of α-particles with different light nuclei, such as H, He, C, O, etc., and have obtained beautiful stereoscopic photographs, which enable one to follow the details of the collision process and deduce accurate results.

Anomalous scattering

If the collision of the α-particle and the light nucleus *takes place within the critical distance,* where the Coulomb law yields to the all-important specifically nuclear forces, considerable deviations from normal scattering may be expected to occur. This is known as *inelastic* collision, where part of the kinetic energy is spent in producing transformations in the internal structure of the nucleus, so that the law of conservation of energy for the collision process will no longer be verified, although the law of conservation of momentum will still be valid. As a result of such an inelastic collision the phenomena of *anomalous scattering* of *α-particle* and even of *actual transmutation of the struck nucleus take* place.

Since according to Rutherford's theory, the critical distance b where the interaction between the α-particle and the nucleus begins to depart from the Coulomb law can be considered as the approximate radius of the nucleus, the condition for normal scattering is

$$4 \, Ze^2 / mv_o^2 > b,$$

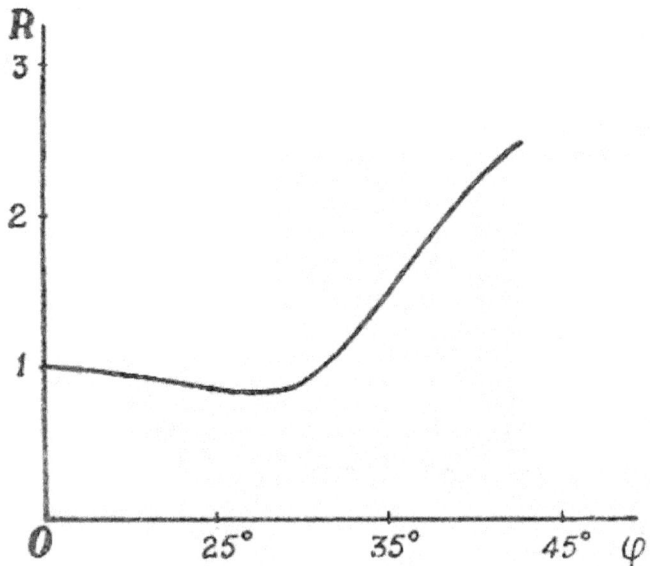

Fig. 330. Anomalous scattering of α-particles by light nuclei.

If this condition fails, then anomalous scattering is to be expected. Experimentally it has been observed that for the fastest α-particles from natural radioactive elements (velocity = 2 x 10⁹ cms /sec) the scattering is normal for all elements heavier than copper (Z = 29). For these nuclei, the above relation sets an upper limit for the nuclear radius. Thus, for gold, silver and copper, b was found to be 3.2, 2 and 1.2 x 10⁻¹² cm, respectively. The actual nuclear radius in these cases must be evidently smaller.

Elements of lower atomic weights show anomalous scattering, since the angular distribution of the scattered α-particles in their case is found to depart appreciably from that predicted by the Rutherford scattering formula. This departure becomes all the more important the lighter the element that scatters and the faster the α-particles, since there is a greater probability that faster particles will penetrate deeper into the region where the Coulomb law loses its importance in the case of lighter nuclei.

The most thoroughly investigated cases of anomalous scattering of α-particles are with the elements H, He, Be, B, C, and Al, all lighter than Cu. When the ratio R of the observed scattering to normal scattering is plotted against the scattering angle, the curve obtained is of the form shown in Fig. 330. It is seen that the ratio R approximates to unity for small scattering angles, but as the latter is increased, R usually decreases first and then increases reaching very high values for large angles. When R is plotted against the speed of α-particle from data obtained with a given element, it is found that it increases very much with increase of velocity. The general aspect of the phenomenon is the same for all the light elements investigated.

Chadwick has examined carefully the case of *scattering of α-particles by helium nuclei* and has determined with great precision the variation of the ratio R as a function of the *R* angle of scattering and of the energy of the α-particle. The curve obtained from these data relative to a scattering angle of 45° is shown in Fig. 331.

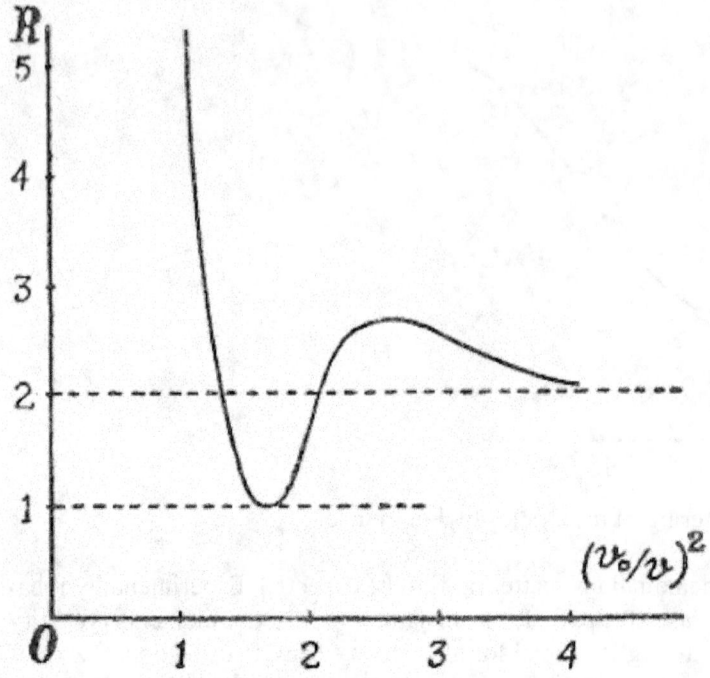

Fig. 331. Anomalous scattering of α-particles in helium at 45°.

For low velocities of the α-particles, [*i.e.,* when $(v_0/v)^2$ has large values] the ratio R does not tend asymptotically to the value 1 as would be expected from the classical theory, but rather to the value 2, which can be explained only on the basis of wave mechanics.

Nevill Francis Mott (30 September 1905 – 8 August 1996), England

Mott, in 1928, worked out the *wave mechanical theory of the scattering of α-particles* and showed that Rutherford's classical scattering formula remains unchanged by the application of wave mechanics in the case of heavy scatterers. The critical distance of the classical theory at which the Coulomb law fails is represented in the wave mechanical theory by the beginning of a rapid exponential decrease of the wave function. According to the classical theory anomalous scattering should occur only for particles of energy higher than the top of the potential barrier, while in wave mechanics there is considerable probability of penetration into the potential barrier, and, consequently, of anomalous scattering for somewhat lower energies, as confirmed by experiment.. The particular result obtained for the anomalous scattering of α-particles by helium is due to *resonance,* arising in the collision between two identical particles, when the incident α-particle has an energy corresponding to a virtual quantum state of the interacting helium nucleus.

Investigation of the anomalous scattering of α-particles by the different light elements has made it possible for the nuclear radius to be evaluated in each case. It is found to increase approximately as the cube root of the atomic number Z. In the case of (α – α) scattering, the nuclear radius is found to be 3.5 x 10^{-13} cm.

The most recent experiments conducted by G.W. Farwell and H.E. Wagner of the University of Washington on the scattering of α-particles of energies varying from 13 to 42 MeV have yielded 1.5 as the value of R_o in the expression for R.

(b) The scattering of protons

Very significant results have been obtained by investigating the scattering of protons by protons, as regards not only the nuclear radius, but also the nature of the intranuclear force between proton and proton. Mott and Massey, in 1933, worked out the wave mechanical theory of the scattering of protons by protons on the assumption of the Coulomb repulsive forces between the two identical particles.

Experiments conducted, in 1936 by White and by Tuve, Hafstad and Heydenburg by projecting protons in hydrogen gas and measuring the intensity of scattering at various angles, as the energy of the incident particle was given increasingly large values, led to results in disagreement with theoretical predictions.

White showed that the theory was valid only for energies less than 600 KeV. At higher energies and for scattering angles greater than 20°, many more particles were scattered than predicted. Tuve and his associates found that for 900 KeV protons the number scattered at 45° was four times greater than that expected from theory.

Breit, Condon and Present were able to show that the experimental results could be reconciled with theory only under the assumption that there existed *an attractive short range nuclear force between the protons at very close contact,* besides the Coulomb repulsive force.

The experimental data also showed that the interacting particles must have approached to a distance less than 5 x 10^{-13} cm which would therefore give the sum of the radii of two protons.

Very recent experiments of D. M. Chase and F. Rohrlich Princeton University on the scattering of 18.6 MeV protons gave R_o = 1.42. It is to be noted that experiments using charged particles as probes are difficult to interpret, because of the electric interaction which is added to the nuclear one.

(c) Neutron interaction

Of the nuclear methods, the most *direct* is undoubtedly that which employs neutrons. In this method, a beam of *fast* neutrons is directed on a sample of the element under study, called the 'target' which partially scatters and absorbs the neutrons, so that the resulting beam is less intense. The fraction of the incident beam that has been thus removed by the target can be experimentally measured. From this, the nuclear radius can be estimated as follows:

If the radius of the nucleus is R, each nucleus presents to the incident beam an effective area of πR^2. Supposing that the target has N atoms per sq. cm, and the cross-sectional area of incident beam is S, the fraction n of the incident beam that has been removed by the target is given by

$n = N S \pi R^2 / S = N \pi R^2$
or
$R = (n / N \pi)^{1/2}$ .... (1)

This relation will be modified if the wave aspect of the neutron is taken into account. For, neutrons which, if they were purely particles, would miss a nucleus, may be deflected by diffraction, in virtue of their wave nature. This means that the effective area which each nucleus presents to an incident neutron is much greater than πR^2. It varies in a complicated manner with the energy of the neutron, having the lower limit of $2 \pi R^2$ for very large energies and increasing for smaller energies. Hence, in order to obtain a simple interpretation of experimental data, it is necessary to use *very fast neutrons*, of energies greater than 50 MeV. But, at such high energies, the nuclei are partially transparent to the incident neutrons, and this reduces the effective area of the nucleus by an unknown amount. It has been found that somewhat lower energies, chiefly between 14 and 25 MeV, lead to the most reliable results. Under these conditions, relation (1), corrected for diffraction effects, becomes

$R = (n / C N \pi)^{1/2}$ .... (2)

where C is a number that depends on the energy of the incident neutrons and decreases towards 2 for very large energies. The dependence of C on the energy of the incident neutrons has been evaluated by V. F. Weisskopf and H. Feshbach of the Massachusetts Institute of Technology. As N is known and n experimentally found, R can be estimated.

We must see now which radius is measured in this method. Since the incident neutron will be affected by a nucleus of the target if it comes within the range of the nuclear force of any proton or neutron, the radius R given by this method should be larger than the radius proper by the above-mentioned range of the nuclear force. Hence one should expect the experimental data to conform to a law of the type

$R = R_o A^{1/3} + b$ .... (3)

where b is the additional range of the nuclear forces, which is roughly equal to 2×10^{-13} cm. From relation (3), we see that if the experimentally measured R $(= (n / C N \pi)^{1/2})$ is plotted against $A^{1/3}$, a straight line graph will be obtained with an intercept on the R-axis, equal to b, and the slope of the straight line will give R_o. Thus the two constants R_o and b can be obtained from experimental data, *i.e.,* by using targets of different elements and measuring R in each case. J. H. Coon, E. R. Graves and H. H. Barschell of the University of California, in 1952, have been able to carry out successfully this type of research and obtain from the graph drawn with their data the values of R_o as 1.4 ± 0.1 and of b in the whereabouts of 1×10^{-13} cm.

The different nuclear methods yield a value of R_o about 1.5, The uncertainty in the results chiefly arises from the present very imperfect knowledge of nuclear forces. *The radius measured refers to* the *outermost limit of the density distribution,* since the nuclear forces, though very strong, area limited to a very short range (about 1 x 10^{-13} cm) which is only a fraction of the nuclear radius.

B. ELECTRIC METHODS

The common feature of these methods, as already pointed out, is that an electrically charged non-nuclear particle is used as the probe, which will interact only with the protons, but not with the neutrons in the nucleus. These methods, therefore, cannot give all the information about nuclear density, but whatever they can give is likely to be more accurate and less equivocal than the nuclear methods, since our knowledge of electric forces is much more complete than of nuclear ones.

The nuclear radius which is measured by all the electric methods is *the average radius of the equivalent nucleus of uniform density.* For, the range of electric forces, unlike that of nuclear forces, is practically infinite and certainly extends to distances hundreds of thousands of times as great as the range of nuclear forces. In consequence, the effect on the electrically charged non-nuclear particles, due to all the protons in the nucleus, is an average one which leads to an average radius only. We shall now indicate the different electric methods actually employed for the estimation of R_o.

(a) Scattering of fast electrons by atomic nuclei

This method is fundamentally equivalent to looking at the nuclei through a microscope. Now, the resolving power of a microscope is limited by the wavelength of the light used, *i.e.,* it is not possible to distinguish detail which is finer than the wavelength used. Hence the only way to obtain wavelengths as short as nuclear dimensions is to use very fast or high energy electrons. The wavelength (λ) of an electron is most conveniently expressed in terms of the so-called Compton wavelength (λ_o = 386 x 10^{-13} cm); likewise the total energy E of the electron, which comprises the rest-mass energy m_oc^2 and the kinetic energy, is measured in MeV. With these units, it can be shown that

$$\lambda = \lambda_o (4 E^2 - 1)^{-1/2}$$

From this relation, the following table can be constructed:

E (MeV) :	1	20	50	100	200
λ (10^{-13} cm.) :	220	10	4	2	1

Since the nuclear radii vary from 3 x 10^{-13} to 8 x 10^{-13} cm, from the above table we see that 1 MeV electrons will show a nucleus as a mere point, 20 MeV electrons will detect that a nucleus has a finite size and electrons of 100 to 200 MeV will give the detailed structure of the charge distribution within the nucleus. R. Hofstadter and his associates of Stanford University, using electrons, accelerated by a linear accelerator, and of energies ranging from 84 MeV to 183 MeV, have succeeded to obtain direct information about the structure of the nucleus. Their experiments yield a value $R_o = 1.20 \pm 0.05$; it has been found that a charge distribution curve with a shallow dip at the center fits better the experimental data; this is to be expected since the electrostatic repulsion between the protons will tend to push each other outwards, leaving a lower density of charge at the center.

(b) Mesic atoms

If negative μ - and π-mesons, which are similar in many respects to electrons, except for their being 200 and 300 times heavier, are captured by atoms, they execute orbital motions, like the electrons; but, on account of their greater mass, the radii of their orbits will be 200 and 300 times smaller than the electronic orbits. This means that the captured mesons revolve very close to the nucleus, so that the energy of the meson in an orbit depends very sensitively on the nuclear radius. In the same way that ordinary atoms emit radiation of definite wavelengths, a mesic atom emits X-rays, whose wavelength determines the nuclear radius.

Experiments performed along these lines by V.L. Fitch and J. Rainwater of Columbia University have given values of R_o close to 1.2. This method is certainly ingenious, though it does not lend itself to very accurate measurements, since mesic atoms are not easy to produce and are extremely short-lived.

(c) Isotope structure

The energies of the electrons in an atom are determined by the distribution of the nuclear charge. Now, if we consider two isotopic atoms whose nuclei have the same number of protons, but one has one or two more neutrons than the other, the corresponding electron energies will differ very slightly. The difference, which manifests itself by a fine structure splitting of a spectral line (isotope structure), though exceedingly small, can be measured by spectroscopic methods, in certain favorable cases at least. From this tiny energy difference it is possible to obtain a measure of the increase in nuclear volume caused by the presence of the extra neutrons and hence of the nuclear radius. P. Brix and H. Kopfermann in Gottingen have been able to deduce, from the analysis of such isotopic shifts, a value for R_o close to 1.2.

(d) Mirror nuclei

Among light nuclei, there are pairs which consist of the same odd number of particles, say $2M + 1$, but one of them has M protons and $M + 1$ neutrons, while the other $M + 1$ protons and M neutrons. Such nuclei are called mirror nuclei, a typical example of which is the pair $_7N^{15}$ and $_8O^{15}$. Of the two mirror nuclei, one is always positron-active and decays into the other. For, *as far as* the first M protons and M neutrons are concerned, two mirror nuclei are identical. But if the last particle is a proton, it is repelled electrostatically by the other protons and is therefore much less firmly bound to the nucleus than a corresponding last neutron. In consequence, the proton-richer of the two mirror nuclei decays into the neutron-richer, the last proton decaying into a neutron with the emission of a positron.

The energy given out in this radioactive decay must be equal to the electrostatic potential energy that the last proton had in the field of the other protons before the decay. On the assumption that the probability of finding this last proton at a particular place in the nucleus is the same as that for any other proton in the nucleus, the potential energy E of the last proton (equal to the decay energy) can be shown to be given by

$$E = 6 \, M \, e^2 / 5 \, R$$

where R is the nuclear radius and e electronic charge. The energy E can be determined experimentally and hence R can be evaluated. Plotting E against $A^{1/3}$ a straight line will be obtained, from the slope of which R_0 can be found.

The experimental points are found to lie on a straight line very accurately, but the value of R_0 turns out to be 1.46. As this result does not agree with those of other electric methods, the mirror nuclei method has been somewhat discredited. But the reason for the discrepancy is at least partly due to the assumption made about the chance of finding the last proton at a particular place. This is evidently wrong, since the one proton that decays is the least firmly bound of all the protons in the nucleus, while the assumption implies that it is just as firmly bound as all the other protons. If this fact is taken into account, the value of R_0 must necessarily have a smaller value than 1.46. Recently, D.C. Peaslee of Columbia University has made a much more exact and fundamental approach to the causes of the discrepancy and has shown that the value $R_0 = 1.2$ is at least not in conflict with the evidence from mirror nuclei.

The different electric methods give a value 1.2 for R_0, with a surprisingly good agreement among themselves, evidently due to the correct knowledge we possess about electric forces. The value of R_0 is less than given by nuclear methods, as is to be expected, since the electric methods measure the average radius of the equivalent nucleus of uniform density, while the nuclear methods the radius of the outer limit of the density distribution.

Theoretical treatment
An interesting theoretical confirmation of the experimental results has recently (1954) been given by M. H. Johnson and E. Teller of the University of California.

In a stable nucleus, the protons and neutrons are held in a potential energy well in the same way in which we described the α-particle as being held in a radioactive nucleus. Since, however, the nucleus is stable in the present case, the total energy of a neutron or a proton in it must be less than the corresponding energy at infinite separation, only in this way there can be no tunneling through the barrier.

Since a neutron is uncharged, it is not affected by a nucleus until it comes within the range of the nuclear forces. In consequence, its potential energy curve is of the form shown in Fig. 332 - the lower of the two curves on the top. For protons, an additional factor is to be taken into account, viz., the electrostatic potential energy which is positive at all distances. Hence the total potential energy curve of the proton will be like the upper of the top curves in the figure.

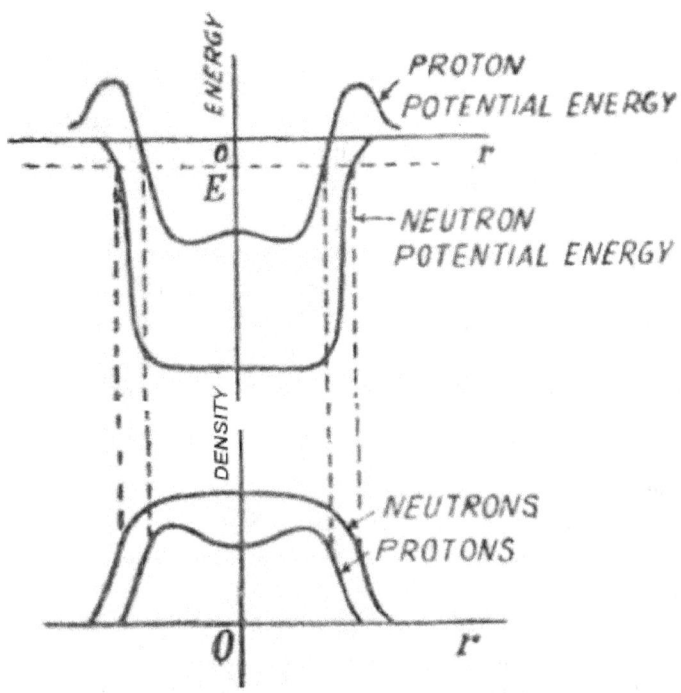

Fig. 332. Potential energy and density curves of protons and neutrons in the nucleus.

According to wave mechanics, there are only certain energies which a particle, held in such a well, can have and there can be only one particle with any given energy in the well. If we imagine that we are filling up such a well, then the first particle goes into the lowest energy level, the second into the next and so on. Because of the sloping walls of the well, particles of higher energy can spread further outwards than those with lower energy.

Since a stable nucleus, by definition, is not β-radioactive, the most energetic neutron and the most energetic proton in it must have very nearly the same energy. For, otherwise, the one with greater energy of the two will change into the other with the emission of electron, positive or negative, which is evidently vetoed in a stable nucleus. The common energy level E of the most energetic proton and neutron is also indicated in the figure. From this it is possible to deduce the variation of the proton and neutron densities, as shown in the lower half of the figure.

We realize at once that (a) there are more neutrons than protons in a nucleus, (b) the neutron density goes further out than the proton density and (c) the proton density has a dip in the middle. All these facts are borne out by experiment. It may be noticed that we have here a slightly different reason for expecting nuclear methods to give a larger radius than electric ones. The nuclear methods measure the outermost limit of the neutron densities. This is important, because the mere difference between the average radius and the radius of the outermost limit of the same distribution is unlikely to be of the order required by experiment.

3. NUCLEAR MASS

Nuclear mass strictly means the mass or weight of the nucleus proper; but it is often called the *atomic mass* or *atomic weight* which would rather mean the mass or weight of the whole atom. Thus, for instance, the mass of an atom of atomic number Z includes besides the weight of the nucleus that of the Z orbital electrons.

On account of the isotopic constitution of elements, the *atomic mass considered here is not the same as the atomic weight ordinarily used in chemistry.* The former is the individual weight of each of the isotopes contained in one and the same element, while the latter is the mean combining weight of an element as found in nature, to which every one of the isotopes in the element makes contribution in proportion to its relative abundance.

Measurement of nuclear masses

From what has been said above, it follows that to measure accurately nuclear masses one should, first of all, separate the different isotopes contained in one and the same element and then determine their masses individually. This cannot be done by the ordinary chemical methods based on physical and chemical properties. But the wonderful high-precision instruments known as the *mass spectrographs* have achieved the task to perfection, as described already in detail in chapter **4.** The weight of the individual isotopes thus measured is ordinarily termed the *atomic mass.* To get at the exact nuclear mass from the above quantity, one must pay attention to the atomic number of the element and to the degree of ionization. Because, positive rays containing atoms singly or multiply ionized are used in mass spectrographs. If Z be the atomic number of the element and singly ionized beam is used, the mass of the (Z - 1) electrons must be deducted from the experimentally measured isotopic mass to obtain the actual nuclear mass. If doubly ionized beam is used, then the weight of (Z - 2) electrons must be subtracted from the atomic mass to get the nuclear mass and so on.

It is to be noted, however, that *in ordinary nuclear reactions no error is introduced if atomic masses are used instead of nuclear masses,* since the nuclear charge must always balance up in every such reaction which means the number of electrons contained in the atoms does not change in any given reaction. The only *exception is positron radioactivity,* where to obtain the correct energy balance of the nuclear reaction, twice the mass of an electron must be deducted from the difference of the atomic masses of the original and product atoms, due to the special nature of the process.

Mass *defect*

As already stated, accurate measurement of isotopic masses with mass spectrographs which are capable of reaching a very high degree of precision, correct up to even the fifth decimal place, has led to the recognition of a very important quantity, closely related to the mass of the nucleus and ordinarily known as the *mass defect.* Mass spectrum analysis reveals the two following facts concerning nuclear masses:

(*i*) *The mass of every nucleus is approximately an integer,* which gives support to the hypothesis that any nucleus is constituted of fundamental particles of unit mass; this when combined with the additional fact of *exactly integral nuclear charge, Which* is always smaller than the *nearly integral nuclear mass,* argues to the proton-neutron constitution of the nucleus, since among the constituent particles of unit mass some should be positively charged (protons) while others neutral (neutrons) and both the proton and the neutron are nearly of unit mass.

(*ii*) At the same time, however, *the masses of the different isotopes are not exactly integers, i.e.,* they depart from the whole number rule systematically. The light nuclei have masses slightly higher than the nearest integer, *e.g.,*

H^1 = 1.00813, He^4 = 4.00408, Li^6= 6.01614;

a very large number of medium nuclei have masses somewhat lower, *e.g.,*

Ne^{20} = 19.9988, Cl^{35} = 34.983, Mo^{98} =97.944, Gd^{157} = 156.976,

while the very heavy ones slightly higher again, *e.g.,*

Hg^{200} = 200.028, Tl^{205} = 205.036, U^{238} = 238.140.

The difference between the measured mass M and the mass number A is known as the mass defect: Δ = M - A. This quantity which is negative for the majority of nuclei, except the very light and very heavy ones, is far outside the experimental error, being for the lightest nuclei about 100 times and for the heaviest ones about 10 times the probable error. Furthermore, the departure from the whole number is too systematic to be disregarded. On the other hand, the magnitude of the mass defect is much too small and depends much too regularly on the mass number A to justify the giving up of the whole number rule.

An adequate solution of this difficulty has been proposed on the basis of Einstein's mass-energy relation.

The grouping together of the elementary particles (*protons and neutrons*) *into a stable nucleus involves a certain interchange of energy.* This is known as the *"binding energy,"* i.e., the energy liberated or absorbed when the constituent particles are *bound* together into a nucleus.

In a real sense, we may think of the phenomenon of *nuclear formation as analogous to chemical reactions* in which energy is released or absorbed. This justifies expressing nuclear reactions by equations very much like those used in chemical reactions. Now, according to Einstein's mass-energy relation (W = mc^2 or m = W/c^2), any energy is equivalent to a certain mass. Hence the binding energy involved in the actual formation of a nucleus can be legitimately assumed to be derived from the mass available, which results in the mass defect. On account of the factor c^2 in Einstein's relation, a considerable amount of energy corresponds to even a very small mass defect. In any nuclear formation, a change in the sum total of the masses involved implies a proportionate change in the sum total of the rest-energies. If rest-mass disappears, a corresponding amount of energy must appear in other forms, as the rest-energy of particles newly created, as the kinetic energy of such particles, or even as electromagnetic radiation (γ-rays). If, on the other hand, the rest-mass increases there must be a corresponding conversion of energy into rest-mass.

We have already seen how the *mass defect* plays a very important role in the study of nuclear reactions by the use of equations of transmutation, appearing as *nuclear reaction energy* Q, and justifying at the same time Einstein's mass-energy relation.

The mass defect is very useful also in considerations of the *stability of nuclei,* defining as it does, to a certain extent, the degree of association between the various constituents of a nucleus, as vie shall see in a later section.

4. NUCLEAR CHARGE
Nuclear charge is the total number of positive charges carried by a nucleus. If Z be the atomic number of a nucleus and e the *elementary* charge, the charge on that nucleus is *Ze,* since the number of positively charged protons in it is Z and e is the same in magnitude whether it refers to the electron or the proton.

From the point of view of *atomic physics,* the nuclear charge is a very important factor, as it determines:
(i) the number of peripheral electrons of an atom in its neutral state,
(*ii*) the energy levels of the atom and
(*iii*) the chemical and physical properties of the atom; all the spectral properties, except for very small corrections such as hyperfine structure, isotopic shift of spectral lines, etc., find an adequate explanation in the nuclear charge.

Measurement of nuclear charge
The nuclear charge has been experimentally determined employing two methods, one based on the *large angle scattering of α-particles* and the other on the *characteristic X-ray spectrum. The* first is an absolute method of historical importance, but less accurate than the second.

(i) *Method of α-particle scattering*

Geiger and Marsden with their apparatus designed for the study of large angle scattering of α-particles were able to get at a rough estimate of the nuclear charge of the scatterer. Chadwick, in 1920, made more accurate measurements with an improved form of apparatus, where the number of observable α-particles scattered through large angles was considerably increased by using the scattering foil in the shape of an *annular ring,* subtending a fairly wide cone at the source of the α-particles.

The *principle* of the method is illustrated in Fig. 333.

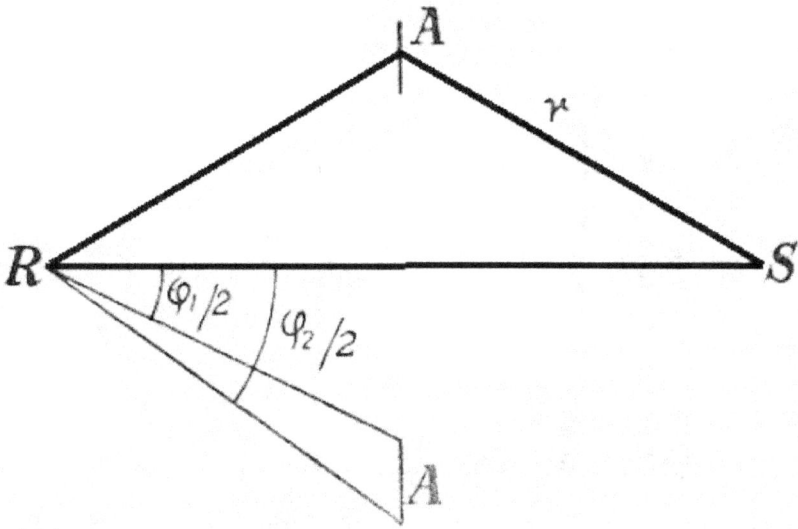

Fig. 333.

Let R be the source of α-particles and AA', the annular ring of the scattering foil with angular limits $\varphi_1/2$ and $\varphi_2/2$, sufficiently large, so that it subtends a wide cone at R. The fluorescent ZnS screen is placed at a point S on the axis of the cone R AA', such that RA = AS = r.

In such a case, it can be shown that the number of scattered α-particles falling upon unit area of the screen per sec is

$$N = \frac{Qntb^2}{64r^2} \left[\log \tan (\varphi_2/4) - \log \tan (\varphi_1/4) \right.$$
$$\left. + \cot (\varphi_1/2) \operatorname{cosec} (\varphi_1/2) - \cot (\varphi_2/2) \operatorname{cosec} (\varphi_2/2) \right]$$

where Q is the total number of α-particles emitted per sec by the source, n the number of atoms per c.c. of the scatterer, t the thickness of the scatterer, b the distance of closest approach of the α-particle towards the nucleus in a *central impact* and r the mean value of the distance of S from the annular ring scattering foil.

With such an arrangement it is found that the number of particles scattered by a suitable foil, even of a heavy element like Pt, is considerable, 1/1000 to 1/500 of the number in the direct beam.

The *apparatus* actually used by Chadwick is diagrammatically represented' in Fig. 334.

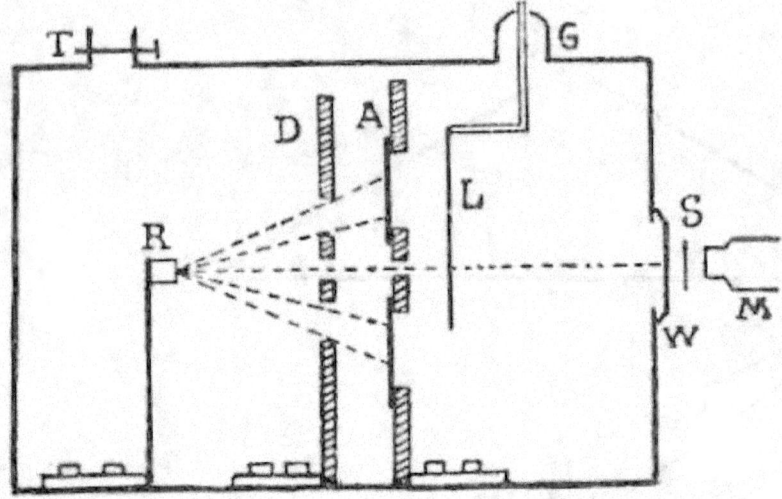

Fig. 334. Chadwick's apparatus for measuring the nuclear charge.

The diaphragm D serves to define the angular limits of the α-particles starting from the source R [a disc 2 or 3 mm. diameter coated on its face with Ra .(B + C)] and striking the scattering foil held on a support A. By suitably disposing D and A, it is secured that no particle is scattered directly to the ZnS screen S from the edges of the diaphragm. The direct pencil of α-particles passes through holes in the central discs of D and A and can be cut off at will by a lead screen L carried by a glass joint G. The different parts, except the screen S and the microscope M, are arranged inside a chamber which is kept under high vacuum by putting it in connection with a vacuum pump at T.

The accuracy of the measurement of the nuclear charge depends chiefly on the total number of α-particles counted in the experiment. If the scattered particles, whose number is considerably increased in this design of the apparatus, are falling on the fluorescent screen at a convenient rate, say about 30 per min, the number in the direct beam will be about 20,000 per min. The counting of so large a number is accomplished by rotating in the path of the direct pencil a wheel W containing a slit and arranged in front of S. By adjusting the ratio of the width of the slit to the circumference of the wheel the number of particles falling on the screen can be reduced to any desired extent. By this device the total number of particles in the direct beam, though great, can be accurately measured.

Experimental procedure and results
Counting the number of α-particles (Q) in the direct beam and the number (N) in the scattered beam, it is possible to evaluate b using the above given expression. Then, from the relation

$$b = 4\,Ze^2\,/mv_0^2\,,$$

Z and Ze can be calculated.

Three series of experiments were carried out, in which foils of Cu, Ag and Pt were used as scatterers. The results of the measurements gave for Z the following values:

Cu 29.3
Ag 46.3
Pt   77.4

with an estimated accuracy of 1 to 2%. As the atomic numbers of these elements are 29, 47 and 78 respectively, these experiments afford a pretty accurate measurement of the nuclear charge which is readily obtained by multiplying Z by the known value of e.

(ii) X-ray spectrum method

There exists a simple linear relation between the frequency v of a characteristic X-ray of an element and its atomic number Z, usually expressed as

$$v = a (Z - b)^2.$$

Measuring the wavelength λ of a given characteristic X-ray line for a few elements, if $v^{1/2} = (c/\lambda)^{1/2}$ be plotted against Z, a straight line graph is obtained. Next determining the wavelength and hence the frequency of the same characteristic X-ray line for the element whose nuclear charge is required, its value of Z can be readily obtained by extrapolation in the above graph. When Z is known, Ze is easily found.

With our present knowledge of the atom, the determination of the nuclear charge of any element is quite an easy affair. The order of an element in the *periodic table* gives at once the value of Z for that element. All elements corresponding to values of Z from 1 to 92 have been actually found in nature except 85 and 87. These last two as well as a few transuranic elements with Z from 93 to 118 have recently been produced in artificial disintegration experiments.

Group

Periodic table:

Group	I	II											III	IV	V	VI	VII	VIII
Period 1	1 H																	2 He
2	3 Li	4 Be											B	C	7 N	8 O	9 F	10 Ne
3	11 Na	12 Mg											13 Al	14 Si	15 P	16 S	17 Cl	18 Ar
4	19 K	20 Ca	21 Sc	22 Ti	23 V	24 Cr	25 Mn	26 Fe	27 Co	28 Ni	29 Cu	30 Zn	31 Ga	32 Ge	33 As	34 Se	35 Br	36 Kr
5	37 Rb	38 Sr	39 Y	40 Zr	41 Nb	42 Mo	43 Tc	44 Ru	45 Rh	46 Pd	47 Ag	48 Cd	49 In	50 Sn	51 Sb	52 Te	53 I	54 Xe
6	55 Cs	56 Ba	*	72 Hf	73 Ta	74 W	75 Re	76 Os	77 Ir	78 Pt	79 Au	80 Hg	81 Tl	82 Pb	83 Bi	84 Po	85 At	86 Rn
7	87 Fr	88 Ra	**	104 Rf	105 Db	106 Sg	107 Bh	108 Hs	109 Mt	110 Ds	111 Rg	112 Cn	113 Uut	114 Uuq	115 Uup	116 Uuh	117 Uus	118 Uuo

Legend:
- Stable elements
- Radioactive elements with half-lives of over four million years
- Half-lives between 800 and 34,000 years
- Half-lives between 1 day and 103 years
- Half-lives ranging between a minute and 1 day
- Half-lives less than a minute

* Lanthanides	57 La	58 Ce	59 Pr	60 Nd	61 Pm	62 Sm	63 Eu	64 Gd	65 Tb	66 Dy	67 Ho	68 Er	69 Tm	70 Yb	71 Lu
** Actinides	89 Ac	90 Th	91 Pa	92 U	93 Np	94 Pu	95 Am	96 Cm	97 Bk	98 Cf	99 Es	100 Fm	101 Md	102 No	103 Lr

Discovery and naming of transuranium elements

The majority of the transuranium elements were produced by three groups:

(a) A group at the University of California, Berkeley, under three different leaders:
o Edwin Mattison McMillan, first to produce a transuranium element:

93. neptunium, Np, named after the planet Neptune, as it follows uranium and Neptune follows Uranus in the planetary sequence (1940).

o Glenn T. Seaborg, next in order, who produced:
94. plutonium, Pu, named after the dwarf planet Pluto, following the same naming rule as it follows neptunium and Pluto follows Neptune in the pre-2006 planetary sequence (1940).

95. americium, Am, named because it is an analog to europium, and so was named after the continent where it was first produced (1944).

96. curium, Cm, named after Pierre and Marie Curie, famous scientists who separated out the first radioactive elements (1944).

97. berkelium, Bk, named after the city of Berkeley, where the University of California, Berkeley is located (1949).

98. californium, Cf, named after the state of California, where the university is located (1950).
o Albert Ghiorso, who had been on Seaborg's team when they produced curium, berkelium, and californium, took over as director to produce:
99. einsteinium, Es, named after the theoretical physicist Albert Einstein (1952).

100. fermium, Fm, named after Enrico Fermi, the physicist who produced the first controlled chain reaction (1952).

101. mendelevium, Md, named after the Russian chemist Dmitri Mendeleev, credited for being the primary creator of the periodic table of the chemical elements (1955).

102. nobelium, No, named after Alfred Nobel (1956).

103. lawrencium, Lr, named after Ernest O. Lawrence, a physicist best known for development of the cyclotron, and the person for whom the Lawrence Livermore National Laboratory and the Lawrence Berkeley National Laboratory (which hosted the creation of these transuranium elements) are named (1961).

A group at the Joint Institute for Nuclear Research in Dubna, Russia (then the Soviet Union) who produced:
o 104. rutherfordium, Rf, named after Ernest Rutherford, who was responsible for the concept of the atomic nucleus (1966).

o 105. dubnium, Db, an element that is named after the city of Dubna, where the JINR is located. Also known in Western circles as "hahnium" in honor of Otto Hahn (1968).

o 106. seaborgium, Sg, named after Glenn T. Seaborg. This name caused controversy because Seaborg was still alive, but eventually became accepted by international chemists (1974).

(b) A group at the Gesellschaft für Schwerionenforschung (Society for Heavy Ion Research) in Darmstadt, Hessen, Germany, under Peter Armbruster, who produced:

o 107. bohrium, Bh, named after the Danish physicist Niels Bohr, important in the elucidation of the structure of the atom (1981).

o 108. hassium, Hs, named after the Latin form of the name of Hessen, the German Bundesland where this work was performed (1984).

o 109. meitnerium, Mt, named after Lise Meitner, an Austrian physicist who was one of the earliest scientists to become involved in the study of nuclear fission (1982).

o 110. darmstadtium, Ds, named after Darmstadt, Germany, the city in which this work was performed (1994).

o 111. roentgenium, Rg, named after Wilhelm Conrad Röntgen, discoverer of X-rays (1994).

o 112. copernicium, Cn, named after astronomer Nicolaus Copernicus (1996).

(c) As of 2011, several elements beyond number 112 have been observed, but none of them were officially named:

o 113. Ununtrium, claimed by RIKEN and by a collaboration between Joint Institute for Nuclear Research (JINR) and Lawrence Livermore National Laboratory (LLNL) (2003 or 2004).

o 114. Ununquadium, by JINR (1998).

o 115. Ununpentium, by JINR/LLNL collaboration (2004).

o 116. Ununhexium, by JINR (2000).

o 117. Ununseptium, by JINR/LLNL/ORNL/Vanderbilt collaboration (2010).

o 118. Ununoctium, by JINR (2002).

5. NUCLEAR QUANTUM STATES

The study of α- and γ-ray spectra and of artificial disintegrations of the resonance type, etc., clearly proves that every nucleus possesses a set of quantum states and a corresponding number of discrete energy levels. Above these actual quantum states there is also an array of virtual states in which the nucleus may exist momentarily (*compound nucleus*) before disintegrating with the emission of particles, γ-rays, etc.

The most important factor about these quantum levels is their *widths,* as already stated. The level-widths are a measure of the probability of disintegration of the nucleus. According to quantum mechanics, if for any reason there exists a finite probability that the nucleus will undergo a transition from one of its quantum states to another, then the energy associated with the initial state is indefinite to an extent that is proportional to the probability of transition. The indefiniteness ΔE in the energy E of any level has the magnitude h/τ, where h is Planck's constant and τ the mean life of the nucleus in that level. The energy level itself is commonly considered as possessing a *certain width* of magnitude ΔE. Hence the *sharpness* of a given level comes into evidence only in proportion to the mean life of that level. Only the normal state of lowest energy has zero width. Other states have widths that vary greatly according to the types of transitions that the nucleus is capable of undergoing.

For many nuclear states transitions are possible only with the emission of electromagnetic radiation (γ-rays) resulting from the electric fields of the protons. The mean-life of a nucleus in such a state is ordinarily short, of the order of 10^{-12} sec. Even the shortest mean lives of such states are, however, long enough to permit a definition of the nuclear energy that thus appears sufficiently sharp. Hence, all γ-rays arise from transitions between relatively sharp levels, so that they form, in practice, monochromatic lines. If the mean life of the state is, for some reason or other (say, because the state is low and differs considerably in angular momentum from the normal ground state) long, perhaps several days, months and even years, the state becomes *metastable,* which may give rise to *isomeric forms* of the nucleus.

Other nuclear states, chiefly *virtual,* are capable of transmutation through the emission of a material particle, viz., a positron, an electron, a neutron, a proton, or a helium nucleus. In many cases the mean lives of such states are extremely short, of the order of 10^{-20} sec. The energy of such states is poorly defined and may be regarded as varying according to the manner in which the state is produced. If a nucleus in such a state results from the absorption of the bombarding particle, its energy increases as the energy of bombardment increases and the kinetic energy of the particle that is subsequently emitted increases to the same degree. It is also possible that the virtual states are long-lived sometimes, in which case the energy of the initial state is relatively sharply defined.

Experimental data indicates that for light nuclei, the lower levels are rather widely spaced, but the spacing decreases as the energy increases. For heavy nuclei, the spacing of level is extremely narrow.

There appears to be quite a good analogy between the quantum states inside the nucleus and those in the extranuclear electronic system in the matter of level widths But a fundamental difference exists between the two as regards/ the nature of the forces acting in their respective regions In the extra-nuclear realm the Coulomb law of inverse squares reigns supreme while inside the nucleus some other law of short range attractive forces even between the charged protons and the uncharged neutrons comes into play The precise nature of this specifically nuclear force will be considered below.

6. NUCLEAR STATISTICS

The properties of the nucleus, considered as a quantum-mechanical system composed of many particles, are determined by the "statistics" of the constituents. There are two alternatives: Fermi- Dirac or Bose-Einstein statistics. It is impossible to give a classical explanation of the significance of the choice of statistics. The consequences of this choice manifest themselves only in the typical quantum phenomena and they disappear in the classical limit.

Let us consider a system consisting of "equal" or indistinguishable particles. From the standpoint of wave mechanics, every particle, a proton included, is described by the specification of its eigenfunction. Let the eigenfunction of one particle (the first) be $\psi_k^{(1)}$, that of the second $\psi_k^{(2)}$, of the third $\psi_k^{(3)}$ and so on; $k, l, m...$ represent the states of the particles in question. The state of the system as a whole can then be described, to a first approximation, (*i.e.,* neglecting mutual action between the individual particles) by a wave function given by the product of the proper functions of the particle:

$$\psi_{k/m\ldots\ldots\ldots} = \psi_k^{(1)}\ \psi_l^{(2)}\ \psi_m^{(3)}\ \ldots\ldots\ldots\ldots$$

If two particles are interchanged, say 1 and 2, another wave function for the system is obtained,

$$\psi'_{k/m\ldots\ldots\ldots} = \psi_k^{(2)}\,\psi_l^{(1)}\,\psi_m^{(3)} \ldots\ldots\ldots\ldots$$

which obviously corresponds to the same value of the energy of the system, viz.

$$E_{k/m\ldots\ldots\ldots} = E_k + E_l + E_m \ldots\ldots\ldots\ldots$$

We can obtain other wave functions for the same value of energy taking an arbitrary linear combination of those wave functions which arise from the one first written down by a permutation of the individual particles among themselves, *i.e.,* by interchange of the arguments 1, 2, 3, of the separate functions.

Hence,

$$\psi_{k/m\ldots\ldots\ldots} = \sum_{P} \beta_{1\,2\,3\ldots}\ \psi_k^{(1)}\,\psi_l^{(2)}\,\psi_m^{(3)} \ldots\ldots\ldots\ldots$$

where the summation is to be made over all the P permutations of the arguments and the factors $\beta_{1\,2\,3\ldots}$ represent arbitrary constant coefficients.

According to classical statistics, these wave functions give as many different states as there are linearly independent wave functions among them. But in new statistics, those cases which arise from one another by mere permutation of particles belong to the same state, owing to the indistinguishability of the particles. Hence the wave function which describes this state cannot change when the particles are interchanged, or at most it can only change in its sign, since only quadratic forms such as $|\psi|^2$ are of any account so far as physical interpretation is concerned.

Now it is readily seen that the only *wave function* of the form specified above, *which does not change* when the particles are interchanged, is the one in which all the coefficients are equal to one, *i.e.,* the *symmetrical wave* r *function:*

$$\psi_S = \sum_{P} \psi_k^{(1)}\,\psi_l^{(2)}\,\psi_m^{(3)} \ldots\ldots\ldots\ldots$$

A second possibility, in which the *sign of the wave function changes,* but not $|\psi|^2$, is the *anti-symmetric wave function:*

$$\psi_a = \sum_{P} \pm\,\psi_k^{(1)}\,\psi_l^{(2)}\,\psi_m^{(3)} \ldots\ldots\ldots\ldots$$

where the plus (+) sign is to be taken for an even permutation of the particles, the minus (-) sign for odd.

This anti-symmetrical form can be obtained by the expansion of the determinant:

$$\psi_a = \begin{vmatrix} \psi_k{}^{(1)} & \psi_k{}^{(2)} & \psi_k{}^{(3)} & \cdots\cdots \\ \psi_l{}^{(1)} & \psi_l{}^{(2)} & \psi_l{}^{(3)} & \cdots\cdots \\ \psi_m{}^{(1)} & \psi_m{}^{(2)} & \psi_m{}^{(3)} & \cdots\cdots \end{vmatrix}$$

No other wave functions exist which satisfy the condition of indistinguishability.

In the case of the anti-symmetric function, there is an additional feature to be noted. According to the theory of determinants, a determinant vanishes if two rows, or two columns are the same; hence if two functions ψ_k and ψ_l are equal, the wave function of this state vanishes, i.e., this state does not exist. This is nothing but Pauli's principle, that two electrons cannot be in the same state.

There are therefore only two possible ways of describing a state by a wave function, viz., by a symmetrical or anti-symmetrical wave function. If we count the possible states on the basis of their wave functions, two different statistics and only two, present themselves. With symmetrical wave functions (without Pauli's principle) we have the Bose-Einstein statistics; with anti-symmetrical wave functions (with Pauli's principle) we get the Fermi-Dirac statistics; which of the two statistics we are to use in a particular case, experiment alone can decide.

Experimental determination of the statistics of nuclei
The type of statistics followed by nuclei has been obtained by investigations on the alternating intensities of the rotational band spectra. A determination of the positional symmetry characteristics for the rotational states allows one to decide the symmetry of those states of greater weight and hence whether the nuclei follow the Fermi or Bose statistics. The Raman spectra of diatomic molecules offer an easy means of achieving this end. It has been found that H^1, Li^7, F^{19}, Na^{23}, P^{31}, Cl^{35} and K^{39} obey the Fermi statistics, whereas D^2, He^4, C^{12}, N^{14}, O^{16}, and S^{32} obey the Bose statistics. Although the statistics is known only in the case of these few *light* nuclei, there appears the *general rule that nuclei of odd mass number follow the Fermi statistics, while those of even mass number the Bose statistics.*

This experimental law is in complete accord with the proton-neutron constitution of nuclei. From the fundamental relation that exists between the type of statistics of a given kind of particles and the wave function of a system composed of those particles, it can be shown that

a system composed of elementary particles, each obeying Fermi statistics, follows Bose or Fermi statistics according to whether the total number of elementary particles in the system is even or odd.

The proof of this statement is readily obtained, if one imagines the exchange of two systems (nuclei) to take place through the *successive exchange of pairs* of elementary particles. Each time a pair of particles which obey the Bose statistics is exchanged, the total wave function remains unchanged; but each time two particles obeying Fermi statistics are exchanged the wave function changes sign. Now, if the total number of particles, each obeying Fermi statistics is even, the total wave function will not change sign and hence the system as a whole will obey Bose statistics. On the other hand, if the total number of particles obeying the same Fermi statistics is *odd,* the total wave function will change sign and the whole system will follow Fermi statistics.

Considering a nucleus of mass number A and atomic number Z, according to the *proton-neutron constitution,* the total number of particles in the system (protons and neutrons) is A, while the number of protons is Z and the number of neutrons A - Z. Using the experimental fact that the *proton* obeys Fermi statistics, and the *deuteron,* made up of a proton and a neutron, Bose statistics, one legitimately concludes that the *neutron also obeys Fermi statistics* from the general principle stated above, viz., a system made up of *two* particles, hence *even* number of particles, both following Fermi statistics, will obey Bose statistics. Since it is the total number which determines the statistics, any nucleus of even mass number A, (*i.e.,* the total number of protons and neutrons is even) will obey Bose statistics, while any nucleus of odd mass number A will obey Fermi statistics, as actually found by experiment. The statistics does not therefore depend on Z but only on **A.**

On the other hand, according to the *proton-electron constitution,* in the same nucleus of mass number A and atomic number Z, the number Of protons is A and the number of electrons A -- Z; hence the total number of particles in the nucleus is (2A - Z) which will be even or odd according to whether Z is even or odd. As electrons, like other elementary particles, obey Fermi statistics, all nuclei with *odd* Z such as $_1D^2$, $_7N^{14}$ should obey Fermi statistics irrespective of the mass number A. This is contrary to experimental results and hence gives another strong argument in favor of the proton-neutron constitution as against the proton-electron theory. Incidentally it may be said that a neutron cannot be considered as a closely bound composite particle made up of a proton and an electron, since such a compound particle would have to obey Bose statistics, whereas the neutrons are known to obey only Fermi statistics.

7. NUCLEAR SPIN AND MAGNETIC MOMENT

Introduction
Although the *nucleus* can be assumed to be a mere point charge-the center of the Coulomb field-as far as an adequate explanation of the main features of atomic and molecular spectra is concerned, yet, as we have seen in the preceding pages, *it has a finite dimensional structure,* consisting of elementary particles such as protons and neutrons. Certain experimentally observed phenomena in atomic spectra, such as the multiplet fine structure, the anomalous Zeeman effect, etc., can be explained satisfactorily only on the assumption that *the peripheral electrons* must be regarded not as mere point charges but as *endowed with dimensional structure, capable of proper spin motion* and in consequence possessing a magnetic moment. These facts naturally suggest a further consideration as to whether the nucleus also, like the peripheral electrons, is endowed with a proper spin motion and a consequent magnetic moment. There exists ample

evidence for admitting such properties of the nucleus, which are of great value in the study of nuclear reactions and in the formulation of a theory of nuclear structure.

Evidences for the existence of nuclear spin and magnetic moment

(*a*) *Indirect evidence*
In general, there are a number of experimentally observed facts, such as the α-ray and γ-ray spectra, resonance disintegrations, chiefly with slow neutrons, etc., which show that the constituent particles inside the nucleus are in incessant motion in discrete quantized orbits, somewhat like the extranuclear electrons. These orbital motions will endow the nuclear particles with mechanical angular momentum and further the electrically charged particles among them will possess magnetic moment. Over and above these orbital motions, one may expect-the nuclear particles of finite dimensions to spin about axes of their own like the spinning electrons and this will give rise to a proper spin angular momentum and an associated magnetic moment. The nucleus as a whole will, therefore, possess an intrinsic angular momentum which is the resultant of the orbital and spin momenta of the different constituent particles and a magnetic moment, the resultant of the moments due to the orbital and spin motions of at least the individual charged particles. Owing to the present imperfect knowledge of the internal structure of the nucleus, the total resultant angular momentum mentioned above is usually called the *nuclear spin* and the total resultant magnetic moment file *nuclear magnetic moment* without referring the two terms to the individual nuclear particles as in the case of peripheral electrons.

(*b*) *Direct evidences*
In particular, there exist several experimental observations which lead directly to and establish beyond doubt the above described *'dynamic'* conception of the nucleus, endowed with spin and magnetic moment. The most important and strikingly convincing among them are:

(*i*) *hyperfine structure in atomic spectra,*
(*ii*) *alternating intensities in band spectra,*
(*iii*) *existence of two types of molecules in ordinary hydrogen ortho and para hydrogen.*

Hyperfine structure of spectral lines
The first piece of evidence for the actual existence of definite nuclear spins was obtained from the observed *hyperfine* structure of each of the components of the multiplet fine structure of lines in atomic spectra. It is now some hundred years since Michelson and others have shown, by means of the interferometer, that some of the lines of the spectra of a number of elements were in reality complex close line groups. Such a complex structure, revealed only by instruments of very high resolving power, was first called "fine structure" but for good reasons was later renamed "hyperfine structure"

The origin of the complex structure of spectral lines remained obscure for a long time, until the multiplet fine structure could be explained adequately as being due to the spin of the orbital electrons. When this was done, since no cause for further splitting was to be found in the extranuclear electronic structure of the atom, it was surmised that the observed hyperfine structure must be a nuclear effect. But it might be due to more than one cause, *i.e.,* more than one nuclear property. *It might be produced, for instance, by the presence of several isotopes in an element.* Since the separate isotopes of an element have identical peripheral electronic system the spectra emitted by each of these are almost the same and the lines of one isotope practically coincide with the lines of another. However, in some cases, a measurable displacement for certain lines have been observed, the displacement being due to the different masses or varying electrical properties or even the motion of the nuclei .of the isotopes. Urey, Schiller and others have demonstrated and analyzed this *isotope effect* in order to study not only the isotopic constitution of elements as an alternative method to the mass spectrograph but also the more important nuclear properties such as *nuclear volumes* (*i.e.,* different isotopes have different nuclear volumes) and *nuclear symmetry* (i.e., whether nuclei are spherically symmetrical or not).

It was soon found, however, that the isotope effect alone could not account for all the observed hyperfine structures, since these occurred even in the case of elements which consisted of only one isotope. For instance, bismuth, though having only one isotope, shows **a** clearly marked hyperfine structure. In such cases, one is led to conclude that the observed hyperfine structure must be due to some other nuclear property, like the spin, as was first suggested, in 1924, by Pauli. If a mechanical angular momentum or spin of the nucleus were assumed, there would, of course, be an associated magnetic moment, whose interaction with the magnetic field due to the motion of the extranuclear electrons could be used to explain the hyperfine setting. In other words, the action of the electrons in the field of the nuclear magnetic dipole will cause the different atomic states have slightly different energies. Thus arises the hyperfine separation of the levels. That the interaction is essentially a magnetic one is readily seen by comparison of hyperfine structure groups with ordinary fine structure multiplets, showing similarities.

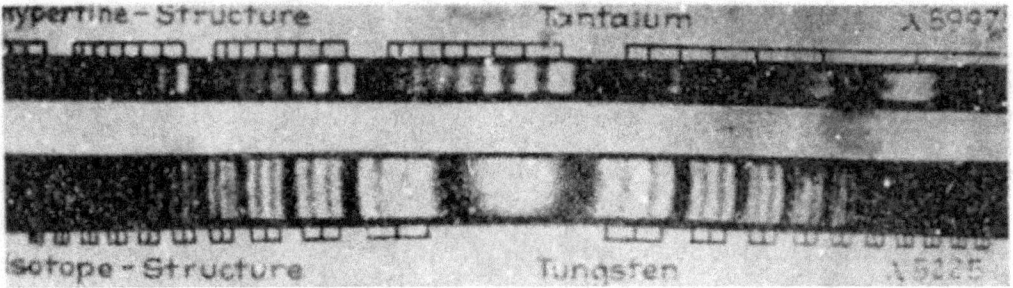

Hyperfine and isotope structures (Grace, Moro, Mac Millan and White)

As the hyperfine structure can arise due to the two different causes stated above, authors prefer to restrict the term *hyperfine structure* to that which is caused by nuclear spin and magnetic moment, the other type being called *isotope effect* or *isotope structure*. The hyperfine structure in the two cases is of the same order of magnitude-a fraction of a wave number. It is, therefore, not always easy to separate the two effects. The hyperfine structure can be clearly observed with instruments of high resolution and dispersion, such as the Fabry-Perot interferometer or etalon, Lummer-Ghercke plate, etc. The photo given above obtained by Grace, More, Mac-Millan and White with a Fabry-Perot etalon illustrates these remarks very well. It is seen that the same pattern is repeated many times in different orders. The separation in the nuclear spin and isotope effects is more or less of the same order of magnitude. The hyperfine structure of a line of tantalum has eight components. The isotope structure of a line of tungsten has three components. Tungsten has four isotopes of masses, 182, 183, 184 and 186 with relative abundances of 22.6, 17.3, 30.1 and 29.8% respectively. The three observed components are due to the three isotopes of even mass number, the component due to the least abundant 183 isotope being supposed to be masked by the others.

Alternating intensities in band spectra

In the spectra of several diatomic molecules, the successive rotational lines of a band *alternate in intensity i.e.,* either the lines are alternately strong and weak or every other line is missing. The phenomenon may be observed in an electronic band and more easily in a pure rotational Raman spectrum. It appears only in the case of *homonuclear* diatomic molecules, where the two atoms are identical, such as H^1, D^2, He^4, C^{12} etc., but not with *heteronuclear* diatomic molecules in which the two atoms are not identical, like CO, NO, $Cl^{35,37}$, $O^{16,18}$, etc.

Heisenberg was the first to correctly ascribe this phenomenon to a *wave mechanical resonance effect of two identical particles- in* the present case, the two identical nuclei of the diatomic molecule. From theoretical considerations based on wave mechanics, he was able to show that if the nuclear spin vanishes alternate lines drop out in the bands, whereas if there is a resultant nuclear spin the band lines alternate in intensity. In the latter case, he derived a simple but important relation between the ratio of the intensities of neighboring associated lines and the nuclear spin, which enables the magnitude of the nuclear spin to be readily estimated. Thus the actually observed alternating intensities in band spectra establish beyond doubt the existence of nuclear spins, nay more, permit a quantitative measurement of their values.

Ortho and para hydrogen

The existence of two kinds of molecules in ordinary hydrogen gas, known as the *ortho* and *para* hydrogen, gives yet another clear evidence for the presence of nuclear spins. The two types of hydrogen were first theoretically predicted by Dennison to account for the anomalous behavior of the specific heat of hydrogen at low temperatures. Attempts made to separate them by Bonhoeffer and Harteck employing catalyzers were successful, so that they could be considered to exist almost independently of each other. It has also been possible to convert *para* into *ortho* by the use of a paramagnetic gas, such as oxygen.

Finally the alternating intensities in the band spectra of hydrogen have been shown to be due to the two types.

All these experimental facts can be adequately explained if the existence of nuclear spin and associated magnetic moment are assumed. The rotational lines exhibit alternating intensities in the ratio 3:1, which gives the nuclear spin of hydrogen nucleus (proton) as

1/2. Since every molecule of hydrogen is composed of two protons, it is possible that some molecules may have the two proton spins parallel, while others have the spins anti-parallel. Using vectorial composition, the resultant nuclear spin in the first case, viz., spins parallel, is 1, while in the second, viz., spins anti-parallel, it is 0.

Assuming that the nucleus on account of its spin I can have $(2I + 1)$ settings with respect to the moment of the rest of the system, in the first case three different states, $+ 1, 0, -1$ occur, while in the second only one state. Hence molecules with resultant nuclear spin 1 will be three times more frequent than those with resultant nuclear spin 0, which accounts for the observed alternating intensity ratio. The first type in which the nuclear *spins* are *parallel* (I = 1) are the *ortho* hydrogen molecules whose rotational states are *anti-symmetrical,* giving rise to the *strong* lines. The second type in which the *spins* are *anti-parallel* (I = 0) are the *para* hydrogen molecules whose rotational states are *symmetrical,* responsible for the *weak,* lines.

Under ordinary conditions, since there are no transitions between the *symmetrical para* states and the *anti-symmetrical ortho* states, the two types exist independently and can be separated. It is, however, possible to induce *para-ortho transitions* by the presence of an inhomogeneous magnetic field such as that due to the paramagnetic oxygen molecule. The rate of conversion has been shown to depend on the magnitude of the nuclear magnetic moment and this dependence has been actually used to determine the ratio of the nuclear magnetic moments of hydrogen and deuterium. As regards the specific heat, at ordinary temperatures kT is large compared to the distance between the rotational states and any possible weighting of these states is irrelevant. But at very low temperatures, this is not the case, so that different statistical weightings of the symmetric and anti-symmetric states acquire significance. Since there can be no transitions between the two states, the gas has to be treated effectively as a *mixture of two gases* and the specific heat determined accordingly, which explains the anomalous behavior of the specific heat at low temperatures. Furthermore, it is found necessary to weight the even and odd states in the ratio 1:3 in order to fit the specific heat.

Measurement of spin and magnetic moment of nuclei
There are a considerable number of methods which are used to determine nuclear spins and magnetic moments. For the greater part, the methods are best applicable to different cases *i.e.,* they are largely supplementary in scope. We shall here limit ourselves to the following important ones:

(i) Analysis of the hyperfine structure in atomic spectra.
(ii) Study of alternating intensities in band spectra.
(iii) Magnetic deflection of molecular and atomic beams.
(iv) Study of magnetic resonance radio frequency spectra.
(v) Analysis of microwave spectra.

(i) ANALYSIS OF THE HYPERFINE STRUCTURE IN ATOMIC SPECTRA

Theoretical considerations

The *general principle* used in the analysis of the hyperfine structure is the suggestion made by Pauli, viz., the hyperfine levels are due to the interaction of the nuclear and electronic magnetic moments. This interaction can be studied in the usual imperfect but simple way of the vector model of the quantum theory.

Let J represent the angular momentum vector of the *peripheral electron* system of an atom, so that J.$(h/2\pi)$ gives the resultant mechanical momentum due to the orbital and spin motions of the electrons.

Let I represent a similar quantity for the *nucleus* of the atom so that I.$(h/2\pi)$ gives the resultant mechanical momentum of the nucleus due to orbital and spin motions of its constituent particles. Hence I, according to our definition, represents the *nuclear spin,* in quantum units, and may be an integer or half-integer like J.

Assuming that I and J combine vectorially the resultant vector may be represented by
F = I + J;
F is called the *hyperfine quantum number* and F.$(h/2\pi)$ gives the total angular momentum of the atom, taking into account the nuclear spin also. Since each of the members can be an integer or half integer, F is also an integer or half integer.

According to the usual rules of addition of quantum vectors, the number of states which F can assume for given values of I and J is (2I +1) if I is less than J, or (2J + 1) if I is greater than J.

Now, if all these suppositions are correct, we may expect to find for each value of J not a single state, but a whole cluster of states. If out of the manifold term system of an atom we select states for which J = 1/2, 3/2 and so on, we may expect, on close scrutiny, to find that these apparent individual states are actually clusters, each cluster comprising a number of sublevels, which for one or two or more of the lowest values of J may be equal to (2J + 1), since in these cases I > J, but for higher values of J reaches a limiting constant value equal to (2I + 1). Thus arise the hyperfine levels of the states.

Since many terms split up in this manner, what is treated as a single line transition between two terms when the nuclear spin is not included, now becomes a hyperfine multiplet. General considerations confirmed by experimental observations show that the *selection rule* which operates here $\Delta F = \pm 1$ or 0 (0 → 0 being excluded).

Considering the *familiar example of the principal series of sodium,* this appears in an ordinary spectroscope as a series of single lines each of which is resolved by a good spectroscope into a *"doublet" fine structure,* which indicates that the P states of sodium are close pairs and thus requires the introduction of *spinning electron.* With an extremely good spectroscope each member of the doublet is, in turn, resolved into a *pair of hyperfine structure,* which therefore demands the introduction of *nuclear spin.* It may be noted that while the fine structure of the principal series lines of sodium implies a splitting of P states only, the hyperfine structure involves something more complex; it is due jointly to the hyperfine splitting of both the P state and the normal S state, the latter being predominant.

A *small magnetic moment* is always associated with the nuclear spin which by analogy with the electronic magnetic moment may be assumed to be given by

$$\mu_N = Eh / 4\pi M_c . F$$

where E is the nuclear charge, h Planck's constant, M the mass of the nucleus and F the hyperfine quantum number.

Supposing that F is of the same order of magnitude as the quantum number L, S or J referring to the electronic system, since M of even the lightest nucleus (hydrogen) is 1,840 times greater than the mass of the electron, μ_N will be small compared to the Bohr magneton, the unit used in the case of the electronic system. The interaction of this small nuclear magnetic moment with the electronic magnetic moment causes the different states to have slightly different energies and thus gives rise to the hyperfine structure.

The *interaction energy* W between the two vectors I and J, in the case of the simple one-electron system, can be shown to be given by the expression

$$W = A \, I \, J \cos (I \, J)$$

which can be further reduced to the form

$$W = (A/2) \, [F \, (F + 1) - J \, (J + 1) - I \, (I + 1)]$$

where A is the *interval factor*. The second relation is known as the *Landé interval rule*, from which the ratio of the intervals in a hyperfine multiplet can be calculated. The first relation shows that the *interaction energy between* I *and* J *is proportional lo the cosine of the angle between them.* These are identical with the formulae referring to the electronic system, J, L and S being replaced by F, J and I.

The interaction energy is dependent upon several factors. First, it depends upon the values of the magnetic moment of the nucleus as distinct from its mechanical momentum, for it is primarily the magnetic fields that couple. Secondly, the nuclear charge also will affect the interaction energy, since the higher the degree of ionization, the greater is the coupling energy between the extranuclear electron momentum J and the nuclear spin I. Thirdly, the eccentricity of the orbit of the optical electron is an important factor to be considered. If the orbit is very eccentric, a high degree of penetration of the inner shells takes place and the optical electron approaches close to the nucleus so that the coupling becomes very strong. The coupling energy of an *s*- electron is, in general, much greater than that of a *p*- electron.

The estimation of the actual magnitude of the hyperfine separation is a very difficult process, where several factors connected with the electrons and the nucleus must be taken into consideration. Only those states which have electrons near the nucleus show the structure. It is well known that *perturbations* between states are very prevalent in atomic spectra, which become frequently very large when the states involved are close together. Deviations from the unit interval rule in the successive hyperfine states have been found, which means that the form of the interaction term must be slightly modified, the deviations being attributable to the probable existence of an *electric quadrupole moment* of the nucleus. In spite of these complications, hyperfine separation formulae have been worked out for a configuration with a deeply penetrating *s* electron by Breit, Wills, Goudsmit, Fermi, Segré and others; but they cannot be generalized to other cases, for instance, for atoms with more than one electron. However, it is frequently the case that the interaction is due to the presence of a single penetrating *s* electron in the group of valence electrons. In such a case, the separation for a given state can be obtained from the formulae worked out. When once the hyperfine separation is determined, it is possible to evaluate from it the nuclear magnetic moment.

NOTE

Nuclear electric quadrupole moments

While the magnetic moment of a nucleus is determined by the distribution of electric currents within the nucleus, the electric moment of a nucleus arises from the distribution of its electric charge. Quantum mechanical consideration of the distribution of nuclear charge shows that *nuclei do not have permanent electric dipole moments, but can have electric quadrupole moments* when their electric charges are not spherically symmetric, which is usually the case if the spin I is not zero or half, *i.e.,* if I is equal to or larger than one. The quadrupole moment (represented by the symbol Q) which is, therefore, a measure of the deviation of the nuclear charge from spherical symmetry, can be expressed, in terms of the nuclear dimensions, by the relations

$$ Q = \frac{1}{e} \int \rho r^2 \, (3 \cos^2 \theta - 1) \, d\tau = Z \, (3z^2 - r^2) A_V $$

where *e* is the fundamental charge, ρ the nuclear charge density, r the distance from the center of gravity of the charge to the element of volume $d\tau$ and θ the angle between *r* and the spin axis *z*.

In the last expression, Z represents the atomic number and the average (A_v) is taken over the nuclear state with $m = I$); *m* is the magnetic quantum number, specifying the orientation of I with respect to the z-axis.

It can be proved that a *prolate* charge distribution with its axis parallel to the z-axis gives rise to *positive* quadrupole moment, while an *oblate* distribution to a *negative* quadrupole moment. Hence a positive value for Q indicates that the nucleus is a prolate spheroid, elongated along the spin axis, while a negative Q means that the nucleus is an oblate spheroid, flattened along the spin axis.

If the nucleus is assumed to be an ellipsoid of revolution with semi-axes c parallel to z and a perpendicular to z, and of uniform charge density, it will produce a quadrupole moment given by

$$Q = (2/5) \, Z \, (c^2 - a^2) = (2/5) \, Z \, (c + a)^2 \, \varepsilon$$

where ε is the eccentricity $= (c - a)/(c + a)$.

If R is the nuclear radius which is approximately equal to $(c + a)/2$ when $\varepsilon \ll 1$,

$$Q = (8/5) \; Z \, R^2 \, \varepsilon \; \dots\dots\dots\dots\dots\dots\dots\dots \; (2)$$

From the observed quadrupole moments, the eccentricities for a number of nuclei have been computed using the above relation and putting $R = 1.5 \, A^{1/3} \times 10^{-13}$ cm. They range from - 0.024 for Cl^{35} to + 0.15 for Lu^{176}, which makes one realize more easily the degree of variation from spherical symmetry than the observed Q values do. Obviously, when ε is positive the nucleus **is** prolate and when ε is negative the nucleus is oblate. The value of Q calculated from relation (1) refers to a bare nucleus in a. definite quantum state, usually the ground state, and is given in square centimeters.

The way in which a nuclear quadrupole moment manifests itself is through the fact that its energy in an external electric field depends not only on its position, but also on its orientation with respect to the gradient of the field. In an atom, this field arises from the electronic charge distribution. In a molecule the field arises from the other nuclei as well as from the electrons. Casimir, in 1936, developed the basic theory of the interaction of nuclear quadrupole moments with extranuclear electrons. This theory has since then been extended to molecules by different workers in the field.

Schuler and Schmidt were able to detect, in 1935, nuclear quadrupole effects in the optical spectra of atoms. Rabi and his associates, in 1940, measured the quadrupole moment of deuteron with the molecular beam magnetic resonance technique. With the advent of the microwave spectroscopy, in 1946, the quadrupole moments of several nuclei have been measured with great precision. We shall speak of these methods of determining the values of Q below.

Quadrupole moments have been found to be related to nuclear shell structure and to nuclear magnetic moments, although exact forms of these relations are not yet available due to want of sufficient and exact experimental data.

Experimental study
Although the greatest amount of information has been obtained from the direct study of .the hyperfine structure of spectral lines, yet this method is beset with very many practical difficulties arising mainly from the *complexity of the patterns* and the *smallness of separations.* It is extremely difficult to measure or even estimate the separations (which are only about 0.1 to 0.5 cm^{-1}), the relative intensities and the mere number of the components forming **a** hyperfine pattern; whence observers of great skill will often

disagree with one another and judgment will often depend on photographs taken with an echelon or an etalon, which look totally different from the pictures obtained with gratings or prisms. Often *several different isotopes of* an element produce different patterns which signify slightly different values of spin and are so nearly superposed on one another as to make analysis superlatively hard.

In spite of all these difficulties, *four different methods of analysis* have been devised which will be briefly described with their relative merits.

(a) Counting the number of hyperfine components

We have seen the number of hyperfine levels into which an atomic state is split by the presence of nuclear spin is $(2I + 1)$ for $I < J$ or $(2J + 1)$ for $I > J$. Hence, if the number of components for any state, which is *less* thin $(2J + 1)$ be counted, we can directly get the value of I by equating the counted number to $(2I + 1)$, since the case considered refers to $I < J$.

Theoretically, this is an *ideal* method. One has simply to investigate the hyperfine line patterns connecting states of as many different values of J as there are. Then applying the selection rules $\Delta F = \pm 1$ or 0, the number N of hyperfine components for state can be derived. Next it is checked that N is equal to $(2J + 1)$ whenever J is less than or equal to some particular value, J_m say and that N is equal to $(2J_m + 1)$ where J is equal to or greater than J_m. All this having been verified the value of I must be J_m.

In practice, one seldom finds such a neat programme as this worked out fully. The difficulties are: (i) a lot of work is involved to analyze the hyperfine structure of even one line, let alone a great number; (*ii*) lines connected with states of certain J values, high ones especially, may not be observed. In many cases, however, the experimental resolution is great enough to compute the number of components, chiefly in patterns known as the *"flag type"*, where the separations decrease uniformly across the pattern. This type arises from the fact that the hyperfine level of one state is very close while they are considerably larger for the other state, so that the resulting pattern will show chiefly the larger separations alone. On account of the interval rule, such a pattern has the characteristic flag appearance.

This method is practically concentrated on such flag patterns. It is useless for cases where I is greater than J, so that it can be used only for the determination of nuclear spin of *small* value.

(b) Measuring the relative separations of the members of a state cluster

Here the relation for the interaction energy W is used. Although all the states of the hyperfine structure group have the same values of I and J, the orientations of I and J will be different leading to different values of W, which results in the separation of the members of a state cluster. Theoretical considerations show that the successive energy differences stand to one another as the successive members of a chain of integers or half integers stepped off at *unit interval* as

$$\triangle W_m : \triangle W_{m-1} : \triangle W_{m-2} \; :: \; (I+J) : (1+J-1) : (I+J-2)$$

This is the interval rule based on the validity of the cosine law. If verified, it enables one to determine the F (= I + J) value for a single cluster with a single value of J. Thus knowing F and J, the unclear spin I can be evaluated.

It is frequently the case in simple spectra of one-electron type that no state of sufficiently large J can be found, which has any appreciable hyperfine pattern. In such a case, this method can be used, even when I is greater than J, unlike the preceding method. Evidently it cannot be used where the cosine law of interaction fails, since the interval rule relies absolutely on the validity of this law. In view of deviations from this law which have been found, probably duo to the presence of an electric quadrupole moment in the nucleus, the method can be considered as safe only for those which would show no quadrupole effects, such as states involving only *s* electrons.

Much use has been made, however, of this method in practice, and there are a few cases, in which a fairly accurate measurement of a chain of intervals has shown that it verifies the interval rule.

It is possible to *evaluate the nuclear magnetic moment* in this method from the measured energy differences between the successive members of a state cluster. But the theory is very complicated and requires a quantum mechanical treatment; even so a lot of approximations have to be made. We may, however, illustrate the theory by a quasi-classical derivation as-follows:

One first visualizes the valency electron as a charged particle revolving round its quantized orbit, equivalent therefore to a steady current running along the orbit, and producing a magnetic field at all points within the orbit and in particular at the point occupied by the nucleus. The nuclear magnetic moment is subjected to this field and when it is shifted from one to another of its permitted orientations a certain amount of work must be done and this constitutes the energy difference in question.

If r is the radius of the orbit, n the orbital frequency, e and m the charge and mass of the electron respectively, p_l the angular momentum is given by

$$p_l = I \, \omega = m \, r^2 \, \omega = m \, r^2 \, (2\pi \, n)$$

According to the quantum theory,

$$p_l = l \, . \, (h/ 2\pi)$$

Therefore,

$$p_l = m \, r^2 \, (2\pi \, n) = l \, . \, (h/ 2\pi) \; \ldots\ldots\ldots\ldots\ldots\ldots \; (1)$$

The strength of the orbital current $= n \, e$ (e. s. u.) $= n \, e/c$ (e. m. u.)

Substituting for n from (1), viz., $n = p_l / (2\pi m r^2)$

The strength of the current $= p_l .e / (2\pi m r^2 c)$

Since the magnetic field produced by a circular current i at the center is $2\pi i / r$, the magnetic field due to the equivalent current of the revolving electron at the center, where the nucleus is situated, is given by

$$(2\pi / r).\, [\, p_l .e / (2\pi m r^2 c)] = \; = l.\, (h/ 2\pi) .(e / m r^3 c) \; \ldots\ldots\ldots\ldots \; (2)$$

If we could ignore the spin of the electron, l could be replaced by J, and the field due to the revolving electron at the nucleus

$$= (e/mc) .r^{-3} . \; J. \; (h / 2\pi)$$

Conceiving the nucleus as having a magnetic moment μ_N parallel to its angular momentum, $I . (h / 2\pi)$, where I is the nuclear spin quantum number, the energy **W** of the nuclear magnet placed in the uniform field of the revolving electron is given by

$$W = \mu_N \; (e/mc) .r^{-3} . \; J. \; (h / 2\pi) \cos (I\, J) \; \ldots\ldots\ldots\ldots\ldots\ldots\ldots\ldots\ldots\ldots \; (3)$$

Remembering that the different members of a state cluster have the same values for I and J, but different orientations permitted between I and J, the above expression will give the energy values of the different members. The only variable quantity in relation (3) is cos(IJ). Hence the differences between the right-hand members of the above equation can be equated to the observed energy differences between the states of the cluster and solved for μ_N.

It is to be noted that this formula when applied to data gives values of μ_N of the same order of magnitude as does the more elaborate quantum mechanical formula.

Though difficult and defective, this method has been used to determine the nuclear magnetic moment in most cases before the advent of the magnetic deflection methods,

(c) Measuring the relative intensities of the components of the hyperfine pattern
The relative intensities of the members of a hyperfine pattern have been shown to obey the same laws which hold for the fine structure multiplet in Russell Saunders coupling, when the quantum members, L, S, J, are replaced by J, I, F, respectively. With J known for both the initial and final states it is possible, from an accurate knowledge of the relative intensities, to deduce I since the F value can be written in terms of I and J. But the formulae are of appalling complexity, while intensity measurements are generally liable to error. Hence this method is probably the *least reliable,* although useful, when the spin cannot be directly determined from the number of hyperfine components as in the first method.

(d) Counting the number of hyperfine components in Zeeman effect

This method readily lends itself to a *direct determination, of the nuclear spin.*

The *theory* is simple in the case where the external field is strong. For, the interaction of I and J with the applied field is much greater than with one another, *i.e.,* the coupling between I and J is broken and, in consequence, each has its quantized projection on the external field H and not their resultant F.

Considering the extranuclear electron system alone, the Zeeman effect may be represented by the relation

$$W = (e h / 4\pi mc) .H. m_j . g \quad ....(1)$$

where W is the energy change of the atomic states due to Zeeman effect, m_j the component of J along H and g the Landé splitting factor for the electron system.

Now, due to the presence of nuclear spin, the above relation will be modified as

$$W = (e h / 4\pi mc) .H. [m_j . g + m_i . g_i] \quad ....(2)$$

where the second term refers to the action of the external field on the nucleus, m_i, being the projection of I along H and g_i the Landé's factor for nuclear motion.

Since the ratio m/M enters in the estimation of g_i, this latter will be very small compared to g, so that the second term in eqn. (2) is negligibly small. Hence, to a first approximation, the Zeeman effect is the same with a nuclear spin as without.

If, however, we include the interaction of the nuclear magnetic moment with the extranuclear electronic magnetic moment, relation (2) will be still further modified as

$$W = (e h / 4\pi mc) .H. [m_j . g + m_i . g_i] + A I J \cos (I J)...........................(3)$$

where the last term represents the above-mentioned interaction.

For a *strong field,* due to the "decoupling" of I and J, the expression

I J cos (I J)

becomes

I cos (I H) x. J cos (J H) = $m_i. m_j.$

Already we have seen that is negligibly small. Hence, in such a case, the expression for the energy change is

$$W = (e h / 4\pi mc) .H. m_j . g + A m_i . m_j \quad ....(4)$$

Here the first term is the Zeeman effect, and 'so with every Zeeman level there will be a hyperfine structure determined by A m_i . m_j.

Now m_i can assume all the $(2I + 1)$ values from - **I...** to + I. Thus every Zeeman level will be split into $(2I + 1)$ equidistant levels with a separation of Am_j; and since m_i does not change during a transition, each of the $(2I + 1)$ levels of one Zeeman state combines with *only one* of the $(2I + 1)$, levels of another Zeeman state.

Thus the Zeeman lines themselves have a very simple pattern, while each Zeeman line is split into a hyperfine structure of $(2I + 1)$ equidistant lines which can be observed with a good spectroscope. Counting therefore the number of hyperfine components of each Zeeman line and equating it to $(2I + 1)$, I is readily obtained.

In *practice,* the usual field strengths which are employed in Zeeman effect fulfill the condition under which eqn. (4) becomes valid. The strength of the field must be such that the ordinary Zeeman effect separation is large compared to the hyperfine separation.

A great number of nuclear spins have been measured in this way and Back's analysis of the hyperfine structure of four Zeeman lines λ = 4722 of Bi is ordinarily cited as a typical example. In this case, each line is split into ten components, which gives the nuclear spin for Bi as 9/2. There exist magnificent photographs of the spectral lines of bismuth exposed to a magnetic field, each line under high resolution exhibiting 10 components. One such photograph obtained by Back and Goudsmit is reproduced here, where the components can be easily counted.

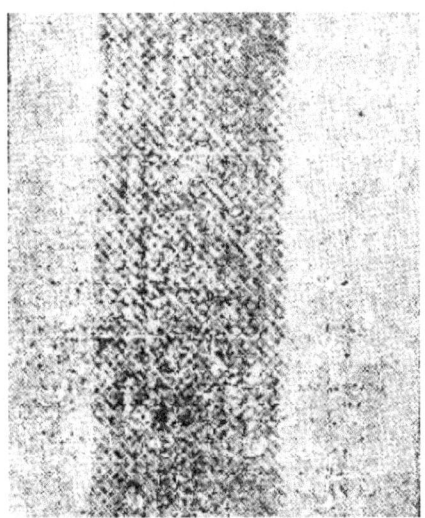

Hyperfine structure in Zeeman effect (Back and Goudsmit)

(ii) METHOD OF ALTERNATING INTENSITIES IN BAND SPECTRA

The *wave mechanical theory* of this method may be briefly stated as follows.

For a diatomic molecule whose two nuclei are identical, two possibilities exist, viz., the wave functions may be symmetric or anti-symmetric in their nuclear spins. Since each of these two possibilities may be realized in several ways, in general, different statistical weights must be assigned to the molecular types.

[The statistical weight refers to the splitting of energy levels caused by "inner" or "outer" precession due to magnetic or electric interaction. For example, in the Stark effect, if a single level is split into five sublevels under the influence of an applied electric field, the statistical weight of the original level is five. In general, to an energy level defined by the quantum number n, the statistical weight $g = (2n + 1)$ must be attributed].

If there were no nuclear spin, the statistical weight of any rotational level for which the total angular momentum is J would be $(2J + 1)$. But the presence of nuclear spin changes the statistical weights of the levels to different extent in the symmetric and anti-symmetric cases. Supposing that the changed statistical weight of the levels in the symmetric case to be

$g_s (2J + 1)$

and that in the anti-symmetric case to be

$g_a (2J + 1)$

where g_s and g_a are the respective weight factors, theoretical considerations show that

$g_a / g_s = (I + 1/2)/ I$

when the nuclei obey Fermi statistics, while

$g_a / g_s = (I + 1)/ I$

for nuclei obeying Bose statistics. In the first case $g_a > g_s$ while in the second $g_a < g_s$.

Now those levels which have the greater weight factor are called *strong levels,* while those having a smaller weight factor *weak.* Hence, for nuclei obeying Fermi statistics those states whose eigenfunctions are anti-symmetrical are strong levels, while those whose eigenfunctions are symmetrical weak. For nuclei obeying Bose statistics the opposite is the case.

Transitions can take place between strong levels or between weak levels, but not between a strong and a weak level. This gives rise to the alternative strong and weak lines actually observed. Thus the intensities of successive lines depend on the weight factors, so that the ratio of the intensities must be equal to the ratio of the weight factors. Hence the ratio of the intensities of the successive lines is equal to $(I + 1)/I$.

In the particular case where g_a or g_s is zero, *i.e.,* where the symmetrical or anti-symmetrical wave function is not present, then obviously $I = 0$, and the alternate rotational lines will be completely missing.

In the case of *heteronuclear* molecules, the nuclear spin values will be different for the two nuclei and consequently resonance phenomenon cannot take place, so that no alternating intensities can be had in their band spectra.

In the *practical determination* of nuclear spin by this method, the intensities of the successive rotational lines in the band spectra of homonuclear diatomic molecules, preferably in the Raman effect, are measured as accurately as possible. The ratio of the alternating intensities is equated to $(I + 1)/I$ and solved for I. If the alternate lines are missing, $I = 0$.

In this way, the nuclear spin of several isotopes, *e.g.,* H^1, D^2, He^4, Li^7, C^{12}, N^{14}, O^{16}, F^{19}, Na^{23}, P^{31} and S^{32} have been measured. The chief advantage of this method over that of hyperfine structure is that nuclear spin can be determined even when its value is zero, while the hyperfine structure methods are unable to distinguish between zero magnetic moment and zero spin. On the other hand, the band spectra method gives no information about the nuclear magnetic moment. Furthermore, on account of the practical difficulty encountered in the accurate measurement of the intensity of the lines, it can be applied satisfactorily only when the spin value is small, as in the case of light elements.

(iii) MAGNETIC DEFLECTION OF MOLECULAR AND ATOMIC BEAMS

Several methods based on the **magnetic deflection of molecular beams,** *in which the magnetic moment of the nucleus is directly determined,* have been developed. They involve, in general, *refined techniques of the Stern-Gerlach method*, in which all possible and more powerful competitors to the nuclear moments are removed.

The first experiments of this character were performed by Stern and Frisch, in 1934, and *the magnetic moments of proton and deuteron* were determined by deflecting molecular beams in nonhomogeneous magnetic fields. The values of the magnetic moments obtained by this method (viz., for proton 2.5 nuclear magnetons with an uncertainty of about 10% and for deuteron between 0.5 and 1 nuclear magneton) agreed fairly well with those obtained by the hyperfine structure method by Rabi, Kellogg and Zacharias (*viz.,* for proton 3.25 nuclear magnetons with 10% uncertainty and, for deuteron 0.75 nuclear magneton with 25% uncertainty).

Rabi and Breit, in 1931, devised a *method of determining both the nuclear-spin and magnetic moment* by employing **magnetic deflection of atomic beams.**

The *theory* of the method as developed by the authors is too long and too complicated to be quoted here. The main line of their argument is as follows.

A beam of atoms flying across a strong magnetic field with gradient perpendicular to the beam shows a separation into $(2J + 1)$ components which is the Stern-Gerlach effect. If a nuclear magnetic moment is present, each of these components consists of $(2I + 1)$ hyperfine structure, which for strong fields all fall together and hence cannot be detected. This is why Stern and Frisch used a molecular and not an atomic beam. If, therefore, one were to use an atomic beam, the problem would be to separate the $(2I + 1)$ components sufficiently by some device and count them individually. Rabi and Breit pointed out *that as the field strength approaches zero, these individual components no longer fall together,* and succeeded to refine the Stern-Gerlach method to such an extent that the above condition could be realized. From the number of components counted the nuclear spin I is readily obtained by equating it to $(2I + 1)$. This method has been used in the determination of nuclear spin of hydrogen and various alkalis such as Li, Na, K, Rb and Cs. With Na, for instance, the beam is split into 8 components, instead of merely 2, proving thereby the value 4 for $(2I + 1)$ and the value 3/2 for nuclear spin. Similarly the nuclear spin for Cs was found to be 7/2.

The magnetic energies of the $(2J +1)$. $(2I + 1)$ states depend not only on the field gradient dH/dx of the inhomogeneous field but also in a rather complicated fashion on H itself. The analysis of the dependence on H shows that certain energy levels go through *zero* at certain values of H. Placing the detector at the position of the undeflected beam at the end of a long, weak, inhomogeneous field, the field is gradually increased and the values of H, at which "zero beam" peaks occur, are noted (method of zero-moments). With these values of H, the hyperfine structure separation can be calculated. From the hyperfine separation, the nuclear magnetic moment also can be evaluated. The regularity or inversion of the hyperfine structure indicates further the sign of the nuclear magnetic moment, whether positive or negative.

The method is advantageously used for the determination of nuclear magnetic moments in cases where the hyperfine separation is far too small to be detected by optical methods. But the chief limitation is the difficulty of producing and detecting a beam of atoms.

(iv) STUDY OF MAGNETIC RESONANCE RADIO FREQUENCY SPECTRA

In 1938, Rabi, Zacharias, Milliman and Kusch developed an ingenious technique for molecular beams, known as the *magnetic resonance method,* which has made possible accurate spectroscopy in the *"radio frequency"* range. The experiment was first announced as a new method for the determination of nuclear magnetic moment, but it became soon apparent that its scope was not limited to the measurement of this quantity alone.

Isidor Isaac Rabi (29 July 1898 – 11 January 1988)

Principle

The principle on which the method is based applies not merely to nuclear magnetic moments, but rather to any system which possesses angular momentum and a magnetic moment. If a particle having an angular momentum J and a magnetic moment μ placed in a magnetic field H, it will execute a precessional motion about the field with the Larmor frequency $v_L = \mu H/J$. The resonance method consists essentially in adjusting the strength of the magnetic field until the precession is in resonance with an impressed oscillating magnetic field H' whose frequency f is in the radiofrequency range. Under this condition, *viz.,*. when $v_L = f$, the particle will suffer reorientations, which can be detected by a suitable device. The smaller the ratio H'/H, the sharper the effect will be on its dependence on the exact agreement between v_L and f. The magnetic moment μ can be evaluated from the known frequency f of the auxiliary oscillating magnetic field and the strength H_v of the primary magnetic field required to produce resonance in the precessional frequency, provided J is also known.

In the case of more complicated systems containing two or more coupled angular. momenta which interact with each other as well as with the external field, it is simpler to view the resonance reorientations as taking place when the frequency of the oscillating field is in resonance with the frequency given by the Bohr relation

$$hv_{nm} = W_n - W_m ,$$

where W_n and W_m represent the energies of the two states of the whole molecular system in the magnetic field. The selection rule which governs these transitions in the case studied is $\Delta m = \pm 1$, where m is the magnetic quantum number of the system. It may be pointed out here that the method detects not only transitions from state m to a, but also the reverse transition n to m. One of these corresponds to absorption of radiation and the other to stimulated emission. As Einstein has pointed out, the two processes are equally probable.

These general principles can be applied to the following particular cases:

(*i*) *Molecular states in which all possible and more powerful competitors to the nuclear moments are removed, as in molecules in the* $^1\sum$ *state* where the energy of the nucleus in the external field depends only on its own orientation, independent of all other molecular interactions. This corresponds to a simple system of angular momentum and associated magnetic moment, in which reorientations will occur if the Larmor frequency of precession v_L (= $\mu H / I$) and the frequency f of the oscillating field are in resonance.

Rabi, Guttinger and Majorana have considered the transition probability in such a case on a quantum mechanical basis and have shown that a maximum number of orientations takes place when $f = v_L$. Under these circumstances, the *nuclear magnetic moment* can be determined, as indicated above, if the value of nuclear spin I is known.

(*ii*) *Molecular states in which the various interactions between the molecular constituents are effective, i.e.,* when the energy of the nucleus in the external magnetic field depends not only on its own orientation in the external magnetic field, but also on the orientation of other nuclei ("shin-spin" interaction) and on the orientation of the rotational angular momentum of the molecule ("spin-orbit" interaction). Analysis of strong field resonance reorientations in such cases has led to the determination of the value and sign of *nuclear electric quadrupole moments* and of *molecular rotational magnetic moments.*

(*iii*) *Atomic ground states.* On account of the existence of nuclear spin, the ground states of many atoms, such as those of the alkali metals, consist of a set of closely spaced energy levels. Each level of this hyperfine structure corresponds to a value of the total angular momentum of the atom. The spacings are caused chiefly by the feeble interactions of the magnetic and electric fields of the electrons with the nuclear magnetic moment and the electric quadrupole moment respectively. Left to themselves, the atoms would radiate this energy in the form of electromagnetic radiation given by the Bohr formula and settle down to their lowest energy state. The region of frequency in which these radiations are emitted lies approximately between 1.5×10^8 and 1.2×10^{10} cycles per sec. Because of these low frequencies, the life-time of a hyperfine structure level is very long and the intensity of spontaneous emission very feeble. In consequence, direct observation of this radiation would be very difficult. But it is possible to irradiate the atom with electromagnetic radiation of the correct frequency and of such intensities (provided by the oscillating field).as to cause it to absorb, or, by the Einstein process of stimulated emission, to emit a quantum of this frequency in a reasonably short time of about 10^{-4} sec. If such a process is detected, it offers a *direct method of measuring hyperfine structure,* which has many advantages over the optical methods. Firstly, the results are simple to interpret since only one atomic energy level is involved; secondly, the accuracy is very high, as only the measurement of a radiofrequency is to be made; thirdly it enables extremely small energy separations to be measured.

Thus, the magnetic resonance method has great potentialities- giving, as it does, precise and direct information about nuclear magnetic moments, nuclear electric quadrupole moments, hyperfine structure separations, molecular rotational magnetic moments, etc., but it involves a knowledge of the nuclear spin.

It is of interest to note that this method reverses the ordinary procedure of spectroscopy and instead of analyzing the radiation emitted by atoms or molecules, analyzes the energy changes produced by the radiation in the atomic system itself.

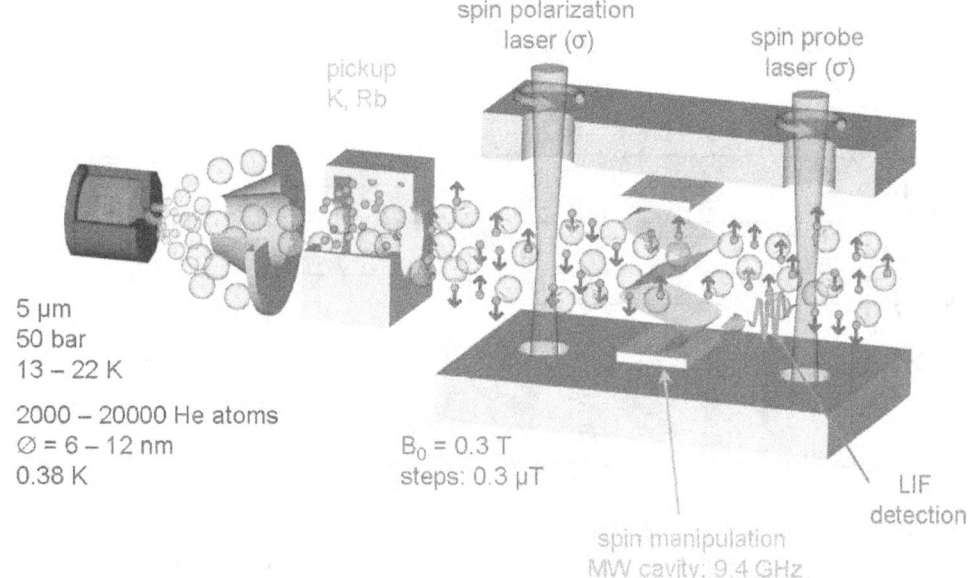

The principle of optically detected magnetic resonance using magnetic circular dichroism.

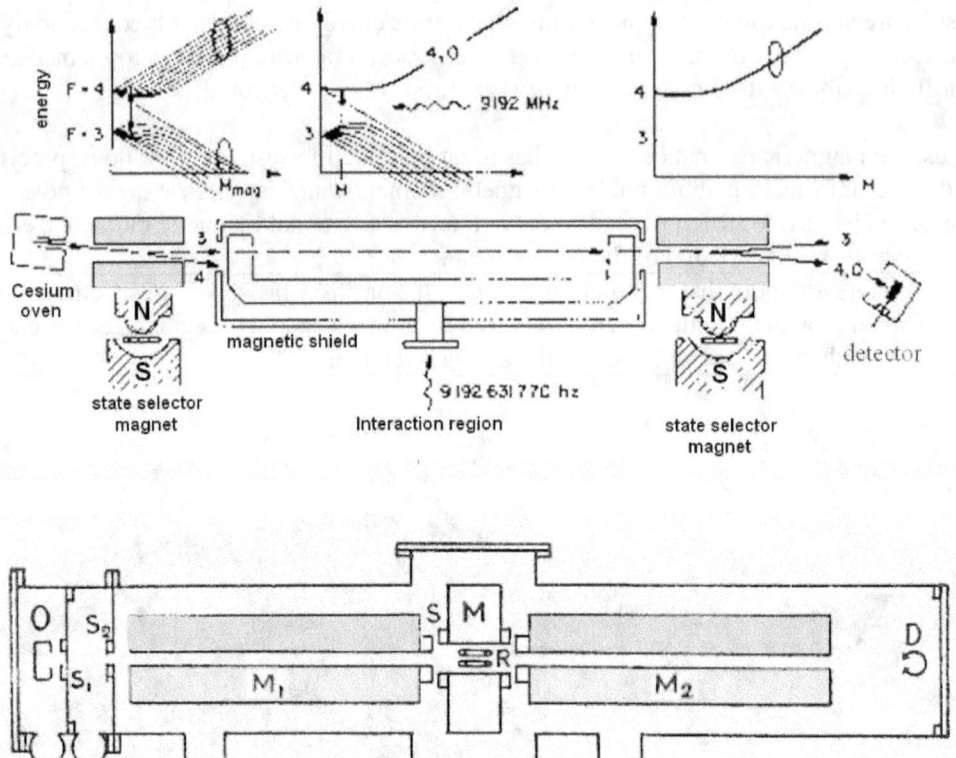

Fig. 335. Apparatus used in the magnetic resonance method.

Apparatus

The experimental arrangement designed by Rabi and his associates is diagrammatically shown in Fig. 335.

The molecules of the substance under test are produced in the oven O of small dimensions (0.01 mm. wide) maintained at a temperature such that the vapor pressure of its contents is a few tenths of a mm of Hg.

Of the molecules that emerge in all directions from such a source, a very small fraction forming a very fine beam passes through a collimating slit S and reaches a detector D.

In the absence of any inhomogeneous magnetic deflecting fields these molecules traverse straight line paths OSD and form the so-called "direct" beam.

The main portion of the apparatus must be highly exhausted (10^{-5} mm. Hg.) in order to render the beam collision-free and two slits S_1 and S_2, several times wider than S; are used near the source to isolate the oven chamber were relatively high pressures are required to secure sufficient intensity of the beam.

In order to obtain reorientations of the molecules and to detect such reorientations the following device is adopted.

Two strong inhomogeneous magnetic fields, whose gradients are in opposite directions, are established using two electromagnets M_1 and M_2 with specially shaped pole pieces, similar to those used in the Stern and Gerlach apparatus.

The first field due to M_1 (which is 25 cms long and begins at 10 cms from the source slit S_2) splits the beam into a number of polarized beams according to different moments and velocities of the molecules.

The second field of opposite gradient due to M_2 (which is 30 cms long and begins at 52 cms from the source) will refocus the divergent beams from the first field at a certain point (92 cms from the source) where the detector is placed.

This refocusing takes place only for those molecules which have remained in the same quantum state through their passage in both the fields. It is found experimentally that when the two fields are properly adjusted the number of molecules reaching the detector is almost the same whether the fields are on or off.

Now a short homogeneous field of variable strength H obtained with an electromagnet M and a superimposed oscillating field, perpendicular to H, produced by a high frequency current flowing through two parallel copper tubes bent in the form of a 'hair pin' and inserted in the gap between the pole-pieces of M are arranged between the two inhomogeneous fields.

If H reaches a value for which the Larmor frequency v_L of the molecules is equal to the frequency f of the oscillating field, resonance sets in, causing reorientations of the molecules which "flop over" into other quantum states. These flopped molecules are no more refocused by the field due to M_2. The resonance, therefore, produces a decrease in intensity at the detector, which thereby offers a means of knowing when the reorientation effect occurs. Each observed minimum indicates a certain Larmor frequency and allows the calculation of the corresponding magnetic moment.

The action of the two inhomogeneous fields on the molecular beam before and after reorientation of the molecules is illustrated in Fig. 336.

manipulation

preparation   probing

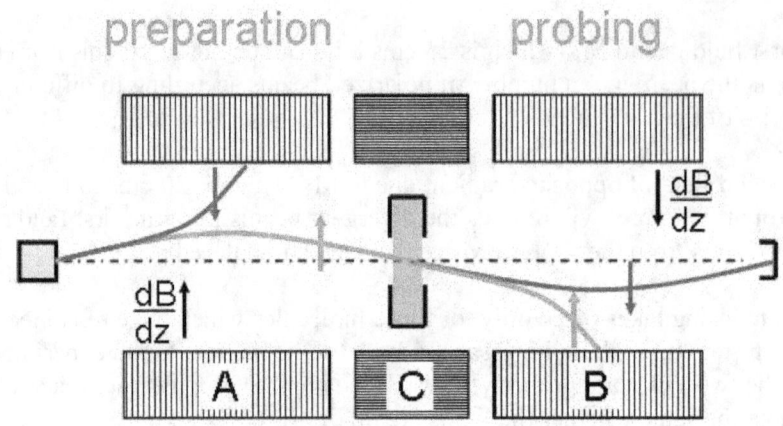

$$\frac{dB}{dz}$$

$$\frac{dB}{dz}$$

A   C   B

Rabi: spatial separation

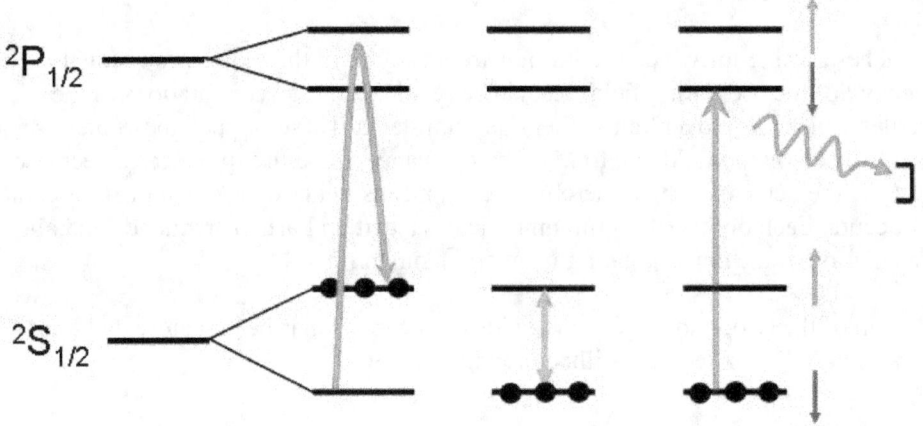

$^2P_{1/2}$

$^2S_{1/2}$

optical detection: Zeeman splitting

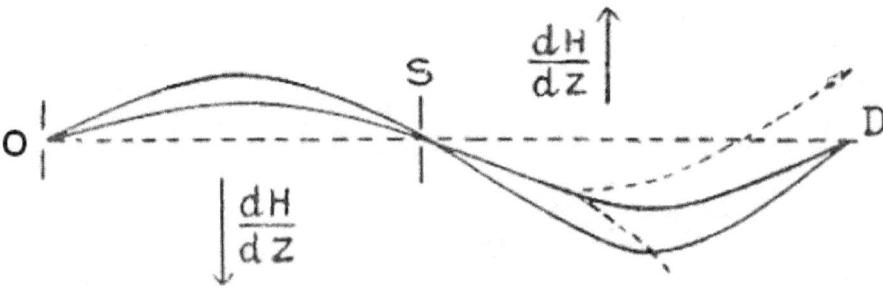

Fig. 336.

The two continuous curves indicate the path of two molecules having different moments and velocities and which have not suffered reorientations and hence get refocused at the detector D. The dotted curves in the region of M_2 indicate the possible changes in path for one of the molecules, which has been reoriented under the combined influence of the homogeneous and oscillatory fields at M and which, in consequence, will not reach D.

The detector must necessarily be a very sensitive device capable of giving a linear response to the intensity of a slender beam of molecules falling on it, as well as of measuring very small changes of intensity. There are, in actual use, three different types of detector, known as the *Pirani gauge,* the *ionization gauge* and the *surface ionization gauge.* This last one is most sensitive and hence commonly used, whenever possible. It consists of a heated thin tungsten filament placed at the refocusing point of the molecular beam. The impinging neutral molecules (or atoms) re-evaporate from the filament as positive ions, each having given up an electron to the surface of the filament. If the tungsten surface is kept sufficiently hot and if the difference between its work function and the ionization potential of the molecule or atom is appreciably greater than kT; practically all the molecules falling on the filament re-emerge as positive ions. These ions are collected by a plate which surrounds the filament and which is kept at a potential of about -10 volts with respect to the filament. The ion current thus obtained is amplified by means of a suitable vacuum tube. With the normal plate current of the tube balanced out, the change in this current is a direct measure of the number of molecules or atoms impinging on the tungsten surface.

The surface ionization device is able to detect an ion current of about 10^{-15} ampere, or about 6,000 molecules or atoms per sec. Measurements are made with it with ease and speed, a complete resonance curve being obtained in ten minutes or less. This, in turn, makes the data more reliable, since changes in experimental conditions are less likely to occur in short-time intervals than in long periods.

The length of the apparatus (about 1 meter) and the use of very narrow slits necessitate a careful alignment of the different parts, which is not easy in practice. But this difficulty is more than compensated by the great precision with which measurements can be made with the arrangement. This is due to the fact that only two quantities, viz., the strength of a given magnetic field and the frequency of an oscillator, have to be measured, which can be done to a high degree of accuracy.

A calibration of the homogeneous magnetic field of the magnet M in terms of the current through the exciting coils may be made with a sensitive fluxmeter, or better still, by the use of atomic transitions of the ground state of alkalis, as was employed by the workers themselves. The frequency of the oscillator may be easily determined to better than 0.01 per cent with a commercial crystal-controlled heterodyne frequency meter.

Experimental procedure
(i) Keeping the frequency of the oscillating field constant, the beam intensity, as indicated by the detector, is measured for different values of the homogeneous magnetic field H. When the data thus obtained are plotted, resonance curves, with one or several minima, depending on the conditions that prevail in the beam analyzed, are obtained.

Such curves are known as *radio frequency spectra.* The minimum in a resonance curve corresponds to the condition $v_L = f$. A typical example of such spectra, obtained by Rabi and his co- workers with deuterium molecules emerging from a source maintained at the temperature of liquid nitrogen is shown in Fig. 337.

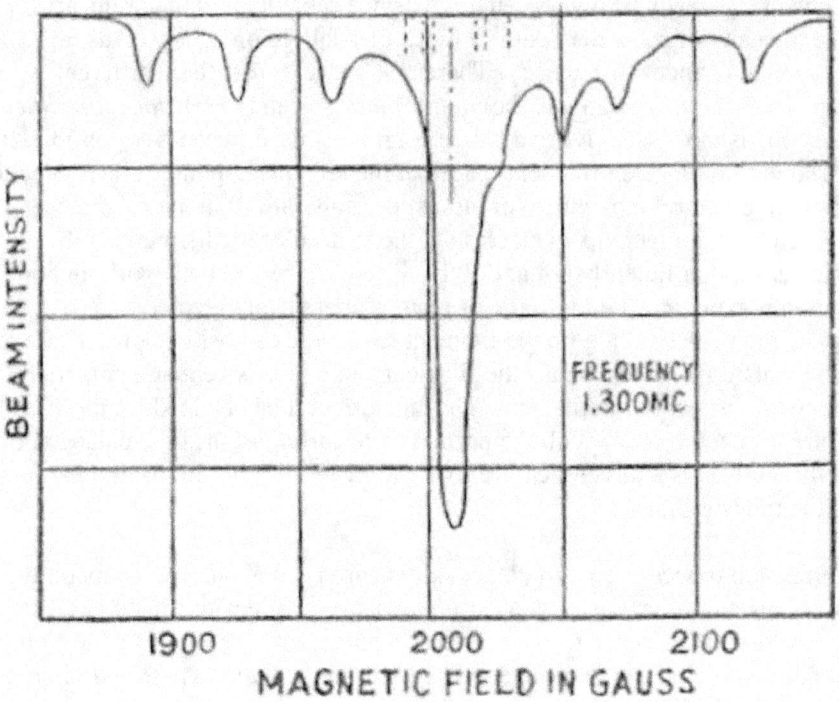

Fig. 337. Radio frequency spectra of deuterium.

The deep central resonance curve arises from reorientation of deuterium molecules for which $J = 0$ and $I = 2$ (case *i*). The value of H corresponding to the resonance condition ($v_L = f$) is readily obtained from the curve, which enables the *nuclear magnetic moment* (μ_N) *of deuteron* to be evaluated as 0.8565 nuclear magnetons.

Further, the asymmetry of the curve with respect to the field value for minimum intensity has been construed to indicate a *positive* moment for deuteron.

The other six resonance curves are associated with reorientations of molecules, for which J = 1 and I = 1 (case *ii*). They are known as *"nuclear spectrum"*, characterized by $\Delta m_J =$ 0, $\Delta m_I = = \pm 1$, in which the nuclei alone suffer reorientations.

Analysis of these curves has led to the determination of the *nuclear electric quadrupole moment Q* of deuteron, whose value is found to be 2.73×10^{-27} cm^2 and is positive.

In addition to the "nuclear spectrum" there exists also another six-line spectrum known as *"molecular rotational spectrum"* (not shown in Fig. 337) arising also from reorientations of molecules in the state J = 1 and I = 1, but characterized by $\Delta m_J = \pm 1$ and $\Delta m_I = 0$. Such a spectrum has been detected and analyzed for the determination of the rotational magnetic moment of molecules. Ramsay has thus measured the rotational moments of H_2, HD and D_2 molecules as 0.879, 0.660 and 0.441 respectively for their first rotational states.

(ii) The complete resolution of molecular radiofrequency spectra, as stated above, is not essential for the determination of nuclear magnetic moments. In fact, the majority of nuclear moments so far determined by the resonance method have been obtained from molecules which yield resonance curves formed by the superposition of many lines-the so-called *unresolved molecular spectrum*. The, particular molecular constituent to which the resonance curve corresponds may be identified by finding the resonance minimum common to two molecules containing the same nucleus. For example, if resonance curves are taken for LiCl, LiF and NaF, the minima common to LiCl and LiF are those of the Li isotopes, while the minimum common to LiF and NaF is that of fluorine. Some of the values obtained in this manner are:

H^1 = + 2.7896   nuclear magnetons,
Li^6 = + 0.8565   "   "   ,
Li^7 = + 3.2532   "   "   ,
F^{19} = + 2.625v,
Na^{23} = + 2.215   "   "   ,
K^{39} = + 0.391   "   "   ,
K^{40} = - 1.290 391   "   "   .

(iii) Using *atomic beams* and obtaining the resonance curves (case *iii*) with weak field (H = 0.05 to 1.5 gauss), the hyperfine separations of the alkali and indium atoms in the ground state have been measured by the magnetic resonance method, which have further enabled to deduce the nuclear spins in some cases.

Alvarez and Bloch, in 1940, by an adaptation of the resonance method to a *beam of neutrons* have been able to measure the magnetic moment of the neutron directly and obtain the value of - 1.945 ± 0 02 nuclear magnetons, which is the moment to be added to that of the proton in order to get that of the deuteron. We shall deal with this curious fact of the uncharged neutron possessing a negative magnetic moment presently.

(v) ANALYSIS OF MICROWAVE SPECTRA

The exceptional resolving power characteristic of microwave spectroscopes readily lends itself to the hyperfine splitting of spectral lines. Analysis of such hyperfine structure in microwave spectra offers a direct means of estimating the spins, magnetic moments and electric quadrupole moments of nuclei. Further, thanks to the high precision of the microwave spectrometers, measurement of the above quantities can be made with great accuracy. Hyperfine structures in microwave spectra have been obtained in the following cases:

A. Atomic spectra

The spectral region from 0.05 cm^{-1} to 5 cm^{-1} now covered by precise microwave spectroscopy includes many transitions between hyperfine levels of free atoms. For example, the ground states of Na^{23}, Rb^{37}, Cs^{112} are split into doublets with separations of 0.059, 0.228 and 0.307 cm^{-1} respectively, due to the interaction of the unpaired electron and the nuclear magnetic moment. Many ionized atoms also have transitions falling in the microwave region. The advantages of microwave methods for the analysis of such transitions are evident. But due to experimental difficulties of getting sufficient amount of free atoms and ions in the apparatus in order to obtain detectable absorption lines, the technique of microwave atomic spectroscopy has not yet been fully developed. All the same, the hyperfine structures of H, Na, and Cs atoms have already been observed. A very important result obtained is what is known as *Lamb-Retherford shift*. Using atomic beam and microwave technique, Lamb and Retherford (1947-1953) were able to make a direct and precise measurement of the separation of the fine structure levels in the first excited state of atomic hydrogen. They found that the $2\ ^2S_{1/2}$ and the $2\ ^2P_{1/2}$ levels do not coincide but that the former level is 1,058 megacycles/sec. or 0.033 cm^{-1} above the latter. This separation, known as the "Lamb-Retherford shift", has allowed a significant modification to be introduced in the theory of quantum electrodynamics. A satisfactory explanation of the observed shift is that it results from the interaction of the electron with its own radiation field. Fluctuations exist in the quantized electric field, and these fluctuations, acting on the electron, cause it to have a very rapidly variable position. This means that the point electron effectively becomes a sphere of radius of about 7 x 10^{-12} cm, and as a consequence, the electron is not so strongly attracted to the nucleus at short distances. As a result, the S states of zero orbital momentum, in which the electron is close to the nucleus, are raised in energy relative to the corresponding P, D, ... states in which the electron has a very small probability of being rear the nucleus.

When Dirac developed his relativistic quantum-mechanical theory of the electron, he omitted the above effect due to the interaction of the electron with its own radiation field. His theoretical conclusions, however, appeared to be confirmed by experimental, results. In contrast to the situation, calculations that attempted to include the field reaction always gave the inertia of the electron's self-field as infinite, corresponding to an electron of infinite mass. Consequently Dirac's theory was favored over the more consistent theory which included the reaction required by Newton's third law. This state of affairs, however, was reversed by the decisive result of the Lamb-Retherford experiment, which proved that a small but certain correction has to be made to the predictions of Dirac's theory. Later Schwinger theoretically showed, from a reformulation of relativistic quantum electrodynamics, that the interaction energy between an electron and an external magnetic field must include a radiative correction term representing the interaction of the electron with the quantized electromagnetic field.

B. Molecular rotational spectra

In the last decade, microwave technique has been chiefly designed for direct observation of the pure rotational absorption spectra of molecules and has met with unprecedented success in the resolution of hyperfine structure of molecular rotational lines. This phenomenon arises through a coupling of the nuclear spin I to the molecular rotational vector J. Vectorially I and J can be regarded as forming a resultant F fixed in space about which J and I precess. The coupling may be electric or magnetic or both. Although both types of couplings can occur in one and the same molecule, frequently one type predominates over the other. We shall now deal with the effects of the two interactions separately for the sake of clearness. It must be borne in mind, however, that if the $(2I + 1)$ degeneracy resulting from a given nucleus is completely lifted by one of these interactions, no further components in the hyperfine structure arising from the same nucleus can be produced by the other type of interaction, but only a small displacement of the existing components.

(i) Hyperfine structure due to quadrupole interaction

A nonspherical distribution of electric charge causes the nucleus to have an electric quadrupole moment which interacts with the gradient of the molecular electric field at the nucleus. This molecular field arises chiefly from the valence-shell electrons which are strongly coupled to the molecular frame and which, in consequence, rotate with the molecule. When the molecular axis changes orientation with respect to the nuclear axis, the components of the molecular field at the nucleus in the direction of the spin axis change values causing a corresponding change in the interaction energy, which must be supplied by the radiation field. It is this change in interaction energy, which is evidently quantized, that gives rise to the hyperfine structure in rotational spectra. To determine the magnitude of the hyperfine splitting of the rotational levels, quantum mechanical perturbation methods have been employed. Casimir gave the basic theory of the interaction of the nuclear quadrupole moments with extranuclear electrons. Nordsieck, Feld and Lamb adapted it to linear molecules. Coles and Good and Van Vleck extended it to symmetric top molecules, while Bragg and Knight and Feld to asymmetric molecules.

A very large number of molecules are found to have observable nuclear quadrupole hyperfine structure in their microwave rotational spectra, W.E. Good was the first, in 1946, to discover such nuclear quadrupole hyperfine structure in the inversion spectrum of ammonia (NH_3), when he found that the individual fine structure lines had satellites. These satellites of one of the ammonia lines, as were seen by Good on the cathode ray screen, are represented in Fig. 338. This beautifully symmetric hyperfine structure about the strong central .undisplaced line arises from the nuclear quadrupole moment of the N^{14} nucleus in the $N^{14}H_3$ molecule.

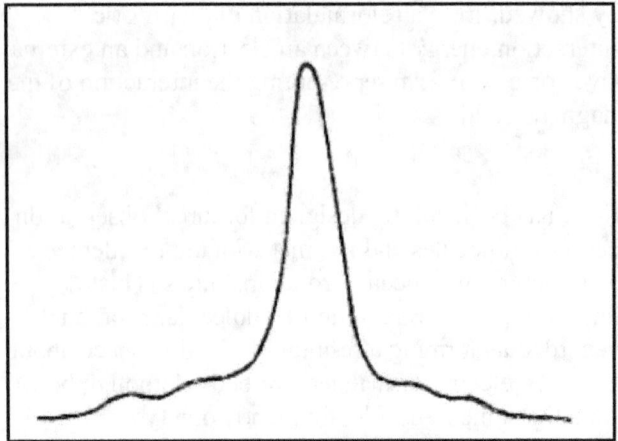

Fig. 338. Quadrupole hyperfine structure of ammonia.

Since the time of this discovery, numerous examples of quadrupole hyperfine, structure in microwave rotational spectra have been observed by different workers in the field. Fig, 339 represents the hyperfine structure of a rotational line of the linear molecule ICN at 0.581 cm wavelength obtained by Gilliam, Edwards and Gordy. The hyperfine splitting has been shown to be due to the quadrupole moment of the I^{127} nucleus.

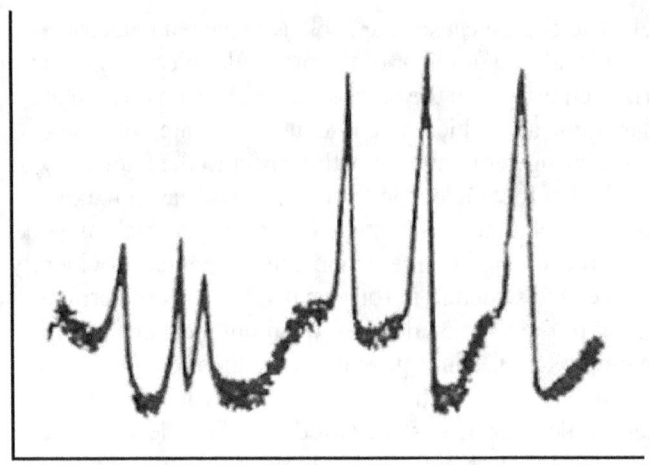

Fig. 339. Hyperfine structure of a rotational line of ICN.

From such microwave quadrupole hyperfine structure nuclear spins and quadrupole moments have been estimated in many cases. Over and above new determinations, results obtained by other methods have been checked. In this connection the analysis of the rotational hyperfine structure of boron carbonyl (BH_3CO) by Gordy, Ring and Burg, in

1949, serves as a very illustrative example. The spin of B^{10}, which was long thought to be one, was shown to be really three; the same B^{10} nucleus was found to have a positive quadrupole moment and hence a prolate (elongated) charge distribution.

The two conditions required for a successful application of this method are:

(a) The nucleus under study must be contained by a molecule which has a dipole moment (essential requisite in rotational spectroscopy) and which is stable in the vapor state as well as simple enough for its spectrum to be easily analyzed.

(b) The nucleus must have a quadrupole moment arising from a non-spherical distribution of its charge, which is possible only when the spin is greater than 1/2. These conditions indicate also the limitations of the method.

(ii) *Hyperfine structure due to magnetic interaction*

In molecular rotational spectra the magnetic interaction leading to hyperfine structure is, in general, not so prominent as the quadrupole interaction. (The reverse is the case in atomic spectra.) The reason for this is that most stable molecules have $^1\Sigma$ electronic ground states and hence extremely small molecular magnetic moments of the order of a nuclear magneton or less. In consequence the coupling of the nuclear magnetic moment to the molecular axis is ordinarily too weak to give a magnetic hyperfine structure sufficiently resolved to allow determination of nuclear moments. The theory of magnetic hyperfine structure applicable to molecules which have no observable nuclear quadrupole interactions (*i.e.,* when I = 0 and 1/2) has been worked out by Gordy and others. A few molecules such as NO_2 and ClO_2 are paramagnetic and the magnetic hyperfine structure of their pure rotational spectra have been observed in the microwave - region and studied. In such cases, to obtain nuclear magnetic moments from the I - J coupling, one must know the value of the molecular magnetic field at the nucleus in question. This, however, cannot be determined with accuracy due to the complexity of molecular moments.

(*iii*) *Zeeman effect in hyperfine structure due to both types of interaction*

The moments of most of the molecules in $^1\Sigma$ state, though very small, are sufficient to cause observable Zeeman effects in the hyperfine structure of their microwave rotational spectra, provided the nuclear spin is coupled to the molecular axis through quadrupole moment. Strong magnetic fields of several kilogausses are also required to .produce the effect. Such a Zeeman splitting of hyperfine structure was first observed in the inversion spectrum of by Coles and Good and has been fully verified by Jen. Nuclear magnetic moments can be estimated from the Zeeman splitting of the hyperfine structure. The relevant theory has been worked out in a manner analogous to the Zeeman effect in atomic hyperfine spectra. in the *molecular beam magnetic resonance method,* here also the value of the spin must be known for the evaluation of the magnetic moment; but the spin can be readily determined from the hyperfine structure which is present. The accuracy of result depends on the strength of the quadrupole coupling. The nuclear magnetic moments of I^{129}, S^{33}, and S^{35}, previously unknown, have been determined in this way. The great importance of this method lies in its application to the determination of *nuclear magnetic moments of rare or radioactive nuclei,* where extremely small samples must be used.

BIBLIOGRAPHY

General
Max Planck, *Survey of Physics*, 1925.
F. H. Newman. *Recent Advances in Physics*, 1934.
E. N. da C. - Andrade, The *Structure of the Atom*, 1934.
A. E. Ruark and H. C. Urey, *Atoms, Molecules and Quanta*, 1930.
G. Castelfranehi, *Recent Advances in Atomic Physics*, (2 Vole.), 1932.
A. Hass, *Theoretical Physics*, (2 Vols.), 1929.
Harnwell and Livingood, *Experimental Atomic Physics, 1933.*
F. K. Richtmyer, *Introduction to Modern Physics*, 1934.
H. A. Wilson, *Modern Physics*, 1944.
Max Born, *Atomic Physics*, 1954.
G. F. M. Jauneey, *Modern Physics*, 1946.
G. Joos, *Theoretical Physics*, 1950.
J. D. Stranathan, *The Particles of Modern Physics*, 1942.
R. A. Milliken, *Electron (+ and -), Protons, Photons, Neutrons, Mesotrons and Cosmic Rays*, 1946.
E. Grimsehl, *Physics of the Atom*, 1949.
H. Semat, *Introduction to Atomic Physics*, 1947.
C. T. Chase, *The Evolution of Modern Physics*, 1947.
S. Tolansky, *introduction. to Atomic Physics*, 1948.
F. K. Richtmyer and E. H. Kennard, *Introduction to Modern Physics*, 1946.
F. K. Richtmyer, E. H. Kennard and T. Lauritsen, *Introduction to Modern Physics*, 1956.
Oldenberg, *Introduction to Atomic Physics*, 1949.
W. Finklenburg, *Atomic Physics*, 1950.
S. Dushman, *Fundamentals of Atomic Physics*, 1951.
F. W. Van Name, *Modern Physics*, 1952.
C. Kittol, *Introduction to Solid State Physics*, 1954.
R. S. Shankland, *Atomic and Nuclear Physics*, 1955.

Chapter 1 -Discharge of Electricity through Gases
J. S. Townsend, *Electricity do Gases*, 1914.
J. J. Thomson and G. P. Thomson, *Conduction of Electricity through Oases*, Vol. I (1928). Vol, II (1932)..
K. G. Emedeus, *Conduction of Electricity through Oases*, 1928.
K. K. Darrow, *Electrical Phenomena in Oases*, 1932.
A. AL Tyndall, Mobilit*y of Positive Ions in Gases*, 1938.

Chapter 2 and 3-The Electron
R. A. Millikan, *The Electron*, 1926.
0. W. Richardson, *The Electron Theory of Matter*, 1916,
D. Grimes, *Meet the Electron*, 1944.
E. C. Stoner, *Magnetism*, 1930.
J. H. Van Vleck, *The Theory of Electric and Magnetic Susceptibilities*, 1932.

411

F. Bitter, *Introduction, to Ferromagnetism,* 1937.

O. W. Richardson, *The Emission of Electricity from Hot Bodies,* 1921.

A. L. Riemann, *Thromionic Emission,* 1934.

H. S. Allen, *Photoelectricity,* 1925.

A. L. Hughes and L. A. du Bridge, *Photoelectric Phenomena,* 1932.

K. H. Spring, *Photons and Electrons,* 1950.

J. Stokley, *Electrons in Action,* 1946,

L. R. Koller, *Physics of Electron Tubes,* 1937.

F. E. Terman, *Fundamentals of Radio,* 1938.

W. L. Everitt, *Fundamentals of Radio and Electronics,* 1958

A. L. Albert, *Fundamental Electronics and Vacuum Tubes,* 19-18.

L. B. Arguimban, *Vacuum tube circuits and transistors,* 1956

Admiralty, *Handbook. of Wireless Telegraphy,* (2 Vols.) 1939.

A. A. Ghirardi, *Radio Physics* Course, 1933.

J. H. Rayner, *Modern Radio Communication,* 1947

W. F. Lovering, *Radio Communication,* 1958

J. L. Hornung, *Radar Primer,* 1948.

Radar School, *Principle of Radar,* 1946.

E. C. Pollard and J. M. Sturtevant, *Microwaves and Radar Electronics,* 1918.

W. Gordy, W. V. Smith & R. F. Trambaru o, *Microwave Spectroscopy,* 1953.

AL W. P. Strand berg. *Microwave Spectroscopy,* 1954.

H. Motz, *Electromagnetic problems of Microwave theory,* 1951.

Campbell and Ritchie, *Photocells,* 1934.

Zworykin and Wilson, *Photocells,* 1934.

-0. A. Briggs, *Sound Reproduction,* 1950.

G. F. Jones, *Sound Film Reproduction,* 1936.

V. K. Zworkyin and E. G. Ramberg, *Photoelectricity and its applications,* 1949.

H. K. Henisch, *Metal Rectifiers,* 1949.

J. T. Mac Gregor Morris and J. A. Henly, *Cathode Ray Oscillography,* 1936.

J. H. Rayner, *Cathode Ray Oscillography,* 1945.

W. C. Eddy, *Television The Eyes of Tomorrow),* 1945.

J. H Boyner, *Television,* 1934.

M. G. Scroggie, *Television,* 1935.

A. Dinsdale, *First Principles of Television,* 1932.

G. V. Dowding, *Practical Television,* 1935.

E. F. Burton and W. H. Kohl, *The Electron Microscope,* 1946.

Chapter 4 - Positive-Rays

F. W. Aston, *Mass Spectra and Isotopes,* 1933,

J. J. Thomson, *Rays of Positive Electricity,* 1923.

Reviews of Modern Physics, *Nuclear Physics* C., July 1937.

Chapter 5-X-Rays

A. H. Compton and S. K. Allison, *X-Rays in Theory and Experiment,* 1935.

Maurice de Broglie, *X-Rays,* 1925.

0. W. G. Kaye, *X-Rays,* 1926.

W. H. Bragg and W. L. Bragg, *X-Rays* and *Crystal Structure.* 1918.

M. J. Buerger, *X-Ray Crystallography,* 1949.

B. L. Worsnop, *X-Rays,* 1930.

W. H. Zachariassen, *Theory of X-Ray Diffraction in Crystals,* 1915.

R. W. James, *X-Ray Crystallography,* 1930.

K. Lonsdale, *Crystals* and *X-Rays,* 1948.

M. Siegbahn, *Spectroscopy of X-Rays,* 1925.

A. J. C. Wilson, *X-Ray Optics,* 1949.

G. L. Clark, *Applied X-Rays,* 1940.

Chapter 6 -Radioactivity

E. Rutherford, *Radioactivity,* 1905,

W. H. Bragg, *Studies in Radioactivity,* 1912.

E. Rutherford, J. Chadwick and C. D. Ellis, *Radiations from Radioactive Substance,* 1930.

G, Newsy and F. A. Paneth, *A Manual of Radioactivity, 1938.*

K. K. Darrow, *Bell Telephone, Some Contemporary Advances in Physics-XII Radioactivity,* 1927.

Madame Pierre Curie, *Radioactivity,* 1935.

Chapter 7-Relativity

A. Einstein, *Relativity : Special and General Theory,* 1922.

L, Silberstein, *The Theory of Relativity, 1914.*

H. Schmidt, *Relativity and the Universe,* 1921.

A. Hass, *Introduction to Theoretical Physics.* (Vol. II), 1929.

Sir James Jean?, *Through Space and Time,* 1934.

W. H. M. C. Crea, *Relativity Physics,* 1935.

A. S. Eddington, *Space, Time and Gravitation,* 1921,

E. Cunningham, *The Principle of Relativit ,* 1921.

Max Born, *Einstein's Theory of Relativity,* 1922.

H. Dingle, *The Special Theory of Relativity,* 1940.

G. J. *Whitrow, The Structure of the Universe, 1949.*

G. Y. Rainich, *Mathematics of Relativity,* 1950.

A. Einstein, *The Meaning of Relativity,* 1950.

L Barnett, *The Universe and Dr. Einstein,* 1952.

Chapter 8-Quantum Theory of Radiation

F. Reiche, *The Quantum Theory,* 1930.

W. Heisenberg, *The Physical Principles of the* Quantum *Theory,* 1930.

L. Infeld, *The World in Modern Science, Matter and Quanta,* 1934.

W. Heitler, *Quantum Theory of Radiation,* 1936.

G. Temple, *An Introduction to Quantum Theory,* 1931.

Louis de Broglie, *Matter and Light, 1939.*

J. Jeans, *The New Background of Science,* 1947,

Chapter 9-Wave Nature of Matter

G. Birtwistle, *New Quantum Mechanics,* 1928.

D. Bohm, *Quantum Theory,* 1952.

J. Frenkel, W*ave Mechanics,* 1934.

E. Schrödinger, *Four Lectures on Nave Mechanics,* 1929.

N. F. Mott, *Wave Mechanics,* 1930.

A. Sommerfeld, *Wove Mechanics,* 1930.

Louis de Broglie, An *Introduction to the Study of Wave Mechanics,* 1930.

H T. Flint. *W woe Mechanics, 1931.*

F. G. Kemble, *The Fundamental Principles or Quantum, Mechanics,* 1937.

P. A. M. Dirac, *The principles of Quantum Mechanics,* 1947.

W. Wilson, *Relativity and Quantum Dynamics,* 1040.

V. Rojansky, *Introduction to Quantum Mechanics,* 1950.

K. R. Dixit, *The Elements of Wave Mechanics and Quantum Mechanics,* 1953.

G. K. T. Conn, *The Wave Nature of the Electron,* 1944.

R. Beeching, *Electron Diffraction,* 1946.

G. T. Thomson and W. Cochrane, *Theory and Practice of Electron Diffraction,*1939.

Chapter 10-Quantum Statistics

J. Rice, *introduction to Statistical Mechanics J or Students of Physics and Physical Chemistry,* 1930.

E. H. Kennard, *Kinetic Theory of Gases,* 1938.

M. N. Saha and B. *N.* Srivastava, *A Treatise on Heat,* 1953.

Chapter 11-Peripheral Electronic Structure of the Atom

N. Bohr, *The Theory or Spectra and Atomic Constitution,* '1924.

E. C. Stoner, *Magnetism and Atomic Structure,•* 1926.

A. Haas, *Atomic. Theory,* 1927.

A. Sommerfeld, *Atomic Structure and Spectral Lines,* 1934.

G. Herzberg, *Atomic Spectra and Atomic Structure,* 1937.

H. E. White, *introduction to Atomic Spectra,* 1934.

F. K. Richtmyer, *Introduction to Modern Physics,* 1934.

0. L. Padding and S. Goudsmit, *The Structure of Line Spectra,* 1930.

G. K. T. Conn, *The Nature of the Atom,* 1944.

F. 0. Rice and E. Teller, *The Structure of Matter,* 1949.

S. Tolansky, *High Resolution Spectroscopy,* 1947.

Chapter 12-Molecular Spectra

L. Kroenig, *Bond Spectra and Molecular Spectra,* 1930.

R. C. Johnson, *Introduction to Molecular Spectra,* 1949.

P. Debye, *The Structure of Molecules,* 1932.

K. B. Ramanathan, *Molecular Scattering of Light,* 1923.

8. Bhagavantham, *Scattering of Light and Raman Effect,* 1940.

G. Herzberg, *Infra-red and Raman Spectra of Polyatomic Molecules,* 1945.

G. B. B. Sutherland, *Infra-red* and *Raman Spectra,* 1935.

Chapter 13-Modern Problems of Radioactivity

K. K. Darrow, *Bell Telephone Series XII, Radioactivity,* 1927.

F. Rasetti, *Elements of Nuclear Physics,* 19 7.

G. Gamow, *Atomic Nuclei and Nuclear Transformations,* 1937.

J. M. Cork, *Radioactivity and Nuclear Physics, 1947.*

Chapter 14 -Artificial Transmutation

K. K. Darrow, *Bell Telephone* Series *XXII, Transmutation, 1931.*

N. Feather, *An Introduction to Nuclear Physics,* 1936.

Reviews of Modern Physics, *Nuclear Physics, C, July, 1937,*

W. B. Mann, *The Cyclotron,* 1940.

D. H. Wilkinson. *Ionizations Chambers and Counters,* 1950.

J, B. Birks, *Scintillation Counters,* 1. 954.

S. C. Curran, Luminescence *and the Scintillation Counter,* 1953.

Curran and Craggs, *Counting Tubes : Theory and Applications,* 1949.

J. 0. Wilson, *The Principles of Cloud Chamber Technique, 1951.*

C. F. Powell and G. P. S. Occhialini, *Nuclear Physics in Photographs,* 1947.

H. Yagoda, *Radioactive Measurements with Nuclear Emulsion,* 1949.

J. B. Hoag and S. A. Kor, The *Electron* and *Nuclear Physics,* 1949-

P. B. Moon, *Artificial Radioactivity,* 1949.

E. Pollard and W. L. Davidson, *Applied Nuclear Physics, 1946.*

J. B. Rajam, *Nuclear Isomerism* (Thesis), 1939.

K. Mendelssohn, *What is Atomic Energy ?* 1946.

J. De Ment and H. C. Dake, *Uranium* and *Atomic Power,* 1945.

J. J. O'Neill, *The Almighty' Atom,* 1945.

G. Gamow, *Atomic Energy,* 1947.

W. E. Stephens, *Nuclear Fission and Atomic* Energy, 1948.

J. Cockcroft, *Development and Future of Atomic Energy,* 1950.

W. L. Lawrence, *The Hell Bomb,* 1951.

Chapter 15 Cosmic Rays

R. A. Millkan, *Cosmic Rays,*

L..Janossy, *Cosmic Rays,* 1948.

J. G. Wilson, *About Cosmic Rays,* 1948.

D. J. X. Montgomery, *Cosmic Rays Physics,* 1949.

L. Leprince Ringuet, *Cosmic Rays,* 1950.

B. Rossi, *High-Energy Particles,* 1952.

F. C. Frank and D. R. Rexworthy, *Cosmic Radiation,* 194.9.

A. Dauvillier, *Les Rayons Cosmiques,* 1954.

R. Marshals, *Meson Physics,* 1952.

H. A. Bathe and F. De Hoffman, *Mesons* and *Fields,* 1955.

J. G. Wilson, *Progress* in *Cosmic Ray Physics,* Vols. I (1952), II (1954) and III (1956).

Chapter 16-Structure and Properties of the Nucleus

Reviews of Modern Physics. *Nuclear Physics, A and B,* April, 1936 and April, 1931.

K. K. Darrow, *Bell Telephone Series, The Nucleus-I,* II, III, 1V, 1933-35.

D. Halliday, *Introductory Nuclear Physics,* 1950.

F. Bitter, *Nuclear Physics,* 1950.

S. Tolansky, Hyperfine Structure in Line *Spectra and Nuclear Spin,* 1948.

N. F. Ramsey, *Nuclear Moments,* 1953.

Reviews of Modern Physics, Jul?, 1946.

S. Devons, *Excited States of Nuclei,* 1949.

0. R. Frisch, *Progress in Nuclear Physics,* 1950.

F. Fermi, *Nuclear Physics,* 1951,

W. Heisenberg, *Nuclear Physics,* 1952.

J. M. Blatt and V. F. Weisskopi, *Theoretical Nuclear Physics,* 1952.

H. I. Bathe, *Elementary* Nuclear *Theory,* 19 7.

C. Wentzel, *Quantum Theory of Fields,* 1949.

W. Pauli, *Meson Theory of Nuclear Forces,* 1948.

R. E. Marshak, *Meson Physics,* 1952.

R. D. Evans, *The Atomic Nucleus,* 1955.